"*This is a fascinating and important* .... academic penchant for ignoring bodies, The Roots of Power: Animate Form and Gendered Bodies *offers a thoroughgoing revision — in favor of bodies — of contemporary attention to language and culture as the root of all meaning. Sheets-Johnstone demonstrates that ignoring the roots of power in animate life and placing all attention on cultural dispositions of power results in an inadequate and inaccurate account of power. This book could single-handedly turn around the current constructionist grain of much feminist thought by its persuasive argument that 'what is political is distinctively concrete in both a corporeal and an intercorporeal sense.' It is a 'must-read' for every critical thinker.*"

—Margaret R. Miles
Bussey Professor of Historical
Theology
Harvard University

"*A brilliantly conceived contrarian book that undercuts the textualist temper of our times. By resolutely taking the corporeal turn — in truth, a re-turn to basic bodily resources in their evolutionary and phylogenetic origins — Sheets-Johnstone effectively trims the culturalist sails of leading deconstructionist, feminist, and postmodernist thinkers (among them Derrida and Lacan, Haraway and Foucault). She accomplishes all this in a densely lucid and unfailingly telling text that is as remarkable for its informative detail as for its philosophical sagacity. In the end, the roots of power are seen to be corporeal and intercorporeal archetypal dominants that, once recognized, complicate and delimit phallocentric behaviors of control.*"

—Edward S. Casey
Chairperson and Professor of
Philosophy
State University of New York at Stony
Brook

"*This subtly argued, extensively documented text opens the door to an archetypal understanding of a contemporary ideological scandal, Lacan's strange conservation, within his otherwise iconoclastic theory, of the totems of patriarchal power. In her unyielding defense of the reality of the female body, and of the body of the child which issues forth from it, the author provides a powerful corrective to the many streams of feminist and postmodern thought that persist in valorizing the psychoanalytic phallic as the major prerogative of being human, when to do so is to collude with our culture's anxiety to make the feminine, for all practical purposes, disappear."*

—John Beebe, M.D.
Jung Institute, San Francisco
Author of *Integrity in Depth*

"In this insistently thought-provoking book, Maxine Sheets-Johnstone develops a compelling analysis of the origins of gendered bodies, power relations, sexuality, and desire in the behavioral and morphological possibilities of animate form—of the 'species-specific body with all its distinctive spatial confirmations and attendant gestures, postures, orientational possibilities, modes of locomotion" (p. 33). Her thesis can be stated quite succinctly: there is a 'fundamental semantics of intercorporeal life' into which 'power relations are interwoven in complex ways' (p. 66).*

By demonstrating the origins of the semantics in careful and insightful interpretations of a multitude of examples, Sheets-Johnstone provides an alternative to currently dominant language/text-centered and social-constructionist theories. Her attention to intercorporeally-presented invariants redirects inquiry from abstract considerations of intersubjectivity and focus upon difference to features of the concrete existence of living bodies that demonstrate intercorporeality and sameness. She shows that these features persist as pan-cultural corporeal archetypes that are continually reworked by the variety of cultural exigencies in human history. Her astute critical readings of contemporary theory as expounded by Butler, Derrida, Lacan, and Sartre are more than matched by her perceptive analyses of cultural practices. The result is that rare accomplishment: a theoretical work that engages the reader's attention from begining to end."

—Lenore Langsdorf
Professor, Philosophy of
Communication
Southern Illinois University at
Carbondale

"Sheets-Johnstone advocates a 'corporeal' or body-centered, biological-evolutionary understanding of power relations between men and women. This view is often opposed by feminists who label it as 'essentialism,' and who claim that power differences are solely the product of cultural institutions. . . . Sheets-Johnstone has integrated a very impressive body of data and ideas from anthropology, primatology, psychology, and psychiatry, which makes for a readable, convincing book. . . . Among the topics she discusses in detail are rape,

exhibitionism, and the role of language in exacerbating the gender power struggle. . . . While The Roots of Power *may antagonize readers who accept the notions of Sartre, Lacan, and Foucault on the problems of gender relationship in human societies, Sheets-Johnstone is uncompromising in her view that their position has been pernicious and patriarchal. . . . The conclusions of the author are supported by a formidable scholarly documentation, which can serve as a guide for further research on this highly controversial subject."*

—Gordon W. Hewes
Professor of Anthropology
University of Colorado, Boulder

"Professor Sheets-Johnstone's research is opening up new questions and methods for feminist scholarship. Early feminist reactions to the types of gender oppressions arising out of essentialist positions led to a distrust of biology and of explorations of biological differences in gendered experiences. Although the misuse of 'anatomy equals destiny' arguments has been and continues to be pervasive in Western sciences and humanities, the feminist distrust of the body too often led to an avoidance of biology or the more recent postmodern reduction of lived bodily experience to discourses and 'meanings.' Sheets-Johnstone's The Roots of Power: Animate Form and Gendered Bodies *provides an essential corrective to feminist scholarship by emphasizing the process of recovering the lived body in such a way as to acknowledge its centrality to gendered experience while precluding a return to essentialism.*

Many feminist theorists acknowledge the importance of the body, but few theorists have attempted to discuss the natural body that is the place of cultural inscriptions. To fully map this terrain, feminists need the type of phenomenological studies of lived experience that Maxine's writings are providing. What her Roots of Power *makes clear is that we must carefully examine the evolutionary body in order to understand the cultural transformations of it. Through her concern for lived experience, Sheets-Johnstone's research places her in the company of such important feminist theorists of the body as Susan Bordo and Iris Young."*

—Nancy Tuana
Professor of Philosophy
University of Oregon

THE
ROOTS
OF
POWER

T H E
ROOTS
O F
POWER

*Animate Form
and Gendered
Bodies*

Maxine Sheets-Johnstone

Open ❄ Court

Chicago and La Salle, Illinois

© 1994 by Open Court Publishing Company

First printing 1994

Library of Congress Cataloging-in-Publication Data

Sheets-Johnstone, Maxine.
 The roots of power : animate form and gendered bodies / Maxine Sheets-Johnstone.
  p. cm.
 Includes bibliographical references and index.
 ISBN 0-8126-9257-8.—ISBN 0-8126-9258-6 (pbk.)
 1. Body, Human (Philosophy) 2. Power (Philosophy) 3. Feminist theory.
4. Sex role. 5. Postmodernism. 6. Sexism in sociobiology. 7. Evolution
(Biology)—Philosophy. I. Title.
B105.B64S48 1994   94-27553
303.3—dc20   CIP

To Albertini

CONTENTS

Prologue

The watchwords of the feminist movement in the early 1970s—the personal is political and the political is personal—had a tremendous impact.[1] They marked a moment of critical awakening and raising of consciousness: women are oppressed, yes, but they can work toward their own empowerment.

This book, an interdisciplinary inquiry into the roots of power, may be seen as an attempt to go back to those formidable and mobilizing watchwords and to mine them further, indeed, to reinvigorate them with both broader and deeper meanings. While the connection between the personal–political equation and the philosophical investigation in chapter 1 of this book is readily apparent, the equation is in fact not a local topic but thematic of the book as a whole. What the book demonstrates from beginning to end is that in corporeal matters of fact lie dimensions of ourselves that are at once both personal and political. It does this basically by taking account of our evolutionary history, of the fact that whatever our color, culture, national origin, class, sex, political affiliation, age, height, weight, religious preference, or what have you, we are all products of a common natural history. We are all Darwinian bodies.[2] That we are such bodies means precisely that there are corporeal matters of fact to be discovered. Moreover, it demands that we recognize the ties that bind us in a common humanity. Our basic differences—like our fundamental human beliefs and practices[3]—are not merely social constructions; they are social constructions that are through and through rooted in our being the animate forms we are. In other words, cultures rework aspects of our communal primate heritage in multiple and intricate ways. Indeed, the difficult task that lies before us, perhaps particularly now, at the tag-end of a fractious and fractionating twentieth century, is to delineate the ways in which cultures differentially rework the heritage that is our common evolutionary heritage. The rewards of this difficult and patient work will be to understand in the most fundamental senses what is pan-cultural and what is idiosyncratically cultural, not in order to have the differences between the two identified as some abstract bits of knowledge to add to our lore, but to appreciate in our bones and behaviors what it is to be the particular animate form and gendered bodies that we are.

The task of this book is to elucidate a particular dimension of our animateness—the roots of power and of power relations—and in the process to sketch out in a preliminary manner how cultures (and individuals), through their exaggerations, suppressions, distortions, and elaborations of animate form and bodily practices, formulate in idiosyncratic ways a certain concept of power and of power relations. It is reasonable to think that by tracing out the roots of power in this way, we can discover the forces driving our particular culture's overarching conception of power as control, and in turn enlighten ourselves, at least in beginning ways, about alternative formulations. In the completion of this task, the personal-political equation clearly takes on its broader and deeper corporeal significance. What is personal is not abstract; it is distinctively concrete in a bodily sense. Similarly with the political. Catherine MacKinnon once wrote that the connection between the personal and political "is not a simile, not a metaphor, and not an analogy." She said that "the personal as political . . . means that women's distinctive experience as women occurs within that sphere that has been socially lived as the personal—private, emotional, interiorized, particular, individuated, intimate—so that what it is to know the politics of woman's situation is to know women's personal lives."[4] While this is one possible meaning of the personal–political equation, it does not coincide with the meanings elucidated here. These meanings emanate not from the situational aspect of women's personal lives, but from an all-pervasive human bodily personal that has a history, which is to say that the political is at root a corporeal built-in, a dimension of our primate heritage that is expressed in that repertoire of 'I can's' that define us both as creatures of a natural history and as culturally and individually groomed bearers of meaning and agents of power. As this book will show, what is political is distinctively concrete in both a corporeal and intercorporeal sense. It is at root part and parcel of animate form.

Before proceeding to specify the progressive way in which the chapters of this book explicate the roots of power and of power relations, it will be instructive to consider the personal–political equation from the viewpoint of both postmodernism and sociobiology, not only because both doctrines offer major twentieth-century viewpoints on power and "the human condition" and therefore require consideration, but because in chapters that follow, both are in fact critically scrutinized and reproached for their neglect of animate form.

That the broader and deeper corporeal significance of the personal–political equation never fully surfaced in the elaboration of the watchwords of the feminist movement can be viewed as a postmodern sign of the times. The corporeal dimensions of the equation were too soon swamped by overpowering concerns with language to be seen as uncharted frontiers in need of exploration. Indeed, the body in postmodern thought is often little more than an accoutrement of language, one of the many possible trappings of discourse, another text to be read as any other. In the skillful hands of some postmodernists, the living body may even be trashed and wasted, and this because, as the title of that hallmark book—*Our Bodies, Ourselves*[5]—implicitly indicates, subjecthood can be denied only if the living body is denied. On the other hand, where subjecthood is celebrated through the body, as in some postmodern French feminist writings, it is celebrated as a certain kind of anatomy. It is not recognized and elucidated in the basic and at the same time far broader context of *animate form*. This fact explains why postmodern feminist writings linking power to language (or language to power) have validity only in either a purely personal or highly abstract theoretical sense: they do not comprehend the living human body as a cultural universal. Blinded by the beguilements of language, such writings either fail to ask the most basic question—where does our concept of power come from?—or, if they do ask the question, they answer it in terms of language itself, averring that the road to empowerment lies in how we use and what we make of language, and correlatively, in our understanding of how language uses us. Because they lack pan-cultural understandings of the body, understandings rooted in an appreciation of our evolutionary and even ontogenetic heritage, linguistically focused writers cannot see those formal and comportmental aspects of bodily life that in the most fundamental sense shape our concept of power and of power relations. They do not see that gendered bodies are related to animate form, that animate form is the common denominator of cultures,[6] and that that common denominator is the spawning ground of corporeal archetypes more ancient than we.

Given the rallying hold of postmodernism and its sign of the times—that in essence, the *social* is political and the political is *social*—it is little wonder that the broader and deeper corporeal significances of the personal-political equation consistently run aground. What these significances would otherwise attest to is not an idiosyncratically personal or theoretical subject, but a concrete locus of *'I can's.'* They would furthermore attest to the fact that this *I* who *can* is part of a natural history, and is thus unequivocally linked to others in a common creaturehood, a common humanity. In its most doctrinaire formulations, postmodernism denies the ties that bind us in this common humanity. As we shall see, it denies the living body in devious ways which redefine it and, in so doing, distort or occlude its evolutionary heritage and concomitant repertoire of 'I can's.' The redefinitions preclude the semantics of animate form and the intercorporeal semantics that anchors socio-political life from coming to the fore. As successive chapters of this book will show, corporeal archetypes of power and of power relations are at the heart of these semantics. They articulate in precise ways how the personal is political and the political is personal. They clearly challenge the thoroughly relativistic, constructionist tenets of postmodern thought. They show that there are meanings to be fathomed outside of language, meanings that are indeed more ancient than language and more ancient than we. Whether by veiling over the body or by rewriting its history, the words of leading postmodernists consistently and effectively shut out these archetypes.

What the postmodernist does with words—and cultural relativism— the sociobiologist does with adaptations—and biological absolutisms. More generally put in terms of academic practice, what the human sciences do to the living body in the name of sociobiology is the reverse of what the humanities do to the living body in the name of postmodernism: not biology—that is, evolutionary history—but culture is denied. Whatever might be the sexual practice or aggressive act in question, for example, the sociobiologist consistently explains it as "adaptive." In essence, all behavior is squeezed through an adaptive filter to emerge at the other end as "good," a benefit to the individual and/or his/her kin. Of course the personal is nowhere to be seen in all of this for the personal is at bottom simply a degenerate form of a genotype; that is, animate forms—living bodies—are simply instruments of behavior, and behavior is simply the outward form taken by a particular constellation of genes. The significance of animate form is in effect nowhere appreciated. What matters—*all* that matters—is adaptation. This is why a major sociobiological text on the evolution of human sexuality can render female sexuality a "service" to males and not once mention, much less elucidate, aspects of the hominid living body as it evolved, neither in terms of primate genitalic modifications nor in terms of the relationship between bipedality and sexual signalling behavior.[7] Instead, an adaptational view is put forth, a specifically *male* adaptational view as the text quickly shows in such chapter headings as "The Female Orgasm:

Adaptation or Artifact?," "The Desire for Sexual Variety," and "Copulation as a Female Service." Given the standard sociobiological devotional stance toward the adaptiveness of male promiscuity, it is surprising that the full range of human male sexuality was not explored and marshalled into sociobiological service. It is indeed remarkable that early morning penile erection was not turned into both a sociobiological justification for males to start each day with a song, as it were, and a sociobiological proof of "the male desire for sexual variety."[8] Adaptively explained in this way, the phenomenon might even be affirmed to coincide with the mission of females to provide a "service" to males, especially insofar as this servicing is linked to the human female ability to provide ever-ready servicing, i.e., to the fact that human females are "continuously copulable."[9]

Though seemingly neutral in the sense of being beyond moral reproach, this radical use of adaptation—to signify what is ultimately for the common good, or more rudely, "passing on one's genes and may the best man win"—has considerable political weight. The constellation of genes that one is, in virtue of the behaviors that constellation specifies, destines one to a certain kind of life. With respect to power, the constellation is unequivocally determinative. Thus, while one might wish to think that providing a service is an act of power—including the power to choose to provide the service or not—that power is an illusion. As several sections of the ensuing chapters of this book will show in detail, adaptive accounts of human behavior do indeed differentially empower and disempower, and in this respect are most definitely *not* beyond moral reproach. In fact, insofar as a sociobiological script of and for human behavior can be endlessly creative in justifying whatever bodily events, sexual practices, or aggressive acts a person might personally find congenial and worth championing, it can make whatever subserves those bodily events, sexual practices, or aggressive acts appear mandatory. At its more sinister, then, the script is not merely suggestive of a certain morality; it carries a moral dictum of sorts since, if a behavior is deemed adaptive, *it is of value to perform it.* Accordingly, some persons might believe that they are here for a certain performance and that with each completion of that performance, they are positively to commend themselves, anointing what they do with an aplomb worthy of a God's "And it was good." Whatever the costs—to themselves or to others—by disseminating (even if not passing on their genes), they give their all to the broader aims of humanity and in so doing succeed in carrying out their adaptive role in society.

By whatever tenets animate form is ignored, whether postmodernist or sociobiological, the personal is thereby weakened if not dissolved. In the end humans conceive themselves as either cultural artifacts or biological robots: *à chacun son goût*. With respect to feminism, however, postmodernism has a further dimension that compounds its oversight. Postmodernism brings with it a preoccupation with theory. For many feminists this means—and has meant—a preoccupation with the construction of a viable and sound ideological framework for addressing the oppression of women and for devising programs for their liberation. The ideological preoccupation, especially several years back, was a near frantic one. At each turn one read about the *need* for theory. Chris Weedon, for example, wrote, "We need a theory of the relation between language, subjectivity, social organization and power";[10] Joan Scott wrote, "That feminism needs theory goes without saying (perhaps because it has been said so often)."[11] At the other end of the spectrum, however, were dissenting voices, particularly from women of color. Barbara Christian, for example, deplored the obsessive emphasis on theory.[12] She inveighed against postmodern-poststructuralist thinkers who were revamping literature. She accused them of "hav[ing] re-invented the meaning of theory." She indignantly decried "the race for theory, with its linguistic jargon, its emphasis on quoting its prophets, its tendency towards 'Biblical' exegesis, its refusal even to mention specific works of creative writers, far less contemporary ones, its preoccupations with mechanical analyses of language, graphs, algebraic equations, its gross generalizations about culture." She spoke of how this race for theory "has silenced many of us to the extent that some of us feel we can no longer discuss our own literature, while others have developed intense writing blocks and are puzzled by the incomprehensibility of the language set adrift in literary circles."[13]

The postmodernist dissolution of the subject not only occurred at exactly the time at which a diversity of subjects was beginning to be recognized—the time at which power differentials among people were seen to exist with respect to race and class as well as to sex—but it brought with it overzealous theoretical concerns. The relationship between the two "events" seems obvious enough. If there are no subjects in the sense of constituting agents—or to use the jargon of which Christian speaks, if there is no "center"—then something must be found to hold everything together, precisely as Weedon's remark specifies. Thus the need for theory. In a very general sense, what postmodern feminists did in answer to this need was to dissolve the subject into a theoretical soup whose major ingredient was language and whose subsidiary ingredients became, as Weedon's list would have it, "subjectivity, social organization and power." The subject was thus in essence to be saved by being concocted, concocted like everything else of moment, from theory, theory which was nothing if it was not language. It is understandable that those disenfranchised to begin with would find this further peripheralization of themselves as subjects affronting.

With respect to animate form, what postmodern feminist theory produced was in some instances bizarre: "We must recognize that what (most) women now share is a positional similarity that *masquerades as a natural likeness* . . . we must be willing to give up *the illusory similarity of nature* [italics added] that reinforces binary logic. . . . My argument is that the structural similarity that *pretends to reflect nature* . . ." (italics added).[14] It is as if whatever the theory one wants to proclaim, even if it flies in the face of corporeal matters of fact, it can be made true by word-waving. "[To] make the social construction of (sexed) identities a project of pressing political concern"[15] is undeniably a commendable aim, but surely it does not warrant *defying nature* by twisting words about so that "right thinking" will prevail. As consecutive chapters of this book will show, however mighty the fear of biological essentialism, nature cannot properly on that account be written out of the picture. It can no more be made to disappear by word-waving than it can be made to disappear by prayer, incantation, wishful thinking, or ritual hand-washing.

In sum, the tenets of both postmodernism and sociobiology preclude insights into the essentially corporeal relationship between the personal and political, thus into those corporeal archetypes that subtend power relations, and into those bodily 'I can's' that are the source of our concept of power. In the broadest of terms, this is because postmodernism culturalizes animate form and sociobiology biologizes gender. *Contra* these consumptive and obfuscatory practices, this book is a vindication of the claim that the deepest of insights into the personal-political equation hinges on a realization of animate form, for it is animate form that basically ties the personal to the political and the political to the personal. By bringing forward corporeal matters of fact—matters of fact deriving from our evolutionary and ontogenetic heritages—and by critically assessing the shortcomings and outright distortions of postmodernists and sociobiologists, successive chapters of this book bring forth evidence that demonstrates the corporeal foundations of power. The purpose of the critical assessment, it should be noted, is not to disparage postmodernism or sociobiology per se, but to use critical analyses of their errors and omissions constructively; that is, to show how an occlusion and/or degradation of the living body prevents an appreciation of animate form, and with it, an appreciation of those corporeal archetypes of power that derive from our primate heritage, that are culturally—and individually—reworked in multiple and intricate ways, *and that, by their very reworkings, testify to a mutability at the same time they testify to a pervasive and emphatically rigid notion of power as control.* We are, in effect, not doomed. On the contrary, by gaining insights into the corporeal foundations of power, we realize our power to rework pervasive culture-spawned corporeal archetypes that thrive on oppression.

Inquiry into the roots of power opens with the claim that Foucault's "optics of power" is Sartre's "the Look" writ large—in broad, socio-political script rather than in fine, interpersonal hand. By subsequently demonstrating how Foucault's interpretive analytic of power is conceptually related to Sartre's analytic of the Other, this initial chapter accomplishes two objectives. It lays the conceptual groundwork for a critical examination of the generally acknowledged but corporeally unanalyzed relationship between vision and power, and it situates the many corporeal investigations and analyses of animate form and gendered bodies that follow within the mainstream of contemporary philosophical views of power.

As its title suggests, the succeeding chapter, "An Evolutionary Genealogy," brings the previous analysis of the optics of power and the power of optics into evolutionary perspective. In identifying ways in which primate power relations—human and nonhuman—are an intensification of the natural power of optics, that is, how the visual is intensified in the socio-political relations of primates, the chapter elucidates dimensions of the intercorporeal semantics that informs primate life and specifies the invariants of animate form that are its foundation.

The evidence gathered in "An Evolutionary Genealogy" clearly substantiates the fact that living bodies are products of a natural history. Living bodies comport themselves in ways that reveal evolutionary continuities. The very idea that human living bodies are part of the natural world and that their behaviors manifest certain relationships with nonhuman living bodies may be discomforting to some—on the grounds of human uniqueness—but the idea is positively threatening to many feminists, postmodern and other. The third chapter, "Corporeal Archetypes and Power: Preliminary Clarifications and Considerations of Sex," begins by showing how feminist disavowals of "the natural body" are contravened by corporeal matters of fact, and it goes on to consider our human evolutionary genealogy in the specific perspective of primate socio-sexual behavior. The chapter shows to begin with how our notion of power and of power relations derives from the animate form we are and from the 'I can's' peculiar to that animate form. It then spells out in detail how cultural archetypes are translations of evolutionary ones. The purpose is to show how our evolutionary heritage generates corporeal archetypes and how these archetypes are reworked by our own twentieth-century Western culture—ironically in the name of biology—to oppress females. In particular, it shows how the phallus is a cultural archetype linked both to penile display—the original evolutionary archetype—and to females as "continuously receptive"—a cultural-spawned archetype.

What "Corporeal Archetypes and Power" brings to light in the way of the sociobiologist's failure to consider animate form, "Corporeal Archetypes and Postmodern Theory" brings to light in the way of the postmodernist's failure. The chapter focuses specifically on Derrida's rewriting of evolution by a suspension of the living body. It shows how his rewriting is structured in an emphatically phallocratic philosophy, and further, how his critique of the metaphysics of presence actually rests upon a metaphysics of absence, an absence of the living body. It might be noted that this chapter departs to a degree stylistically from others. In keeping with its subject matter, it is written partly in deconstructive style.

The fifth chapter examines the complex relationship between sex and aggression. It does so by taking up the cultural archetype of the phallus and showing how the equation male threat/female vulnerability derives from the cultural notions of "year-round receptivity" and docile bodies on the one hand, and "perpetual erection"—the myth of the phallus—and the power of penetration on the other. From an examination of these themes emerges the dominant conception of females as, in Sartre's well-known phrase, being "in the form of a hole." The ensuing sections of the chapter are concerned with the ways in which this archetypal formulation of females relates to female fear and male assault, in other words, to rape. In particular, the following claims are examined: the sociobiological claim that rape is an evolutionary male sexual strategy; the claim that aggression is a built-in of the male brain or of male neurohormonal factors; the claim that rape is consistent with male chemistry. In the context of these claims, the critical question of a male's discriminatory abilities is raised. Given the fact that sexual and aggressive behaviors are linked in the corporeal economy of nonhuman primates, and given the variety of studies showing that penile erection *cannot* be taken as an exclusive indicator of sexual inclinations or desire since it occurs in moments of anger and excitement as well, the signal importance of a male's being able to distinguish his feelings cannot be denied. The chapter shows how explanations of rape that fail to take this fact into account not only fail to acknowledge corporeal matters of fact; they absolve rapists of responsibility by showing them to be merely pawns of evolution or of some aspect of their physiology.

"Corporeal Archetypes: Penetration and Being 'in the Form of a Hole'" takes up Sartre's "psychoanalytic of things," laying out to begin with both the patriarchal and archetypal aspects of his psychoanalytic ontology. The purpose of the chapter is to spell out—flesh out—his characterization of females, not only with respect to *their* archetypal psychoanalytic as being "in the form of a hole" but with respect to a male archetypal psychoanalytic as well, thus with respect to being "in the form of a filler of holes"—what Sartre calls "strange flesh." The chapter in this sense answers to the basic challenge implicit in Sartre's psycho-ontology: to identify and describe the consistently unspoken male corollaries of female being. The consideration of cultural variations on the fleshed-out archetypal psychoanalytic gives substantive support to the idea of the archetypes themselves and their strategic place in a culture's elaboration of power.

Just as the chapter on Sartre's psycho-ontology is a psychoanalytical extension of the previous chapter on sex and aggression, so chapters 7 through 10 are a psychoanalytic extension of Sartre's psycho-ontology. The four chapters constitute an extended critical meditation on Lacan's psychoanalytic theory. As noted at the very beginning of these chapters, Lacan's psychoanalytic is a microcosm of twentieth-century Western practices, tendencies, and ideologies that preclude an appreciation of animate form and thereby preclude concerns about, and understandings of, the roots of power. His psychoanalytic is for this reason of critical significance. Like twentieth-century Western practices, tendencies, and ideologies, his psychoanalytic scientizes human life, pedestals human language, and silences the living body. As successive chapters show, each in a distinctive way, it is the silencing of the living body that is foundational; the effacement of animate form clears the way for the scientization of human life and the pedestalling of human language. The roots of power in consequence remain as unexposed in Lacan's psychoanalytic as they do in twentieth-century Western life. Their phallocentric origins are in each case protected, which of course means that the preeminence of the phallus is and can be guaranteed in other ways than by explicit display or ritual veneration.

It is all the more significant, then, that in addition to providing basic insights into the roots of power in twentieth-century Western society, the extended critical examination of Lacanian psychoanalysis exemplifies how culturally elaborated corporeal archetypes deriving from our primate past can be individually elaborated, individually elaborated in ways that entrench even further those archetypal forms already subtending major cultural dispositions, in effect, solidifying even more strongly the reigning conception of power.

Chapter 7 offers beginning perspectives on how the living body is silenced. It specifies substantive socio-cultural linkages, in particular, Lacan's ties to Sartre's ontology, to Lévi-Strauss's structural anthropology, and (perhaps surprisingly) to Jung's analytical psychology. In addition, it raises basic questions about the relationship between the unconscious and the imagination, each construed as a source of power. In the context of these questions, Lacan's psychoanalytic of the unconscious is examined in Kantian perspective. Four topics emerge as central in this chapter: introspection, infancy, the unconscious, and the phallus. Each is tied in a preliminary but specific way to a silencing of the living body. The latter three topics are then treated in turn and in detail in each of the three successive chapters. What emerges in the course of a critique of their meaning and value within Lacan's radical poststructuralist formulations is an ever more consistent and persuasive demonstration of how his psychoanalytic reflects twentieth-century Western culture's equation of power with control, of how Lacan thereby covers over any hint of male fears or deficiencies, and of how simultaneously, by twists and turns, he projects those fears and deficiencies onto females.

As the epilogue makes clear, this exaggerated, one-dimensional, typically Western concept of power is out of all proportion to the living body. It is enshrined at a considerable cost, and for all: the price of control is high for males and females alike. What the epilogue also makes clear is that there are other ways to be human; what our evolutionary history teaches us is that the roots of power lie in other directions as well. Though we see them dimly, we know them with utmost clarity. They are anchored not in vision but in touch; they are structured in intercorporeal invariants no less than are the roots of optics of power. We have the potential to elucidate these other roots—these other 'I can's'—and the concept of power and of power relations they engender. Moreover, though the effort would be enormous, we have the potential to realize these neglected 'I can's' by reworking our dominant cultural archetypes. With their realization, power would no longer be preeminently the power to control. The phallus would lose its mythical proportions. It would not, however, lose its miraculous powers of self-transformation. On the contrary, those powers would be enhanced in the essentially *relational* possibility of male being that lies latent in the shadow of a phallus blown up out of all proportion and exceeding the bounds of animate form.

1

Optics of Power and the Power of Optics

To be looked at is to apprehend oneself as the unknown object of unknowable appraisals . . .

Jean-Paul Sartre[1]

Visibility is a trap.

Michel Foucault[2]

I. Introduction

Foucault's "optics of power" is Sartre's "the Look" writ large—in broad, socio-political script rather than in fine, interpersonal hand. In a similar way, Foucault's "interpretive analytic" of bio-power is conceptually related to Sartre's analysis of the Other. To show these relationships is to lay the conceptual groundwork for a critical examination of the generally acknowledged but corporeally unanalyzed relationship between vision and power, indeed, to show in a beginning way how power is rooted in the very facts of being a body. These facts are latent but unrecognized in Foucault's account of power. In particular, while bodies are everywhere apparent in his work, a flesh-and-bone body is nowhere concretely evident and analyzed. While such a body is on the contrary present and examined in Sartre's account—not of power per se but of freedom—its full elucidation is held in check by a restrictive ontology. Philosopher Hubert Dreyfus and anthropologist Paul Rabinow, in their joint book on Foucault, come very close to recognizing the necessity of identifying and examining a flesh-and-bone body when they write that "specific facts about the body no doubt have influenced those who developed disciplinary techniques," and mention such specific facts as "that the body has a front and a back and can only cope with what is in front of it, that bodies can move forward more easily than backwards, that there is normally a right/left asymmetry, and so on."[3] The body which is the subject of these specific facts is most succinctly defined as the body *simpliciter,*[4] the body in abstraction from any historical investitures of power, that is, considered apart from its grooming in the ways of a particular socio-political cultural heritage. Through a brief but detailed consideration of Dreyfus's and Rabinow's estimation of Foucault's work, it will readily become clear why an understanding of the body *simpliciter*—and in a broader sense, why a corporeal analysis of power—is needed. It will furthermore become clear how Sartre's analysis of the Look and of the Other can elucidate aspects of Foucault's optics of power by bringing to light specific intercorporeal dimensions of flesh-and-bone bodies and in turn of socio-political life.

Dreyfus and Rabinow first of all present a compelling rationale for what they call Foucault's *interpretive analytics:* it is *the* method, they claim, for studying human beings. Foucault's new methodology makes possible an understanding of those cultural practices that have "made us what we are": a society driven by power and by certain relations between power and bodies.[5] Quoting from Foucault's *Discipline and Punish,* they write that "power relations have an immediate hold upon [the body]; they invest it, mark it, train it, torture it, force it to carry out tasks, to perform ceremonies, to emit signs."[6] The central question Dreyfus and Rabinow leave unasked, however, is the same one Foucault leaves unasked: how is it that power has an immediate hold upon the body? How is it that the body is *accessible* to power relations?

The question needs to be addressed because it is immanent in the analytic. If as Dreyfus and Rabinow clearly show, "Foucault . . . is asking how the body can be divided up, reconstituted, and manipulated by society" and answering the question in terms of a proliferation of "disciplinary technologies,"[7] then the question of the body's accessibility is *not* being addressed. In fact, Dreyfus and Rabinow specifically state that something is missing. After writing that "specific facts about the body no doubt have influenced those who developed disciplinary techniques," and giving examples of these specific facts, they go on to note that there are questions concerning both the identification of these facts about the body and "the role [these facts] play in the successful deployment of disciplinary techniques." They insist that "Foucault is uniquely placed to address these questions," and point out, "So far, he [Foucault] has remained silent."[8]

An answer to the question of how it is that power has an immediate hold upon the body requires first a recognition that there *are* basic facts about the body, basic facts yet to be discovered and articulated. The "optics of power" Foucault so deftly identifies and describes, for example—where does it come from? On what does it rest? Each of the disciplinary technologies identified by Foucault clearly operates not simply by "intensifications" of the body, as he affirms, but more precisely by intensifications of the *visual* body: all of the various practices he examines focus on the body as a preeminently visual entity. In writing about public executions, for example, Foucault speaks of "an invincible force" being "deploy[ed] before all eyes'"; in specifying what he means by a *scientia sexualis,* he pinpoints way after way in which the body is self- and other-examined in never-ending rituals of observation.[10] Even the purely verbal confession is a spectacle; it is a detailing of thoughts or acts in such a way as to make them visible happenings; private experiences are *made public.* Indeed, it is in virtue of their publicity—whether actual through verbalization to another, or imagined in the specular confines of one's own reflections—that thoughts and acts are shameful and warrant censure. Close examination of all the disciplinary technologies enumerated and analyzed by Foucault strongly indicates that intensifications of the visual body draw their force from a power of vision itself, that an optics of power is rooted in a concrete bodily *power of optics.* Power, in other words, does not merely appropriate the visual as the means whereby it will be deployed. The visual is already charged with power.

An answer to the question furthermore requires recognition of the fact that Foucault's always abstract body—his generalized, non-delineated body—rests upon a concrete, sexed, and gendered bodily form, one which can be examined and analyzed in detail with respect to conceptions and uses of power. Foucault hints at the significance of sexed and gendered bodies when he states, for example, that "The Greek ethics of pleasure is linked to a virile society, to dissymmetry, exclusion of the other, an obsession with penetration, . . ."[11] and when he notes that in Athenian society, "The rapist violated only the woman's body, while the seducer violated the husband's authority."[12] Implicit in many of his analyses and remarks is a recognition of distinctively sexed and gendered behavioral possibilities, possibilities in virtue of which the very concept of power and the very possibility of power relations inhere. These possibilities and the concepts to which they give rise need to be spelled out. Clearly sex and gender are related to "specific facts about the body." Clearly they play a role in power relations. A sex- and gender-spawned differential ratio among the senses, for example, and, correlatively, a sex- and gender-spawned differential corporeal sense of power are prime aspects of these corporeal facts. Psychoanalytic theory and cross-cultural anthropological studies substantively support the corporeal evidence for distinguishing between a male and female sensorium and a male and female sense of power. The question of how everyday behavioral dispositions toward certain power relations derive from one's being the distinctively sexed and gendered body one is, is thus clearly a question in need of answer.

In sum, while Dreyfus and Rabinow are correct in asserting with Foucault that the body in Western society has not simply been scarred by power relations but has itself shaped the disciplinary technologies which enforce those relations, they have not pressed on specifically to the obvious conclusion, namely, that the body is not a *tabula rasa* on which power makes its marks: the body is already inscribed and potent. The obvious conclusion brings into sharp focus the *two*-way relationship between bodies and power and the *two* intertwined questions which in turn demand answer: on the one hand, there is the question of how it is that the body is accessible to disciplinary technologies; on the other hand, there is the question of "specific facts about the body" which make it a natural source of power relations. If Foucault's interpretive analytic is to fulfill its promising press, it needs considerable empirical deepening. The body which molds power and which power invests must be progressively exposed. To this end, corporeal analyses of power are needed.

This chapter will take up the first of the intertwined questions, and that within a limited context. It will specifically examine and elaborate the notion suggested above: that undergirding optics of power is a power of optics. The incisive relevance of Sartre's original analyses of the Look and of the Other to Foucault's socio-political analyses of optics of power will become readily apparent in this context. One might even say that an examination of Sartre's analyses are mandatory to the needed empirical deepening of Foucault's work.

II. Sartre's the Look and the Other

What Sartre describes at the level of everyday interpersonal encounters, Foucault describes at the level of socio-political practices and institutions. As noted above, Sartre's analyses of the Look (and of the Other) are not actually carried out in the name of power. Sartre's central concern is freedom, not power relations per se. Indeed, and interestingly enough, he uses the word "power" specifically and in passing only in describing what it is like to be the person who looks. In a triangulated optics of looking, where one looks at Others at the same time that one and the Others looked-at may possibly be looked at by a third, Sartre writes that "in looking at them [the Others] I measure my power. But if the Other [a distinctively third Other] sees them and sees me, then my look loses its power; it can not transform those people into objects *for the Other* since they are already the objects of his look."[13] This perspicuous observation on power relations coincides with his earlier observations on being caught peeking through a keyhole. He notes, for example, that "With the Other's look the 'situation' escapes me. . . . I *am no longer master of the situation.*"[14] Clearly, though power goes virtually unnamed as such, it figures centrally in Sartre's account of the Look, as in his account of "The Existence of Others" and of "The Body" generally.[15]

Sartre's analyses of the Look are deft. They ring with a clarity that is consistently true to experience. Sartre points out, for instance, that "a look turned toward me appears on the ground of the destruction of the eyes which 'look at me.' If I apprehend the look, I cease to perceive the eyes; they are there, . . . but . . . they are neutralized, put out of play." He observes furthermore that "It is never when eyes are looking at you that you can find them beautiful or ugly, that you can remark on their color. The Other's look hides his eyes; he seems to go *in front of them.*"[16] When he writes of the difference between seeing the Other-as-object and the Other seeing him-as-object, he brings to the surface facets of everyday experience that go unsaid but that are constitutive, at times even determinative, of situations in which we find ourselves. For example, where it is a question of seeing the Other as Other—as an object but an object wholly distinct from a table or book—the Other becomes a "privileged object"[17] who takes with him all of the immediate environment: everything that surrounds the Other—trees, grass, whatever—is in relation *to him.* As Sartre says, "an object has appeared which has stolen the world from me."[18] Hence he speaks of the *"disintegration* of the relations which I apprehend between the objects of my universe."[19] He goes on to say of this "stolen world" that "We are not dealing here with a flight of the world toward nothingness or outside itself. Rather it appears that the world has a kind of drain hole in the middle of its being and that it is perpetually flowing off through this hole."[20] He speaks later of this flight of *his* world toward the Other-as-object as "an internal hemorrhage."[21] When it is Sartre who becomes object for the Other, as when the Other catches him in the act of peeking through a keyhole, the situation parallels the first in that everything which surrounds him is taken away by the Other, just as earlier the Other-as-object took away everything from him: "The Other's look embraces my being and correlatively the walls, the door, the keyhole. All these instrumental-things in the midst of which I am, now turn toward the Other a face which on principle escapes me."[22] The "hemorrhage" in this case, however, is unbound. In this situation, the world flows off away from Sartre in limitless indeterminacy. In fact, the Look of the Other usurps his freedom; Sartre becomes a given object for the Other. As he himself affirms, "My being . . . is written in and by the Other's freedom."[23] Thus, although it can be equally said of this situation that "an object appears which steals the world from me," in this instance the flight of the world is catastrophic: "The Other's look makes me be beyond my being in this world and puts me in the midst of the world which is at once *this world* and beyond this world."[24] In short, dissolution of oneself into an object through the Look of the Other is a dissolution of one's own possibilities as one's own; those possibilities, whatever they might be, exist only across the freedom of the Other.

Sartre's analyses of everyday human experiences validate a *power of optics.* There is no question but that the Look, though not analyzed in these terms, is subtended by a power of optics inherent in *seeing* itself. Sartre's analyses of the Look bring to light power relations already structured in a pure visuality; various meanings of the Look are laced through and through with these relations. Accordingly, the experiences of seeing and of being seen that Sartre describes are not simply evidence of an optics of power. They are not simply testimony to the visual nature of power relations—any more than they are testimony to facts about visual behavior, or about light patterns falling on the retina and engaging the brain in some way. A tacit but formidable power of optics undergirds these experiences. Not only this, but it is set in play in nonverbal ways. The power of optics implicit in one's experiences of the Look and of the Other is not in the least language-dependent. It unfolds rather from ways of being a body, ways that are at bottom not culturally spawned but present in the intercorporeal phenomenon of vision itself.

Substantive hints, if not outright descriptions of this power and its potential to usurp, to modify, to control, are apparent many times over in Sartre's description of the gaze of the Other. For example:

> To be looked at is to apprehend oneself as the unknown object of unknowable appraisals—in particular, of value judgments. . . . [25]
>
> Being-seen constitutes me as a defenseless being for a freedom which is not my freedom. . . . In so far as I am the instrument of possibilities which are not my possibilities . . . and which deny my transcendence in order to constitute me as a means to ends of which I am ignorant—I am *in danger.* This danger is not an accident but the permanent structure of my being-for others.[26]
>
> In order for me to be what I am [rather than being my possibilities, my freedom], it suffices merely that the Other look at me. . . . If there is an Other, whatever or whoever he may be, whatever may be his relations with

me, and without his acting upon me in any way except by the pure upsurge of his being—then I have an outside, I have a *nature*. My original fall is the existence of the Other.[27]

Through the Other's look I *live* myself as fixed in the midst of the world, as in danger, as irremediable.[28]

Clearly, for Sartre the look of the Other is a *perpetual* threat; it has the ever-present possibility of instantiating an ontological vulnerability at the core of being.[29] The question is, how can vision have this power? How can the mere look of another put me at risk, rob me of my possibilities, deny my freedom? Although in his analysis of the Other and of the Look, Sartre shows us the extraordinary power potentially there in the Look of the Other, although he in fact presents us with many facets of that power, and although he gives us clues as in his allusion to the Look's bringing to the fore "an outside" and "a *nature*,"[30] he leaves such questions basically unasked and unanswered. Thus precisely the *corporeal nature* of the power of optics needs further and finer examination, and even its evolutionary genealogy needs to be acknowledged in the context of *perpetual threat*. The question of its evolutionary genealogy—and with it the question of "specific facts about the body"—will in fact be addressed in the next chapter.

III. The Foucault-Sartre Relationship Concretized

When we compare power relations engendered in the Look with power relations in disciplinary technologies, we find the latter relations situated on a far broader and wholly impersonal terrain. Yet we nevertheless readily discern in them the same power of optics Sartre describes in the Look. Power relations might in fact at this level be described as the socio-political apotheosis of the Look. While we learn nothing from Foucault of how optics of power are driven by the power of optics, we learn a great deal about how power is set in play not by bodies alone, but set in play and kept in place by language.

To begin with, in *Discipline and Punish* Foucault shows us how punishment transposed into spectacle plays off the body. The body is the centerpiece, so to speak, of the corrective treatment. Even with the change from public torture to forced labor, for example, the body retains its focal place in that, while no longer punished by direct assault, it serves as an instrument upon which constraints and privations are applied. Thus Foucault writes that "in our societies, the systems of punishment are to be situated in a certain 'political economy' of the body: even if they do not make use of violent or bloody punishment, even when they use 'lenient' methods involving confinement or correction, it is always the body that is at issue—the body and its forces, their utility and their docility, their distribution and their submission."[31]

Docility is a major theme in Foucault's analytic of power relations. As he shows in his history of Western punitive techniques, the creation of docile bodies is, and has been, central to typical Western socio-political practices. Docile bodies of course do what they are told; hence, their utility. The means to establishing and maintaining docility is through some mode of permanent observation of the bodies in question. At Mettray, for instance, in the mid-1800s, the inmates of what Foucault describes as at once a cloister, prison, school, and regiment were observed day and night by those operating the institution.[32] The practice validates his earlier remark that "The perfect disciplinary apparatus [makes] it possible for a single gaze to see everything constantly."[33] The same unrelenting emphasis on surveillance is shown to have been present with respect to behavior ordained by the government in the event of plague at the close of the seventeenth century. Foucault writes that in the instance of plague, "Inspection functions ceaselessly. The gaze is alert everywhere . . . [to insure that] everyone [is] locked up in his cage, everyone [is] at his window, answering to his name and showing himself when asked—it is the great review of the living and the dead."[34] Moreover this same practice of visual isolation is practiced on the leper, another kind of "abnormal individual," Foucault notes.[35] In fact, "disciplinary partitioning"[36] is linked directly by Foucault to the Panopticon, Jeremy Bentham's eighteenth-century architectural prototype of all disciplinary socio-political institutions. Foucault's descriptive analysis of the Panopticon shows beyond doubt that docile bodies are produced by distinct forms of spatial constraint which not only physically limit but *visually* limit. Indeed, it is the visual rather than the strictly physical limitation that constrains and punishes. In Foucault's sharp words, "Visibility is a trap."[37]

Foucault's analysis of the Panopticon, like his analyses of disciplinary technologies in general, is an elaboration of how an optics of power operates and has operated in Western culture. It should not be surprising, then, that it consistently recalls Sartre's descriptive analyses of the Look: the Look instantiates the intuitively known but unarticulated power of vision itself. It is this same intuitively known power of vision itself that keeps the disciplinary technologies in place and allows their institutionalization to begin with. For example, Foucault writes that in the Panopticon, "each individual . . . is securely confined to a cell from which he is seen from the front by the supervisor; but the side walls prevent him from coming into contact with his companions. He is seen, but he does not see; he is the object of information, never a subject in communication."[38] In consequence of this architectural arrangement, the Panopticon is able to induce "a state of conscious and permanent visibility" in its inmates.[39] Foucault goes on to emphasize in particular the visibility and unverifiability of power in Bentham's hypothetical institution: as to visibility, "the inmate will constantly have before his eyes the tall outline of the central tower from which he is spied upon"; as to unverifiability, "the inmate must never know whether he is being looked at at any one moment; but he must be sure that he may always be so."[40] The closeness of this description to Sartre's statement that "through the Other's look I *live* myself as fixed in the midst of the world, as in danger, as irremediable"[41] is unmistakable. Clearly, Foucault's optics of power—like Sartre's the Look—is rooted in *the power of optics.*

It is in fact revealing to flesh out in more specific, i.e., Sartrean, terms what is denied in Bentham's prototypical disciplinary institution—in his 'inspection house', as Bentham actually termed the Panopticon.[42] In particular, Sartre's affirmation of the reciprocality of the Look provides a rich context for understanding Foucault's observation that "The Panopticon is a machine for dissociating the see/being seen dyad: in the peripheric ring, one is totally seen, without ever seeing; in the central tower, one sees everything without ever being seen."[43] Sartre affirms that "a subject who sees me may be substituted for the object seen by me." In other words, seeing another as an Other and being seen as Other by another are intrinsically related possibilities—or as Sartre more emphatically puts it, " 'Being-seen-by-the-Other' is the *truth* of 'seeing-the-Other'."[44] By the same token of reciprocality, "the Other is a qualified object for me [i.e., a subject] only to the extent that I can be one for him."[45] What panoptical institutions and practices destroy is precisely the possibility of reciprocality that these basic dyadic relationships engender. In virtue of an uncoupling of possibilities, a disruption of common potential reciprocalities, docile bodies are produced: on the one side, an absolute subject is created; on the other side, an absolute object is created. This pathological state of affairs highlights the extraordinary impact of a pure immutable visibility. It attests incontrovertibly to the fact that panoptical vision is vision already charged with power, that subtending an optics of power is a power of optics by which human life can be reduced on the one side to a pure gaze and on the other side to a pure spectacle. The infrastructure of the power relations Foucault describes is by this insight cast in the closer terms of actual intercorporeal experience. Specific facts about bodies begin in rudimentary, inchoate ways to come to light. But the troubling question remains, how does vision come to have this power? How is it that simply *being seen* by an Other can constrain and punish? The disciplinary technologies Foucault analyzes and discusses give us deep understandings of how power is visually deployed, yet even more urgently than with Sartre, further and finer corporeal analyses are needed to elucidate the formidable power of optics that consistently undergirds optics of power.

The need is perhaps most dramatically illustrated by a consideration of Foucault's account of sexuality where power relations are expressly shown to be set in play and held in place not directly by the body alone but by language. In the first chapter of his first volume on sexuality titled "The Incitement to Discourse," Foucault speaks of the transformation of sex into discourse about sex.[46] He goes on to show specifically in the book as a whole how "power's hold on sex is maintained through language."[47] The disciplinary technologies of Western sexuality, he says, have a "juridico-discursive character" precisely in virtue of their being structured in language. By this Foucault means that language and behavioral rules are intrinsically related; power controls sexuality "through the act of discourse that creates, from the very fact that it is articulated, a rule of law."[48] But while there is what might be termed *a power of discourse* operative here, there is here too both an optics of power and a power of optics that subtends it. Indeed, the discourse can only be formulated on the basis of the visible since it articulates a law that applies everywhere to everyone; it is in the public domain. In his trenchant analysis of a *scientia sexualis* in which the truth of sex is *told,* and in which his thesis of the intrinsic relation between power and knowledge is clearly spelled out, Foucault makes the relationship between discourse and visuality readily apparent. For example, he writes of the confession that it "still remains . . . the general standard governing the production of the true discourse on sex," and this because it is the model for those other practices in which it is a question not simply of telling "what was done—the sexual act—and how it was done; but of reconstructing, in and around the act, the thoughts that recapitulated it, the obsessions that accompanied it, the images, desires, modulations, and quality of the pleasure that animated it. For the first time no doubt, a society has taken upon itself to solicit and hear the imparting of individual pleasures."[49] In short, the transformation of sex into discourse about sex rests on "going public." Either one makes oneself visible, or one has before one—whether in plain print or in plain speech—all the details of another's sexual experience; either one creates or recreates one's own sexual scene as a display for others, or a scene is recreated as if happening before one's very eyes. This endless gathering of information about sex is part of the optics of power that controls sexuality not simply with respect to the act itself, but with respect to desire and pleasure—what will be done, with whom, what intensity the experience will have, and so on. A critical and extended exposition of Foucault's and Sartre's accounts of desire will bring to the fore the power of optics that subtends this *scientia sexualis*. It will show how corporeal analyses can flesh out the power of optics by providing insights into the sense ratios that empower it, that is, that give vision power to begin with.

IV. A Corporeal Fleshing Out of Foucault's and Sartre's Accounts of Desire

A. Foucault's Scientia Sexualis *and* Ars Erotica

Foucault speaks of two major sexual traditions, a *scientia sexualis* and an *ars erotica,* and points out two major differences between them.[50] The first difference centers on a distinction between "sex-desire" and "bodies and pleasures"; the second on a distinction between the truth of pleasure transmogrified by disciplinary technologies, i.e., by the production of discourse about sex, and the truth of pleasure experienced first-hand as pleasure itself. At its core, each distinction attests to an artifactual schism at the heart of corporeal life; each attests to a separation of a purely visual body from a whole-bodied existence. While Foucault describes aspects of the distinction in both cases, he acknowledges the schism in neither. This is because Foucault's identification of the body as the target of disciplinary technologies and his repeated emphasis upon its intensification center consistently on an abstract, essentially *undefined* and *undescribed* body. In consequence, Foucault is never driven to a consideration of the concrete corporeal distinctions that subtend the more abstract ones he makes between a *scientia sexualis* and an *ars erotica.* An *ars erotica* is in fact an *ars* of the *felt body;* a *scientia sexualis* is in fact a science of the *voyeur* in the double sense of a body put on display for others, and both self- and other-examined in never-ending rituals of observation. Foucault nowhere hints at an identification of, and distinction between, felt and visual bodies. He nowhere intimates the possibility of plumbing the corporeal depths of the distinctively visual body at the center of a *scientia sexualis,* inquiring into its origins or into the basis of its cultural refinements and elaborations. As for the distinctively felt body at the center of an *ars erotica,* it remains equally unidentified and unanalyzed by Foucault—perhaps precisely because it arises originally outside of historical investitures of power, indeed, judging from Foucault's descriptions, perhaps because it seems to arise at the level of the body *simpliciter*. Granted that an *ars erotica*—and thus the felt body and the original truth of pleasure—falls outside of Foucault's essentially Western socio-political concerns (Foucault associates an *ars erotica* with "numerous" societies: "China, Japan, India, Rome, the Arabo-Moslem societies"),[51] what is on display, observed, and told in the one instance, and what is treasured for itself in the other instance are nevertheless clearly distinguishable on corporeal grounds. That Foucault's focus is on socio-political practices and institutions—on disciplinary technologies rather than on interpersonal relations, or in more pointedly contrastive terms, *on behavior rather than experience*—is not sufficient reason for hewing to an always abstract body. Corporeal matters of fact subtend socio-political practices, particularly where bodies themselves are the target of the practices. Specific consideration of what Foucault calls "sex-desire" will demonstrate the essential relevance of corporeal matters of fact to his account of power and power relations.

When Foucault writes that "The rallying point for the counterattack against the deployment of sexuality ought not to be sex-desire, but bodies and pleasures,"[52] he only implicitly urges a turning away from the merely visual toward the felt; "bodies and pleasures" resound sweetly but remain merely vague "counter-products" to sex-desire. By sex-desire, it should be noted, Foucault means not merely the desire to have sex, but the desire "to have access to it, to discover it, to liberate it, to articulate it in discourse, to formulate it in truth."[53] Desire for sex is thus linked to juridical and discursive powers which at the same time they expose sex-desire, they contain it; while they impose on it certain restrictions, they search it out and illuminate it in finer and finer detail. In other words, disciplinary technologies which are everywhere on the trail of sex and which attempt to fix it on a juridical-discursive map—all the way from birth to death—are in Foucault's account part of the very practice they relentlessly pursue and investigate. Sex-desire is a socio-political industry saturated in complex power relationships.

However it is had, scrutinized, contained, exposed, documented—even as a confessional—sex-desire is clearly at its core the visual fastening of one body upon another, including the visual fastening of one's own body upon itself. It is an emphatically observed, second-hand sexuality. A pleasure so radically tied to the visual is no longer a pleasure of the felt body; it no longer resonates with an erotic corporeality. Being appropriated to an optics of power, it is celebrated as something shown, discovered, displayed, ogled, and so on. Foucault notes briefly the pleasure of "showing off, scandalizing, or resisting," but does not probe the particular ratio among the senses that founds such bodily acts. Equally, in invoking bodies and pleasure over against sex-desire, Foucault invokes but does not examine a different kind of body and a different kind of pleasure from that which the power of discourse has vitiated. What he overlooks in his otherwise compelling analysis is any indication both of what desire is prior to its corruption into sex-desire, and of what the nature of the corporeal visual matrix is which anchors sex-desire.

Now if the rallying point for a counterattack against the deployment of sexuality ought to be bodies and pleasures as Foucault claims, then precisely desire within what he pictures to be the uncorrupted domain of an *ars erotica* needs to be uncovered and spelled out. Clearly there is an "un-denatured" desire that is present in an *ars erotica* and that needs explicit exposition. Indeed, only in this way can the optics of power that produces sex-desire and that transforms it into discourse be fully understood. A power of optics, after all, also generates an *ars erotica*. The intensification of the body of which Foucault speaks is in actuality an intensification of the body's natural power of optics, a power rooted in its essentially inescapable visibility, its "seen-ness." Thus the power of optics that fuels sex-desire does not arise *ex nihilo* with sex-desire itself but rises to the juridico-discursive occasion of a *scientia sexualis* inflated with overpowering importance. To trace its radical transformation will give us insight into the corporeal infrastructure of the optics of power that constitutes a *scientia sexualis*. In effect, an analysis of *erotica-desire* is not idle filler with respect to Foucault's socio-political analysis of sexuality but a critical necessity. If un-denatured desire is the foundation of those ultimately corrupted sexual practices Foucault finds everywhere rampant in Western society, then an understanding of an un-denatured power of optics is foundational to their understanding.

Sartre's account of desire offers an excellent point of departure both for balancing Foucault's account and for grasping that original animate totality of which the abstracted visual body is but "an impoverished image."[54] More specifically, Sartre's descriptions provide a beginning framework for a deepening of Foucault's interpretive analytics through their implicit recognition of the sensory-kinetic dimensions of desire, that is, through their elucidation of the nature of desire as corporeally experienced. A brief look at these dimensions will show that in the separation and ascendancy of a visual body, a resonant corporeality is indeed eroded and compromised. It will also show that although Sartre renders the body concretely, as Foucault does not, Sartre's rendition also falls short of doing full justice to the body *simpliciter,* and this not because of his descriptions which are never less than perspicuous, but because of a blindered sense of the body, the result of the ontological prism through which he sees it.

B. Sartre's "Desire-Proper"

In its initial awakenings, sexual desire is typically a preeminently visual phenomenon, a longing after something seen. Sartre's descriptions of desire bear out this yearning attraction to something visible; desire is born in *a look,* but a look substantively different from those that he has described earlier and that we have already considered. In fact, Sartre, by carefully distinguishing an abstracted visual body from the concretely resonant body of the Other, is describing something patently other than an observed sexuality. Desire, says Sartre, "is addressed not to a sum of physiological elements but to a total form"[55]; "the arm or the uncovered breast [is desired] only on the ground of the presence of the whole body as an organic totality."[56] Desire here, while visually tethered, is clearly not oriented toward a merely visual thing. Moreover it is not oriented toward a certain act or satisfaction. Desire is not "a desire of *doing*," says Sartre;[57] though the act of intercourse ordinarily terminates desire, "coitus . . . is not its essential goal."[58] The goal of desire is rather a matter of flesh: to incarnate the Other and to incarnate oneself; to make the Other live his/her body as flesh and equally to live one's own body as flesh.

Desire, as Sartre describes it, is clearly not sex-desire, either in the sense of a desire "to have sex"[59] or in the extended Foucaultian sense to produce discourse about sex. For Sartre, there is no such phenomenon: what is desired is always a transcendent object, not the having of a certain state of affairs nor the having of certain pleasures or sexual satisfactions, whether through some discourse about sex or not. This is why Sartre can say that "desire is expressed by the caress."[60] It is Other-oriented, and its thematic is the feel of flesh by and against flesh. Moreover even here, in the caress, it is not a question of partial bodies. What characterizes amorous gestures as desire is not "taking hold of a part of the Other's body but placing one's own body against the Other's body."[61] As Sartre indicates, if it were otherwise, desire might be expressed by an erection, or by areolar tumescence.[62] The contrast is not only apt but quintessentially telling. Where desire is so expressed, then clearly sex has already been transformed by, and into, discourse about sex: tumescences here are not felt but provoked, measured, displayed, compared, and so on, all in the name of the truth of sex. Though confounded by a cumbersome ontology—a duality of con-

sciousness and body, and a consciousness *"making itself body"* in desire[63]—
Sartre's distinction between body parts and their behaviors on the one hand,
and the erotics of flesh on flesh on the other, allows us a decidedly rich if
beginning glimpse of the corporeal nature of the distinction between
sex-desire and erotica-desire.

Still, Sartre's encumbering ontology cannot be summarily discounted.
A consciousness that *"makes itself* body" suggests either an abdication or a
transformation. In either case, there is the question of how consciousness, in
a reversal of some kind, can make itself consciousness again after dissolving
into flesh; or can a mere body *"make itself* consciousness"? Clearly, in
Sartre's ontology consciousness and flesh live essentially either at two
different addresses or, if at the same address, share the same facilities with a
degree of discomfort. There is moreover no doubt as to which of the two is
the more elite—or in control. Indeed, from the perspective of Sartre's
ontology, to enjoy the pleasures of the flesh is akin to a temporary loss of
autonomy. It is in fact in his words "to be swallowed up in the body,"[64] to
be " 'absorbed by my body as ink is by a blotter'."[65] In effect, however
trenchant his analysis in terms of concrete sensory-kinetic bodily experiences, by his ontological orderings, his consciousness which makes itself
body is akin to a hand that dips into a box of chocolates: it acknowledges
and yields momentarily to the joys of the flesh, but then immediately and
dutifully retracts itself, reassuming control and behaving in ways proper to
its stature in the world.

Insofar as the desire Sartre describes clearly coincides with desire prior to
its corruption into sex-desire, it is as appropriately called *desire-proper* as
erotica-desire. Its goal is grounded not in power relations but in purely
sensual ones. Its aim, in Sartre's words, is "to be revealed as flesh by means
of and for another flesh."[66] This is why, according to Sartre, "desire is
expressed by the caress." Nothing in the discourse of sex-desire remotely
suggests a caress. There is nothing *tactile* about sex-desire—except of course
in the once-removed-from-life terms of cutaneous stimulation, and understandably so since cutaneous stimulation, like sex-desire, stays on the surface
of things like a look. A caress, on the contrary, is a tactile-kinesthetic gesture
which knows nothing of visuality; it is attentively consumed in the feel of
flesh on flesh. Whatever might be the gestures in sex-desire, they are part of
the *regime* dedicated to enacting and pursuing a *scientia sexualis*, the truth of
sex. Foucault at one point questions whether this regime is not perhaps "an
extraordinarily subtle form of *ars erotica*,"[67] the pleasures of telling sex being
"something like the errant fragments of an erotic art."[68] He cites as
evidence, "the learned volumes, written and read; the consultations and
examinations; . . . the delights of having one's words interpreted; all the

stories told to oneself and to others, so much curiosity, . . . the profusion of secret fantasies and the dearly paid right to whisper them to whoever is able to hear them, . . ." and so on.[69] But he does not seem to take the question seriously, for he proceeds to underscore once more the emphatically visual nature of sex-desire when he writes that a *scientia sexualis* is "a process that spreads [sex] over the surface of things and bodies, arouses it, draws it out and bids it speak, implants it in reality and enjoins it to tell the truth."[70]

Though not described in the finer corporeal terms of tactility I have indicated, Sartre's descriptions show that in desire-proper, what begins in vision passes over into touch and is fulfilled by touch: desire is expressed by the caress. In an *ars erotica,* the visual is thus transcended. In effect, what fundamentally distinguishes sex-desire from desire-proper is a shift in the ratio among the senses. Earlier remarks about the difference between tumescences as incarnations of desire and caresses as incarnations of desire corroborate this shift from an incisively visual matrix to a tactile one.[71] Elucidating facets of this shift will enable us to see how Sartre's analyses set Foucault's in relief. In particular, we will see how the power of optics that undergirds a *scientia sexualis* refuses either to be deposed by touch or to make room for touch by accommodating it into its structure, and how in consequence the natural power of optics is denatured. Now swollen with singular importance, it gives rise to that discourse of the visible called "sex-desire."

V.   Desire and the Ratio among the Senses

To begin with, the natural power of optics in desire is akin to the natural power of optics in both the everyday and artistic worlds where things seen can draw us to them simply on the basis of their form, their colors, their movement. Our attraction to a painting, for example, and our longing to see the ocean are not desires for mere *sights.* In drawing us near, painting and ocean alike invite us to touch them and/or to be touched by them in more complex and deeper ways than simple seeing. This tactile power of the visual explains why the eminent art critic and historian Bernard Berenson speaks of the *tactile values* of painting.[72] A similar visual embodiment of tactile values is found in our attraction to a rose garden or to a bountiful buffet. Aristotle long ago suggested that smell was like taste and taste was "a sort of touch."[73] The natural power of optics is thus not strictly a *visual* lure; whatever attracts us visually or whatever we long to see is not a purely visual datum but something that encompasses or spills over into other sense modalities, most specifically, *touch.*[74] That desire is expressed by the caress is in effect not an anomaly with respect to the natural power of optics. A longing for the Other is clearly not a longing after a merely visual body. In terms of the ratio among the senses, what is initially a preeminence of the visual gives way to a preeminence of the tactile. As Sartre might say, initially borrowing an image from Foucault, desire-proper does not spread sex over the surface of things and bodies; it incarnates them.

A second facet of the shift in the ratio among the senses highlights a near uncanny presentiment by Sartre of Foucault's work. In elucidating the nature of the object of desire and of that object's radical transformation once "we have learned that the sexual act suppresses desire," Sartre writes that "the average man through mental sluggishness and desire to conform can conceive of no other goal for his desire than ejaculation."[75] In other words, by adding on "a little bit of knowledge" to desire, Sartre says, and further, by adding on the idea of "pleasure as desire's normal satisfaction," men come to believe that ejaculation is the due and proper end of desire. Moreover Sartre goes on to say that "This is what has allowed people to conceive of desire as an instinct whose origin and end are strictly physiological since in man, for example, it would have as its cause the erection and as its final limit the ejaculation."[76] Here in essence is both sex-desire and a *scientia sexualis*. Foucault's thematic of the power/knowledge relationships that generate and maintain the great Western truth of sex is adumbrated in the travestied object of desire Sartre describes. Foucault would not of course agree that sluggishness and the desire to conform explain such a conception of desire, but he would likely agree that the "orientation" of such attitudes is linked to that historically conditioned socio-political economy of bodies and sex that has transformed sex into discourse about sex, or in brief, that "adding a little bit of knowledge" goes a long way toward instantiating certain laws of sex. Certainly where the object of desire is a matter of body parts and their behaviors, or where it is a matter simply of one's own pleasure, desire falls short of the "reciprocity of incarnation" that is its goal.[77] In such a situation, what is missing is the *incarnate* body of the Other. On the one hand, like one's own body, the body of the Other exists only as so many body parts behaving in so many ways. On the other hand, where the pleasure of caressing dissolves into the pure pleasure of being caressed, desire evaporates: there is no longer an Other who is desired.[78] Sartre's description of "the object to which desire *is addressed*," though sounding oddly like postal directions, clearly emphasizes not only a whole-bodied existence over a visual body, but a reciprocity of being in the flesh: the object to which desire is addressed is "a living body as an organic totality in situation with consciousness at the horizon."[79] It is not that touch is actually missing, then, in the travesty of desire Sartre describes, but that it exists in the degenerate form of cutaneous stimulation and thereby cannot but fail to incarnate. At the same time, a preoccupation with body parts and their behaviors aligns desire with visual desiderata that transform sex into discourse about sex.

A third facet of the shift in the ratio among the senses is an elaboration of an aspect of the second, namely, the nature of a caress which "does not want simple *contact*."[80] Sartre's point here is both to show that the meaning of the caress is not a *superficial* matter, and to draw a quite subtle distinction: "if caresses were only a stroking or brushing of the surface, there could be no relation between them and the powerful desire which they claim to fulfill; they would remain on the surface like looks and could not *appropriate* the Other for me."[81] There are two related ways of fleshing out the distinction that warrant mention in this context. First, as Sartre expressly puts it, "the caress is not a simple stroking; it is a *shaping*."[82] A caress is thus a creative act and not a bland if particular kind of tactile gesture. It is creative insofar as it incarnates the Other; it causes the Other to become flesh, and equally, it causes the caresser to become flesh: "The caress by *realizing* the Other's incarnation reveals to me my own incarnation."[83] As a creative act, a caress thus clearly has power. The natural power of optics that opens desire is transformed through the caress into *a power of the tangible*. The change in the ratio among the senses is in consequence not simply a shift from one sensory modality to another, a visual preeminence giving way to a tactual preeminence, but a modification of possibilities. Indeed, insofar as the caress which "does not want simple *contact*" has power to awaken corporeally, it can, like the Look, make things happen.

What it makes happen has to do with the second way of fleshing out the distinction. When the body is experienced as flesh and not as instrument, the world it inhabits correlatively becomes a world of flesh. Objects that it contacts matter *as matter:* they are soft, gritty, hot, jagged, and so on, rather than things to use or means to such-and-such ends. Thus with desire-proper comes a world different from the everyday one in which things have primarily a utilitarian or instrumental value. The change in valuation is in fact a change in the way objects, including the privileged objects which are Others, are experienced. As Sartre puts this change in valuation with respect to everyday objects, "I discover something like a *flesh* of objects. My shirt rubs against my skin, and I feel it. What is ordinarily for me an object most remote becomes the immediately sensible; the warmth of air, the breath of the wind, the rays of sunshine, etc.; all are present to me in a certain way, . . . revealing my flesh by means of their flesh."[84] The change in valuation with respect to Others is most succinctly illustrated by juxtaposing the inflated power of optics that operates on the surface of things in sex-desire and the power of the caress to cast consciousness "at the horizon" of flesh—or in basically corporeal rather than ontological terms, the power of the caress to awaken the deeply sentient and affective tactile-kinesthetic body. In particular, in sex-desire, the power of optics is unsoftened by the power of reflection, the latter not only in the sense that one blindly follows the rules—the *scientia sexualis* goes unquestioned in a way recalling attitudes of "sluggishness" and "a desire to conform"—but that the object of one's desire remains a bare object—no reciprocal flesh is called forth. Lacking reflective illumination through the power of the caress, the Other as Other remains unrecognized and unknown. A mere body, an unilluminated Other, is present. Sartre has described this kind of bald apprehension of Others not in his analysis of desire, but earlier, in the context of his analysis of fundamental concrete relations with others. There he writes, "I brush against 'people' as I brush against a wall. . . . I do not even imagine that they can *look* at me. . . . [they] are functions: the ticket-collector is only the function of collecting tickets; the cafe waiter is nothing but the function of serving the patrons. In this capacity they will be most useful if I know their *keys* and those 'master-words' which can release their mechanisms."[85] Sartre terms this mode of being-for-others *indifference.* It is this kind of relation that closes the Other off to reflection and that precludes a consciousness at the horizon of flesh. Indeed, Sartre initially describes indifference in terms of *blindness,* a look that sees but registers nothing in the way of a living body. In indifference, the Other remains part of an instrumental complex completely subservient to one's needs; he/she is never engaged as flesh. In contrast, the caress which "does not want simple *contact*" is a gesture having the power of reflection through the reflexivity of touch. Given back through the reflexivity of touch is not just a body but a living body latent with the sentient possibilities of its own tactile-kinesthetic life.

In testifying in detail to the power of optics undergirding optics of power, the preceding corporeal analysis of desire has provided a preliminary answer to the question of how the body is accessible to power relations and pointed us in the direction of answering how, correlatively, "specific facts about the body" undergird disciplinary technologies. Within the natural ratio among the senses, the visual is already charged with power. This natural power of optics exists because Others are already on the scene, *not* in the relativist's sense that "the world is already there," but in the sense that from the moment we are born, Others dominate the landscape. To see is to see Others. We cannot in fact readily escape *seeing* Others; we can only readily escape acknowledging them and this through deformations of the visible in which indifference or other attitudes are taken up which reduce them to merely visual objects. Thus, at the same time that vision necessarily imposes relations with Others, the simple act of looking has the power to qualify those relations, to structure the nature and dynamics of what is a virtually inescapable intercorporeality. The Look is indeed a powerful force in human relations. A power of optics swollen in importance distorts the natural ratio among the senses. It has the power to reduce bodies to pure exteriorities. Disciplinary technologies thrive on such bodies.

2

An Evolutionary Genealogy

Man in his arrogance thinks himself a great work worthy the interposition of a deity. More humble and I think truer to consider him created from animals.

Charles Darwin[1]

I. Introduction

Corporeal analyses of power must uncover not only the fundamental modes in which power is articulated in the everyday human world; they must also take into account an evolutionary history. Modes of power derive from bodies and bodies are animate forms that have evolved. Accordingly, animate form, in the broad sense of a species-specific body with all its distinctive spatial conformations and attendant gestures, postures, orientational possibilities, modes of locomotion, and so on, must be reckoned with. Moreover its phylogenetic relationships must be considered, and this because whatever its fundamental modes of power and whatever their possible intercorporeal elaborations, those modes and elaborations also have an evolutionary history; they evolved along with the animate form itself—hence the necessity of considering phylogenetic similarities and dissimilarities in an inquiry into the roots of power.

An example will demonstrate concretely the nature of this evolutionary dimension. In particular, an extended examination of optics of power and the power of optics within the living context of a natural history will carry forward the corporeal insights of the last chapter by exposing the evolutionary ground in which an "intensification of the body" has its roots. Indeed, we shall see that optics of power and the power of optics that makes an emphatically visual body possible have a long evolutionary as well as cultural history.

II.   Optics of Power and the Power of Optics in Evolutionary Perspective

A.   *An Illustration from Natural History*

The point of departure is a popular but by no means wanting account of dominance and submission in macaque, baboon, and monkey societies generally, by writer Sarel Eimerl and primatologist Irven De Vore:

> There is no mistaking a dominant male macaque. These are superbly muscled monkeys. Their hair is sleek and carefully groomed, their walk calm, assured and majestic. They move in apparent disregard of the lesser monkeys who scatter at their approach. For to obstruct the path of a dominant male or even to venture, when unwelcome, too near to him is an act of defiance, and macaques learn young that such a challenge will draw a heavy punishment.
>
> . . . A dominant animal controls the space around it; a dominant baboon occupies the best site when a group is resting, and asserts an exclusive right to more space than its inferiors. It can invade an inferior's space as a right, whereas no inferior would dare to venture into its space without first making a gesture of appeasement.
>
> . . . Suppose that a dominant male [baboon] is annoyed by a squabble. Its first reaction will be to stare at the offenders. The stare is long and steady, with the animal's whole attention concentrated behind it. If the stare is not enough to quell the trouble, it pulls back the skin on the top of its scalp, drawing back its ears and opening its eyes wide. . . . If the facial threat is still not enough to impose order, the male stands erect, with its body tensed and the fur on its mane stiffened. A baboon may bark, take a few steps forward, slap the ground threateningly and take a few more steps. Finally, if it still feels defied, it will give chase.
>
> . . . On being threatened by a definitely dominant monkey, a subordinate is likely to display submission. Confronted with a fixed stare, it will look away. Faced with a possible charge, it is likely to crouch close to the ground, its head turned away. And if it flees and is chased, it will cringe away from the threatened bite or try to avoid punishment by presenting its hindquarters.[2]

Displays of dominance and of submission are living testimony of optics of power and of the power of optics. Though not described as such, they are recognized by primatologists, anthropologists, and ethologists as being both integral to many nonhuman animal behavioral repertoires and a common kind of behavior. As the etymology of the term suggests (*to display* is to unfold, to expand, to spread out), a display is primarily a visual phenomenon. Vocal displays are common to many animals of course, especially avian ones, but visual displays in primate species are frequently more complex, more detailed, and more numerous than vocal displays. For example, visual displays commonly alter the size and/or shape of an animal's body—as in the piloerection of a male baboon or chimpanzee, or in a male orangutan's inflation of his laryngeal pouches—or they center on a gesture or orientation of certain bodily parts—the genital rump area in nonhuman primate presenting, for instance, or the mouth area in the lip-rolling gesture of a gelada baboon—or they constitute a certain attitudinal bearing of the whole body—as in a dominant primate male's confident, unhurried walk. Displays are clearly an *intensification—in Foucault's unspecified but implicit sense*—of a naturally visible body. It is significant to point out that one primatologist in fact differentiates a signal from a display in the following way: "Some signals [stereotyped communicative acts] have been affected by natural selection, i.e., *ritualized,* so that the signal is exaggerated and incorporates several elements that make it more complex. We call such a signal a *display.*"[3]

Intensifications are not always in the service of dominant/submissive power relations. There are other forms of display—in courtship and certain greeting behaviors, for example. When intensification is in the service of aggressive power relations, however, as in the above illustrations, it is strikingly similar to those human intensifications Foucault describes in terms of "a micro-physics of power,"[4] that infrastructure of tensile force relations that dominates and has dominated Western human socio-political relationships. For instance, it can as easily be said of the nonhuman animal to whom the dominant one addresses its displays as of the human animal upon which disciplinary technologies operate, that "the power exercised on the body is conceived not as a property, but as a strategy, that its effects of domination are attributed not to 'appropriation', but to dispositions, manoeuvres, tactics, techniques, functionings; that one would decipher in it a network of relations, constantly in tension, in activity, rather than a privilege that one might possess; that one should take as its model a perpetual battle rather than a contract regulating a transaction or the conquest of a territory."[5]

In both dominance displays and disciplinary technologies, power is a setting in motion of a certain intercorporeal relationship. It is not a static proprietary stance of one individual toward another, but a dynamic ritual of intercorporeal behaviors repeated over and over again. Moreover when Foucault writes that "power is not exercised simply as an obligation or a prohibition on those who 'do not have it'; it invests them, is transmitted by them and through them, it exerts pressure upon them, just as they themselves, in their struggle against it, resist the grip it has on them,"[6] it is equally clear that the power of optics, insofar as it is an intensification of the naturally visible body, is perpetuated as much by the seer as by the seen. Indeed, from the perspective of evolutionary theory, the behavioral fixation of a display in a species' repertoire is contingent on meaning to the seer, the displayed-to animal. Staring, for example, until understood as a warning or a threat, and until consistently and predictively understood as such, would not otherwise "come under selection pressure as a signal"[7] and be inscribed in the behavioral repertoire of the displayer. The responding animal solidifies and perpetuates the power of optics by acknowledging and acquiescing to its intended meanings. Whether human or nonhuman, it becomes a docile body.

There are further similarities. Foucault's descriptions of a king by whose measure a crime is punished by public execution is remarkably parallel to descriptions of a dominant male whose sovereignty has been transgressed. In the first place, "over and above the crime that has placed the sovereign in contempt," Foucault writes, "[the public execution] deploys before all eyes an invincible force."[8] There is no public execution in nonhuman animal societies. In fact, there is relatively little physical contact in which fatal or near-fatal exchanges take place. But there are threats and appeasements, public avowals of transgression and repentance. There is unquestionably "an invincible force" which is "deployed before all eyes." That "there is no mistaking a dominant male macaque" makes the point unequivocally. Whatever the infringement on the domain of the male sovereign, it is similarly a matter dealt with not privately but publicly. The visual enactment of power draws attention to itself. Warnings and admonishments on the one side, and submissions and appeasements on the other, are there for all to see.

In the second place, the aim of the public execution according to Foucault is "not so much to reestablish a balance as to bring into play, at its extreme point, the dissymmetry between the subject who has dared to violate the law and the all-powerful sovereign who displays his strength."[9] Here too, the deployment of power in nonhuman primate societies is similar to that in human ones: the actions of the sovereign are not so much to keep subjects in line or to restore calm as they are to reassert inequities in status. In each case there are sovereign bodies for whom all other bodies are docile ones.

The parallel between nonhuman primate power relations and those power relations in Western human societies described by Foucault is indeed striking. The relations are woven of the same visual cloth; they are public; they are consolidated in ritualized behaviors; they are enacted intercorporeally; they are fixed in an individual's behavioral repertoire in virtue of a particular and consistently grasped meaning. Clearly there are *optics of power* in nonhuman as well as human primate societies. Just as clearly, those optics of power are nourished and supported by the *power of optics*. As emphasized and spelled out in the last chapter, power does not merely appropriate the visual as the sense through which it will be deployed. The visual is already charged with power. Though there are radical differences between nonhuman primate dominance displays and human disciplinary technologies, and radical differences as well in the mode and degrees of intensification—all of which differences need explicit elucidation precisely because they constitute *dis*-analogies—power relations are structured and enacted in analogous ways in human and nonhuman primate societies. We shall look at five major primate modes of intensification in turn in order to exemplify the quintessential relationship between primate vision and primate power, or in broader terms, how the roots of power are grounded in animate form.

B. The Look: Its Intensifications and Variations

The act of staring is a particularly rich example of how the visual is already charged with power, of how that natural power is intensified, and of how both the natural power of vision and its accentuation are rooted in a common primate heritage. In the Western world, staring is considered rude. Fixing one's gaze on another person is akin to pinning him or her to the wall. The power of such a gaze is an intensification of the natural capacity of the eyes to see and to behold, the natural power of sight being transformed into a look which now gapes, glares, leers, pierces, pries, bores into, and so on. Such acts transcend mere sight in the sense of carrying to an extreme a beginning interest, inquisitiveness, curiosity, or questioning. They are not simple noticings nor more aroused inquiries. They are acts willfully committed against another person. In Sartrean terms, they encroach upon the Other's freedom, robbing him/her of his/her possibilities. Hence, in like manner, they are an intensification of the seen. Eyes which pierce, gape, glare, bore into, and so on, transform the person seen into an object, a mere outside. As a particular expression of the Look, staring may even be devastating, as when the intention is not merely to make another uncomfortable or to keep him or her in line, but to do him/her in: casting an evil eye on someone has its intercorporeal or "socio-etymological" roots in staring, just as staring has its socio-etymological roots in the simple act of looking. I have elsewhere traced out this progression in intensifications of the power of vision.[10] Clearly, a fixed, overly inquisitive, or sinister gaze on someone is an aggressive act.

The following observations of staring in different primate societies document the extraordinary similarity in intensifications of the visual body between human staring and primate staring generally. In their descriptions, primatologists unambiguously if tacitly affirm that the power of the eyes to see others is transformed into *an intercorporeal power* to intimidate or to threaten. Thus K. R. L. Hall and Irven DeVore write that "Facial expressions observed in threatening animals [baboons] consist of 'staring,' sometimes accompanied by a quick jerking of the head down and then up, in the direction of the opponent."[11] Judith Shirek-Ellefson states that "Every species of Old World monkey and ape has at least two distinct threat patterns. One is some form of open-mouth threat, . . . the eyes stare at the opponent, the brows are either lowered or raised, . . . depending . . . on the species. . . . Open-mouth threats are given by the dominant member of an interacting pair."[12] Generalizing on the basis of her lengthy study of chimpanzee society, Jane van Lawick-Goodall writes that "Glaring . . . was frequently associated with other threat patterns; a fixed stare is a form of threat in other primate species and also in man."[13] John H. Kaufmann writes that "Threat [in rhesus monkey society] usually meant simply staring or running toward another monkey, often with the mouth open and with a threat call."[14] Donald Stone Sade states that "In its mildest form the charge [by a rhesus monkey] is expressed simply by staring at the victim. . . ."[15] J. A. R. A. M. van Hooff, well known for his comprehensive comparative studies of facial expressions in nonhuman primates, observes that "The Tense-mouth Face," "The Staring Open-mouth Face," and "The Staring Bared-teeth Face" are expressions common to numerous species of monkeys and to chimpanzees. He describes them as acts in which the eyes "are staring fixedly towards a partner" or "towards the opponent."[16] In the context of describing common primate acts, Eugene Linden succinctly points out that "Humans still display ape behaviors such as bowing, begging, and aggressive staring."[17]

The intensification of the natural power of optics is conclusively attested to not only by *the stare,* but by its opposite. A primate's typical response to staring is to avert the eyes. Primatologist Stuart Altmann, who has written extensively on primate social communication, has remarked that "Among primates—and doubtless among many other animals—facing and looking at the addressee is probably the most common means by which social messages are directed. . . . Thus, one interpretation of avoiding visual contact . . . is that it is a means of avoiding interactions." He goes on to say, "Not surprisingly, this behavior is usually given by the subordinate member of a pair, and its converse, direct staring, is usually a form of threat."[18] Similarly, Thelma Rowell, a specialist in monkey social behavior, writes that "'Looking away' can be itself a visual signal: since a direct stare at another's face is a threat for all monkeys, looking away indicates non-aggressiveness—either mild fear or conciliation—depending on whether the animal looking away is lower or higher ranking that [sic] the other." With respect to her subsequent parenthetical remark—"(Remember this when looking at caged monkeys and try not to stare directly at them. It is also true of our own species, which is why children are taught that 'staring is rude'—rudeness is a form of aggression)"[19]—the distinction that Rowell does not draw and that should be drawn is between the natural inquisitiveness of infants and children which impels them to look fixedly at others and the act of staring. Children are not "taught that 'staring is rude'"; they are taught to distinguish between their natural curiosity and an archetypally aggressive act.

Clearly, there is a relationship in primate societies—human as well as nonhuman ones—between visual contact and dominant/submissive interactions. Scales listing increasingly aggressive and increasingly submissive gestures in North Indian langurs document the relationship further. They show the former gestures to begin with "staring," the latter to begin with "avoid visual contact."[20] The same relationship is noted in the behavior of rhesus monkeys. A table of the "Components of Fights" in rhesus society shows that *attack* begins with a stare and *flight* begins with a "glance away."[21] An understanding of the power of optics is indeed enhanced by a consideration of the sequential components of a rhesus monkey's act—really *acts*—of submission. The monkey first glances away, then cowers; then it moves aside and/or hops aside and then flees, or instead, it presents and then flees, or rather than doing either, it flees directly after cowering.[22] Thus, submission is not simply a matter of averting the eyes but *a whole body gesture:* the reaction to staring is not simply a looking away, but a more generalized corporeal attitude of drawing back and shrinking. Definitive action—moving aside/hopping aside, presenting, or outright fleeing—is taken consequent to this whole body gesture. This means that the power grip of a stare is not broken simply by refusing to meet it, by ducking it visually, so to speak. Otherwise stated, merely averting the eyes will not do because aversion by itself only acknowledges the visual intrusion; it does not subdue it. To break the stare, one must abase oneself bodily in some way, and only then ultimately remove oneself in some equally whole-bodied way from the situation—by moving aside, for example, by presenting, and/or by fleeing.

Presenting is of course a standard nonhuman primate way of both acknowledging the threats of another and calming them. It is a common gesture of submission. In Gombe Stream chimpanzees, for example, Jane van Lawick-Goodall observed presenting to be "one of the most frequently observed elements in submissive behavior patterns." Furthermore, she writes that "After a particularly severe attack the subordinate . . . may back toward the aggressor from a distance of eight feet or more, looking over its shoulder. It then adopts the 'extreme crouch', with all its limbs completely flexed and its head almost on the ground."[23] The backward orientation of the body, together with the radical lowering of the body and particularly the radical lowering of the head, is a thoroughly self-abasing act. By itself, the backward orientation may in fact be viewed as an enhancement or intensification of the act of cowering. Although the submissive animal does not actually remove itself bodily from the situation, it effectively breaks the stare by figuratively removing itself. In virtue of its backward orientation, the fixed gaze can no longer connect with its target; the target itself has virtually disappeared in that the submissive animal has subverted and denied the socially strongest surface of its body. In consequence, for the aggressor, not just the eyes of the victim but the victim itself cannot be conclusively reached. Other whole body behaviors such as moving aside and outright fleeing also mute social presence, but in comparison with presenting, they are less acts directed *toward* the aggressor than acts directed away from the aggressor—hence they are less self-abjuring solutions to a threatening stare. Indeed, compared to presenting, they fall far short of being the fine and unequivocal act of a subservient, docile body.

A further aspect of presenting behavior is worth noting in this context. In both less and more severe aggressive interactions, an animal who presents also frequently looks over its shoulder at the animal to whom it presents. In view of the fact that "presenting [in nonsexual contexts] . . . is often accompanied by nervous, even fearful, behavior on the part of the presenting animal,"[24] the head-turn gesture is not surprising. As suggested above, a presenting posture is a vulnerable one; the presenting animal is unable to see easily what the other's reaction is or what the other is doing, and this visual inability, coupled with the fact that the animal is exposing the least defendable part of its body to the other, makes the situation a considerably dangerous one.[25] But the optical significance of the head-turn gesture in conjunction with presenting extends further. The presenting posture-*cum*-head-turn is a spatio-temporal miniature of panoptical life. Here too the observed is an object for an Other, its submissive gestures part of a disciplinary technology which, through rituals of surveillance, instantiates a situation of indeterminacy. The look over the shoulder is like a glance at the tower. One does not know how, or if, one is being monitored, but one cannot help looking in the direction of the Other's eyes to see if one is, and what the situation might be. At the same time, the look, like the glance, intensifies submissiveness because it implicitly acknowledges being trapped and held in a certain *exposed* position by a commanding Other.[26]

From an optical viewpoint, it is interesting to compare a presenting posture-*cum*-head-turn with what van Hooff describes as "The Frowning Bared-teeth Scream Face," a face common to many species of primates. In this facial expression, van Hooff writes, *"The eyes* are closed or opened only to a small degree. When not closed the eyes are never directed straight towards the opponent; the animal looks away, and often moreover 'faces away'."* The circumstances in which this face is observed is an agonistic one:

> It is shown, without exception, by the subordinate animal involved in the encounter. It occurs as a reaction to an imminent or actual attack and, in the performer, is frequently alternating [sic] with flight. The attacker will stop its action fairly soon in most cases and turn away. It now depends on the beaten animal, whether the agonistic encounter continues. If the next movements of the subordinate are very conspicuous (for instance, if it starts running away immediately after the dominant has turned its back towards it) this may release a new attack or pursuit.[27]

The facial expression and attendant behaviors van Hooff describes highlight the complexity of intercorporeal life with respect to docile bodies. The dynamics of an optics of power as of an intensified power of optics can take many forms. By not only averting the eyes, but by virtually shutting out vision entirely by closing them and "facing away"—by not looking in a *double* sense—a subordinate animal can hope to begin defusing an attack if not to end one. But the timing of its subsequent movements are critical with respect to the dominant animal's field of awareness. Should the dominant animal *sense* the subordinate running away—one infers from the above description that the dominant animal does not actually witness the fleeing action of the subordinate—a new attack will be generated. Clearly what one does with one's own eyes and what one does in the eyes of another, both figuratively and literally, are key constituents of the intercorporeal behaviors that define power relations. Those behaviors are, of course, species-overlapping as well as species-specific. It is our own human experience that allows us to understand so readily the optics of power and the power of optics in nonhuman primate societies.

Taken together, all of the above examples leave no doubt but that a natural power of optics—a capacity to see, in particular, a capacity to see others and observe what they are doing, together with its corollary, the possibility of being seen by others—can be and is intensified in primate life in multiple intercorporeal ways, ways defined by certain common denominators. Thus, staring is neither a cultural invention nor a peculiarly human act. Neither is the act of averting the eyes in answer to a stare. Both acts are built-in possibilities of being human. Both are quintessentially primate possibilities. The possibilities are differentially expressed in human cultures in the same way they are differentially expressed in nonhuman primate species. There are variations on themes, but the themes themselves remain intact. Averting the eyes, for example, is the accepted mode of social discourse in Kaluli (Papuan New Guinea) society. In this society, gazing into the eyes of another in the context of social discourse is considered intrusive.[28] Even the mere act of seeing can be taboo, as in Wik-mungkan society where there is a restriction against seeing a man, though not a woman, defecate.[29] The general human proscription against looking in the direction of, or specifically at, other people's genitals, even in cultures in which people are nude, is in fact an excellent example of a *pan*-cultural recognition of the archetypal power of vision to intrude, intimidate, or threaten. Rather than being suppressed on the grounds of its intrusive or threatening powers, the archetypal power of vision may in contrast be dramatically exaggerated; the Look amply testifies to this fact. So also, of course, does the evil eye, a belief in which, it should be noted, is extensive throughout Indo-European and Semitic cultures; an evil eye can strike and enter another person with destructive force.[30] Clearly, what one does with one's eyes is significant, and this both because eyes in and of themselves have power and because the intercorporeality of vision is an inescapable fact of intercorporeal life.

In sum, at the root of optics of power is a power of optics which makes intensifications of the visible body possible, intensifications that are structured in corporeal archetypes, primate invariants, thus *pan-hominid invariants.* Just as van Lawick-Goodall's observation that threat gestures are common to human and nonhuman primates obliquely attests to these invariants—as does Rowell's parenthetical remark cited earlier—so also if primatologically extended would Sartre's observation that "My original fall is the existence of the Other" or that "L'enfer, c'est les Autres"[31] and Foucault's remark that "a body is docile that may be subjected, used, transformed and improved. . . . it [is] a question of exercising upon it a subtle coercion, of obtaining holds upon it at the level of the mechanism itself—movements, gestures, attitudes, . . . an infinitesimal power over the active body."[32]

C.   *Invariants of Animate Form*

Other fundamental aspects of optics of power derive from invariants at the heart of the power of optics. These invariants are implicit in some of the above descriptions of dominant and submissive nonhuman animal behaviors. This is because what is being described in many instances are basic features or dynamic possibilities of animate form such as body size, posture, attitude, orientation, movement, and/or gesture. These basic features and dynamic possibilities articulate power relations no less definitively and forcibly than staring and averting the eyes. Like the latter explicitly optical gestures, they are an intensification of the natural power of optics. While differentially expressed, they too are thematically invariant.

It is important to point out that the omission of these features and dynamic possibilities of animate form in present-day cultural analyses of power results in critical oversights. Lacking anchorage in animate form, present-day accounts of power assume power and power relations to be both a peculiarly human idea—power is something a human mind conceives (whether at the generalized bio-socio-political level of Foucault or at the individualized 'for-itself' level of Sartre) and a human body executes—and a peculiarly human problem—an epiphenomenon of culture. In consequence, the accounts nowhere come to grips with the question of the genesis of power. They necessarily bypass inquiries into the origin of the concept of power and into the origin of concepts about how power is and may be wielded. Moreover where a near-exclusive concern with relative aspects of culture prevents a hominid evolutionary heritage from coming into view and with it those primate pan-cultural invariants that bespeak commonalities rather than differences, basic phylogenetic relationships are ignored as are those corporeal facts leading to understandings of how power relationships are built into living bodies, built into the very fabric of bodily life. Indeed, since a living body is undeniably the site of power relations, both its form and its history must be elucidated. In sum, the essential centrality of animate form to concepts and expressions of power must be acknowledged if present-day accounts of power are to be righted.

Let us begin by considering a basic aspect of animate form that plays not only a well-known but decisive role in power relations—*size*. In keeping with an evolutionary perspective, we will first consider a description of the value of size from the viewpoint of chimpanzee society:

> [The] habit of making the body look deceptively large and heavy is characteristic of the alpha male. . . . The fact of being in a position of power makes a male physically impressive, hence the assumption that he occupies the position which fits his appearance. The impression of a connection between physical size and social rank is further strengthened by a special form of behaviour which is the most reliable indicator of the social order, both in the natural habitat and in [captivity]: the *submissive greeting*. Strictly speaking, a "greeting" is no more than a sequence of short, panting grunts

known as pant-grunting or rapid-ohoh. While he utters such sounds the subordinate assumes a position whereby he looks up at the individual he is greeting. In most cases he makes a series of deep bows which are repeated so quickly one after the other that this action is known as bobbing. Sometimes the "greeters" bring objects with them (a leaf, a stick), stretch out a hand to their superior or kiss his feet, neck or chest. The dominant chimpanzee reacts to this "greeting" by stretching himself up to a greater height and making his hair stand on end. The result is a marked contrast between the two apes, even if they are in reality the same size. The one almost grovels in the dust, the other regally receives the "greeting." Among adult males this giant/dwarf relationship can be still further accentuated by histrionics such as the dominant ape stepping or leaping over the "greeter." . . . At the same time the submissive ape ducks and puts his arms up to protect his head. This kind of stuntwork is less common in relation to female "greeters." The female usually presents her backside to the dominant ape to be inspected and sniffed.[33]

The last comment regarding female 'greeters' aside for the present, this testimony to the value of size hardly needs comment, except perhaps to note that *size* has been consistently recognized as a central factor in the social relations of animals from the time Darwin called attention to its significance in natural and sexual selection.[34] Many animals, by making themselves larger than they actually are, are able to ward off predators—as, for example, the common European toad which inflates its body and stands on tip-toe when it meets a snake—or to gain a mate—as, for example, the male sea-elephant whose nose, being potentially both a weapon for fighting other males and a display for attracting females, "is greatly elongated during the breeding-season, and can then be erected."[35] The significance and value of size are clearly written into the social relations of many vertebrates. Humans themselves are by no means unaware of the potential advantage either of their given size or of giving the illusion of being larger than they actually are. Fashions document this fact. So also do the social problems of growing children, adolescents, and even adults. What warrants extended comment is a particular spatial aspect in the social life of at least some primates that is related to the significance of size. This aspect is implicit in the above account of chimpanzee greeting behavior. With respect to power relations, size is not merely a measure of comparative bulk and/or height; it is central to a *relational* invariant in the spatial semantics of intercorporeal life, an invariant carrying for each individual a specific spatial valence depending upon the particular *position* each individual holds with reference to the other. This spatial invariant is exactingly depicted in the following description of a matrix behavioral code among the Tikopia (Melanesian peoples):

In the vertical plane degree of elevation is a very important Tikopia status index. In Tikopia language there is a direct correlation between physical elevation and social elevation. The term *mau runga* indicates either higher above the ground or superior socially or both. Consequently, in bodily posture in an immediate personal context standing is ordinarily superior to sitting, squatting, crouching or kneeling. Standing children are continually told "Sit down" in the presence of adults who are sitting. Sitting is superior to crouching, squatting, kneeling or crawling on hands and knees. When a chief is seated in a house people will crawl over the floor in his vicinity. When it is necessary for a person to stand up in order to take down some fishing-line or a bowl stored in the rafters, an apology is made to the man of rank or head of the house if he is sitting near by. In Tikopia the sitting and kneeling postures of men and women differ considerably . . . senior men, but not women, sometimes use coconut-grating stools or other objects as seats which raise them off the ground. On public occasions such seats are reserved for chiefs.[36]

Again, we will put aside gender distinctions for the present and consider the spatial invariant itself. Raymond Firth, the anthropologist who studied Tikopia culture and wrote the above description as part of his study of the postural and gestural ways of the people in that culture, compared his foreign findings with his own British experience. He mentions situations in which degree of elevation is not a mark of status in present-day British society, but rather an acknowledgment or salutation, as when one briefly inclines one's head toward another in greeting. But he notes too that degree of elevation *can* mark a status differential in British society—as when one *bows* on formal occasions. *Kneeling* may also serve as a power index on ritual occasions and in ceremonials as when dignitaries of a church kneel in homage before their seated sovereign or when a man is knighted by his sovereign.[37] Moreover as Firth notes, "Further abasement of the body is still possible," i.e., one can *prostrate* oneself in some way before another.[38] He points out first that bending over and touching the feet of a superior is a mode of salutation practiced in some Asian societies, and then remarks that although lowering the head in this extreme way is not a British practice, prostrating oneself before another is not "a completely unfamiliar procedure in Britain." He gives as example its use in Catholic ordination rites.[39] All the same, Firth affirms that in Western societies degree of elevation plays a highly restricted social role; it is maintained primarily as a formal rather than everyday measure of status.

Firth also speaks of outright differences between Tikopia and British cultures—for example, of the fact that an act of apology in Britain involves "a lowering of the *idea* of the self, not of the *physical* self . . . it is a verbal process, not as it is in Tikopia a bodily process."[40] He describes how in Tikopia culture, an apology is at the nether end of a series of possible spatial relationships with others. To grasp the distinctive intercorporeal relationships underlying this series, it is necessary first to spell out how the *nose* rather than the hands (or lips, or language) is the organ by which respect is shown. In Tikopia greetings between equals, noses are pressed together with no change in elevation on either side; in greetings between a senior and a junior individual, the nose of the latter is pressed to the wrist of the former at the same time that the latter genuflexes or kneels; on certain formal occasions with respect to a commoner and his chief, or in the context of marriage when the bridegroom's kin apologize to the bride's kin for taking away a woman of their family, the person of lesser status, who is on all-fours, presses his/her nose to the knee of the person of greater status, the latter person being seated; in an acknowledgment of loyalty (or again, in the context of apology), noses are pressed, but in this instance, the one of inferior status is on all-fours before the other who is seated. Thus, in greeting and in apologizing, that is, in situations engendering a show of respect, the height of one's nose is all important; depending upon the interpersonal relationship and the situation, lowering oneself to a certain degree before another may be mandatory. The lower the degree of elevation, the lower one's status.

Firth's account clearly demonstrates how body position, conceived as a variable along a vertical axis of low to high, can be a measure of intercorporeal power relations no less than the given or amended gross size of naturally standing bodies. It is not of course just in Tikopia, or in Britain, or in chimpanzee society, that one's physical position relative to another plays a decisive role in supporting and even determining the particular power relationship obtaining between oneself and that other. What primatologist Frans de Waal calls a "giant/dwarf relationship" is not a rare but altogether commonplace corporeal expression of power in the animate world. A vanquished wolf, for example, may either twist and lower its head, offering its neck to the victor who is above it, or it may lie flat in front of the victor.[41] That domestic dogs do the same as gestures of submission is common knowledge. This spatial positioning of one's body below another is a natural marker of power relations because in such a position, one's resources are compromised. To be "on top" has superior value for the very reason that one's full powers, whether a matter of display or a matter of bodily weapons such as claws or teeth, remain uncompromised. But looking up someone and looking down on someone are not on that account to be construed merely as acts having pragmatic value. Spatial positioning *tout court* is a natural marker of power relations. In other words, at the most basic level, looking up to someone and looking down on someone are naturally meaningful acts not because of weaponry or resources but because the spatiality of one's body is integral—one might even say, indigenous—to being the individual one is, and because, in view of that indigenous spatiality and its ever-present possibilities of social meaning, a spatial semantics is a built-in of intercorporeal life. From this *power-of-optics* perspective, *esse est percipi*—to be is to be seen—that is, to be seen is, *inter alia,* to have a certain spatial position relative to others. Accordingly, by making oneself vertically smaller, one makes the other appear larger; one acknowledges a dominant Other by bowing, groveling, kneeling, or lying before him or her. By the same token, by maintaining one's full height or aggrandizing oneself in some physical way relative to another, one maintains a superordinate position. As to why in a semantic sense *down* is not *up*, and *up* is not *down*, the natural power to see is progressively intensified, proportionally augmented, with respect to greater and greater elevations. To be *at the very top* or *above it all,* for example, is to be in visual command of a situation, to behold an entire panorama, to have maximum powers of surveillance, to hold sway over those below. It is in this *optical* sense that elevation is a relational invariant in the spatial semantics of intercorporeal life. As indicated, the invariant derives first of all from being a body oneself and in consequence being open to a range of spatial possibilities; it derives secondly from living among other such bodies which are open to the same (or to a similar) range of spatial possibilities. Hence it is clear why one's physical standing along a vertical axis can be indicative of social status, and why where one stands in relation to another along that axis is neither a capricious nor trivial matter when it comes to power relations. Where the power plot thickens is in creatures for whom the vertical axis is a consistent fact of life. Consistent bipedality brings with it certain elaborations and certain ramifying values vis-à-vis the spatial dimensions of the body. A more complex vertical spatial semantics is evident. A clue as to why the semantics is more complex is suggested in de Waal's final two sentences about chimpanzees' "giant/dwarf greeting": the giant/dwarf relationship is a common greeting between two adult males, not between an adult male and an adult female. An adult female greets an adult male by presenting. The latter behavior does not appear in the behavioral repertoire of consistently bipedal primates. A fundamental primate positional marker of power relations, one that is strongly represented in nonhuman primate social relations, is thus absent in hominids. But what is absent behaviorally is not absent morphologically. Insight into the corporeal facts of presenting will bring out the morphological-semantic concordance.

Discussed earlier only as a mode of submissive behavior, presenting actually exemplifies a third fundamental aspect of primate animate form with respect to optics of power and the power of optics. As described previously, a presenting animal (*male or female*) in agonistic situations turns its back on the animal it is addressing. Given quadrupedal animate form, this means that *the horizontal axis* of nonhuman primate bodies is as significant as the vertical axis when it comes to power relations. In each case, intercorporeal meanings are generated by the very nature of the body itself; distinctive corporeal dispositions toward certain intercorporeal meanings are the direct result of being the quadrupedal body one is. These distinctive dispositions, insofar as they are species-specific invariants, may be called corporeal archetypes: front/back and high/low are spatial corporeal archetypes in the power semantics of primate intercorporeal life. They carry with them a fundamentally invariant spatial valence: front and high are positive; back and low are negative. Although hominids, through consistent bipedality, differ markedly from nonhominid primates in having lost a horizontal axis and with that axis, a behavioral pattern of both socio-sexual and socio-aggressive significance, the primate corporeal archetype *front/back* is semantically and axiologically unchanged. This is because planar surfaces are the equivalent of a horizontal axis. With consistent bipedality, in other words, primate front and back *ends* became fully exposed front and back *sides*. Insofar as back-ends and back-sides are equally vulnerable to attack, and insofar as frontal facings, whether in the form of ends or sides, are, in contrast, the power locus of the body—face-to-face, frontal bodily postures are typically in the direction of movement, in the direction of activity, and in the direction of concordant social interactions as well as aggressive encounters—the difference between hominid and nonhominid primates with respect to front/back as a natural marker of power relations is a difference in behavior only: the typical primate behavior expressive of the corporeal archetype—presenting—has disappeared, but the corporeal archetype itself has not. It is for this reason that the invariant of front/back, like other invariants, can be designated an order-, i.e., a primate-, specific invariant.

As suggested, in primate social relations, to be in a position of power is ordinarily to orient oneself face-on toward others. But with consistent bipedality, particularly complex variations are apparent, and this because virtually the entire body is exposed. Front and back surfaces are consistently on exhibit, as it were, open to view, subject to attack, and so on. Compound spatial relationships are in turn possible with respect to intercorporeal life. For example, *leading the way* puts one in front of others at the same time that it puts one's back to their front. Since taking someone from behind is always a possibility, leading the way theoretically puts one at an orientational disadvantage. Yet the basic positive value of frontality commonly remains: a leader leads the way. Firth notes an interesting general rule in Tikopia society with respect to this spatial *precedence,* as he calls it. "The general Tikopia proposition," he states, "is 'My front to the rear of a person indicates he is of higher status than I. My rear to his front indicates that I am of higher status than he'."[42] Given spatial precedence, the superior position of a leader remains uncompromised. But as Firth points out, *control* of spatial relations may take precedence over *actual* spatial precedence, as when a chief asks or directs a follower to lead the way. The general rule is in such circumstances reversed. Taking a cue from Firth, we can see that in instances of the kind, turning one's back on another can in fact be a sign of great power precisely because the act is a sign of fearlessness, of undauntability. People attending a bullfight intuitively understand the power of this act when they see a matador, after a pass at the bull, flair his cape, turn his back on the bull, and walk away from it. Moreover in Western as in Tikopia society, turning one's back on someone can be a movement of consummate disrespect toward another or of aversion toward an inferior as well as a straightforward display of power over a subordinate. By such an act, one affirms one's power over others by essentially denying relations with them, by dismissing them from one's world. The optics of power in these situations is such that one dominates others by not addressing them. To present oneself frontally toward them would be to recognize them not only as worthy of respect (if not as equals), but as part of one's immediate world. In short, what one faces and addresses visually is what is important; what one turns away from and refuses to see lacks value. Expressed specifically in terms of a complex intercorporeal spatial semantics, turning one's back on others can intensify the fundamental positive value of one's frontality.

D.   A Behavioral Invariant

The final major form of intensification is essentially different from the previous ones. It has to do not with the three-dimensionality of animate form per se nor with eyes themselves but with an individual's possibilities for animation, in particular, *its capacity to make a visual spectacle of itself.* The illustration at the beginning of this chapter aptly suggests this aspect of behavior. It began, "There is no mistaking a dominant male macaque," and moved on to describe the effect this "superbly muscled," "carefully groomed" animal and his "assured and majestic" walk had on other macaques. A description of increasingly agonistic behavior in chimpanzees will further illustrate the capacity for intensified whole body display.

In his research report on Ugandan chimpanzees, Japanese primatologist

Yukimaru Sugiyama spells out four modes of graduated aggressive action: first, there is the "Confident gesture" ("A large male, who may be dominant, sometimes walks, with an exaggerated composure, near or comes up to a small individual. The latter moves to another branch, crouches on the spot, or sways back grinning. . . ."); then there is the "Light attack" ("A large male stares at a small male or juvenile who had inadvertently approached him heedlessly. Sometimes he threatens the smaller animal by bobbing and weaving or by slapping at a branch of a tree, or the ground. . . ."); next comes "Chasing" ("A large male will often stamp violently on the ground, bark, swing a branch exaggeratedly, and chase a certain individual, sometimes beating the buttresses of trees. . . ."); finally, there is "Grabbing and biting" ("A violent attack accompanied by grabbing and biting was rarely observed. Even if a large animal grabbed a smaller animal, the former did not bite the latter. . . .").[43]

Sugiyama's description of what might be called "generic" display behavior is similarly instructive: first, there is "Slapping" ("Staring, . . . sticking its head out, and bobbing, an excited chimpanzee slapped the ground or a branch."); next, there is "Branch throwing" ("Then, taking a small branch or twig, the chimpanzee put it into its mouth or brought it up to shoulder height and threw it down. . . ."); then comes "Exaggerated eating" ("Sometimes excited chimpanzees ate leaves whenever they could lay their hands on them." [Sugiyama makes reference here to primatologist George Schaller's similar observations of 'symbolic feeding' in gorillas.]); then there is "Buttress beating and branch shaking" ("Beating and drumming on the trunks or buttresses of trees or shaking branches violently, an excited chimpanzee uttered a kind of growl, and opening its mouth for a little while, displayed a threatening expression."); finally, there is "Hooping display" ("Uttering a violent high-pitched cry, an excited male ran about on the tree and brachiated from one branch to another. . . .").

The above examples give dramatic evidence of a primate's capacity to make a visual spectacle of himself, a spectacle which, though it may begin with a relatively low-key intensification of presence—e.g., an unhurried, confident walk, or a slapping of the ground—can easily accelerate and swell into a tumultuous visual performance. Here, the intensification of the visual is such that there is a profusion of goings-on, a commotion far beyond the ordinary. The power of the visible is intensified beyond the body *tout court* to all the things a body can do not only in the way of boisterous movements and gestures, but in the way of sounds it can generate to swell the general visual bustlings and agitations, and in the way of objects it can appropriate —tree branches, stones, missiles—to add to the visual fanfare.

In nonhuman primate society, hitting, stamping, slapping, throwing, brandishing—all are components of threat behavior; lifting, slamming, dragging, pulling, scratching—all are components of attacking behavior. It is of interest to note moreover that van Lawick-Goodall writes that "temper tantrums are a characteristic performance of the infant and young juvenile [chimpanzee]. . . . The animal, screaming loudly, either leaps into the air with its arms above its head or hurls itself to the ground, writhing about and often hitting itself against surrounding objects."[44] This nonhuman primate way of calling attention to oneself—of intimidating, threatening, or defying others by making a spectacle of oneself—is in every way similar to a human primate's way of calling attention to itself. The natural power of the visible to attract attention is indeed augmented by movement and gesture, which in and of themselves regularly attract attention. By intensifying movement and gesture, in fact, by making them extraordinary, one insures that visual attention is not simply drawn; it is riveted. Temper tantrums in both nonhuman and human primates are a clear demonstration of this accentuated visual-kinetic relationship. It is not surprising then that hypervisualism—an intensification of the visual—and hyperanimation—an intensification of movement—go hand in hand in the formation and preservation of power relations. A show of political force in human affairs, for example, is not ordinarily a static assemblage of weapons and machinery for all to see; it is street demonstrations of rolling tanks, marching soldiers, waving flags, and much more. Moreover, sound is drawn into the service of the visual to intensify the spectacle further as when planes scream overhead and loudspeakers blare. The appropriation of objects is an even richer possibility in human primate displays because a manifold of cultural artifacts is available. Where power is personified, for example, objects such as crowns, jewels, red carpets, pillars, and gonfalons all call attention to or celebrate a certain personage along with aural accouterments such as trumpets that further accentuate the person's visual presence. By festooning the visual scene in multiple ways, one magnifies the natural power of the visible: what is seen is incisively intensified beyond the ordinary.

Of course, individuals in contemporary American society readily understand—even if only implicitly—the power of making a spectacle of oneself. They understand, that is, that one can gain power, that one can rivet the attention of others, by committing violent acts. One thereby makes the headlines, is seen on TV, or is interviewed on talk shows. One is, in effect, "the top story." Such a manner of making a spectacle of oneself is painfully evident in present-day American life—a chilling actualization of Andy Warhol's idea that in the future, "everyone will be famous for fifteen minutes."

Philosopher Judith Butler, in her book, *Gender Trouble*,[45] also implicitly makes use of this archetype. The power to shock others by calling attention to oneself in some way, to disturb others from their familiar judgments and categorizations, is well subsumed by Butler in her example of a performative act, namely, dressing in drag. However heavily such an idea might fall on postmodernist ears, there is no doubt but that such performative acts—acts intended to jar, threaten, or trouble others in some way—are archetypally anchored in our primate heritage. In fact, anyone who would use such acts knows this intercorporeal power—the power of making a spectacle of oneself—in his/her bones. This is why Butler, for example, can draw on it to advantage as a way of making a public statement about what she believes to be the social construction of gender and sex. The instance is ironic in that the belief that such performative acts are acts demonstrating the social construction of bodies is contravened by corporeal matters of fact which show them to be bona fide primate power plays, a fundamental part of our phylogenetic heritage.

III. Related Morphological Considerations

Swiss biologist Adolph Portmann's book, *Animal Forms and Patterns*,[46] is of considerable interest with respect to an elucidation of what may properly be called an *intercorporeal social semantics*. His concern is to show how visible morphology structures and affects recognition and communication among animals. The insights he brings into the visible, and by extension, into the power of optics, are remarkable and merit elaboration.

To begin with, an early discussion in his first chapter titled "The Outside and The Inside" calls attention to the fact that the insides of related animals are fairly uniform. "Let us try to distinguish the more familiar animal species merely by examining their viscera, such as the shape of the heart, or the position of the intestinal coils," Portmann suggests, and immediately remarks on the difficulty of arriving at distinctions.[47] As he points out, in related species, internal organs are monotonously the same, so much so that highly advanced knowledge is necessary to distinguish them properly. Portmann contrasts this uniform similarity of insides with the visible body, noting first by way of example what a difference there is between a lion and a tiger in spite of their close relationship. His point is that what impresses us—what we *see* directly and forcefully—is very different with respect to insides and outsides, and that "as regards forms, it is our eyes and the memory co-ordinated with them" that is dominant.[48] We are, in short, geared to seeing and remembering *formal* distinctions. Indeed, Portmann states that "what is presented to the eye is formed according to different laws from what is invisible. Only when we remain conscious of this twofold type of structure can we really grasp the special character of the visible animal body."[49]

At the very beginning of his analysis of "form production" and "form value" across a very broad range of creatures, Portmann develops his notion of optical patterns as organs equivalent to neurological organs. Structures that are seen—markings, colorations, designs, conformations—all are *organs* in the sense that they have a certain function, namely, to be looked at. When so perceived, they become meaningful forms. They inform those who see them of the age, sex, spatial orientation, and so on, of the individual concerned; they can as well inform of danger or of friendly or hostile feelings; they can equally deceive. At the level of mere appearance, the seen clearly has semantic value. The distinctiveness of that value is in each case related to a particular manner of life. In a broad sense this explains why the coloration and patterns on fish are generally found on their bodies, not on their heads as such. A fish's head, as Portmann points out, "is simply the anterior pole of a body built for swimming."[50] Accordingly, one fish is recognized by another not in terms of its head and anal end but by its overall bodily markings, by a certain bodily patterned design. It is quite otherwise with many mammals and birds, who have, for instance, distinctive heads in the form of manes, horns, erectile crests, crowns, hair, and so on, and distinctive anal ends in the form of tails, feathers, scrotal sacs, and so on. The seen here is not only a matter of form as design but of form as spatial placement. In other words, meaningful bodily conformations, markings, and the like do not appear just anywhere but are found in places where they are accessible to sight, either straight off or through the performance of certain behaviors. Insofar as they exist to be looked at—in Portmann's sense, to be meaningful—strategic location with respect to bodily form as a whole only makes sense. In effect, there is not only formal *arrangement* but formal *placement* which controls social communication—or what we may more precisely designate, "intercorporeal meanings."

As suggested above, form values may be effectively solidified by behavior. At a very basic level, for example, Portmann observes that "optical form production is reinforced in its function by special ways in which the animal behaves; these either ensure that the sematic [sic] parts show up, or else guarantee the disruptive effect of a pattern."[51] By the latter, Portmann means that an animal may behave in such a way that its actual form is made "cryptic" in contrast to being "semantically" displayed, as when a male peacock opens its train and in so doing exhibits its ocelli to a female. Through cryptic-producing behavior, a creature's coloration, for instance, may meld into the environment, thus "disrupting" attention to form value as an aspect of the living creature itself. At more sophisticated semantic levels, as we shall see in more detail momentarily, optical form production is in the service of expressive values; visible form is the organ of psychical processes.

If we follow Portmann, we find that a richly documented and irrefutable case is made for the impact and meaning of outsides, and in consequence, of the natural power of the visible and its propensity for accentuation. Portmann shows in detail how the visible has "intrinsic value"[52]—not only how designs and colorations can warn, invite, or deceive, or inform as to sex, spatial orientation, and so on, but how they can indicate mood and feelings. Portmann speaks of these latter form values as "expressions of inwardness." He insists that if we are led to acknowledge form value, then we "shall not allow it to be degraded to a mere shell which hides the essential from our glance."[53] What Portmann ultimately means by this is that in those creatures for whom form values play complex strategic roles, social life also plays a strategic role, that is, social life for these creatures is not a matter of *swarms* or *schools,* for example, but a matter of one individual seeing another as a distinct visual form. Interestingly enough, Portmann speaks of this kind of individualized social life as "an *intensive* communal life," as "*intensified* forms of social relationship," as being of "increased vital *intensity*" (italics added).[54] Over and over again, there is an emphasis on an *intensification* of the visible in the evolution of individualized sociality. It is this intensification of the visual that culminates in what Portmann calls "the expression of inwardness." Not only can patterns, designs, and colorations inform others of the presence, gender, and status of a certain individual, they can also inform others of the mood of that individual; thus, form values can be expressive values, as when human skin turns rosy or pale, according to emotional life. In fact, as Portmann points out, "A similar *intensification of meaning* is given to the play of the pupil of the eye, which as its elementary purpose controls the admission of light; but in the higher animals as well as in man special nerve tracks cause it to follow the changes of the inwardly felt emotions" (italics added).[55] Moreover expressive values attach to visible behavior in the sense of postures, movements, gestures, facial modulations, and bodily shapes. Such behaviors also readily inform others of an individual's mood or "inwardly felt emotions," as when bodily position indicates whether an individual is resting or alarmed. In Portmann's sense, visible form, through rich and various intensifications, is clearly the organ by which a felt life becomes part of an intercorporeal social semantics.

An important aspect of Portmann's thesis is that formal manifestations of psychical processes—"intensified forms of social relationship"—are not the result of higher organization "but a constituent part of it."[56] This notion, that a creature who leads "an intensive communal life" leads it in virtue of its *form,* was suggested by the earlier quotation to the effect that outsides must not be considered simply a covering for "the essential." Portmann's point, of course, is that higher organization is not to be thought of and measured exclusively in terms of *insides,* i.e., brains, as is common, but in terms of animate form itself. By this very token, inwardness or expressive values in an evolutionary sense are not epiphenomena of higher organization but coeval with it. One might paraphrase this point by saying that what distinguishes individual forms of life is not hidden away inside them; what distinguishes individual forms of life are the *forms* of life themselves, all the way from functional organs to manifest "inwardness." Portmann's insistence upon an appreciation and understanding of these forms is ultimately an insistence upon the visible as a key to distinct individuality and the psychically intensified sociality that goes with it. Where there is "mutual expression of moods so that the being together is raised to a richer relationship," where there is "a true meeting of independent creatures," there are not merely crowds, but individuals.[57] For such individuals, higher organization is higher organization of the visible, which means an intensification of both form production and form value, or in short, an intensified power of optics.

Portmann's theme of "intensification" clearly dovetails and complements the already developed theme of the power of optics to intensify naturally visible bodies. The intensifications of which Portmann speaks, however, are less in the service of Foucaultian-type power relations than in the service of what might in Sartrean terms be designated "reciprocal incarnations." The Other is not reduced to its outside as a thing; rather, its outside is apprehended as a *form;* the body is thus not a pure object for a pure subject, in essence a docile body, but a veritable *individual,* a subject recognized by another subject. The difference strongly recalls the conclusions of the last chapter. Where form values are not recognized and appreciated as such, whether in particular a matter of inwardness, for example, or of display, age, or species, neither is the creature itself. Where vision stops short of form values, it stops short of recognizing and *valuing* a fellow life. The Other, be it human or nonhuman, is simply something susceptible to being used, manipulated, coerced, inspected, and so on, in a word, susceptible to conscription in a Foucaultian socio-political sense. In such a situation, a living body is stripped of its autonomous meanings; certain objectifying power relations already obtain. If form values intercede at all, they likely do so only in the way of inciting to violence or augmenting violence: where weakness, fright, naiveté, or terror is formally expressed, for example, so also is an unmitigated vulnerability that can be inviting. As Sartre says of the sadist, "[his] effort is to ensnare the Other in his flesh by means of violence and pain. . . . If the victim resists and refuses to beg for mercy, the game is only that much more pleasing. . . ."[58] Put specifically in terms of the conclusions of the last chapter, Portmann's insights are subsumed in the observation that where form values of creaturely life go unattended, or where, in his words, "a true meeting of independent creatures" is lacking, the natural ratio among the senses is distorted. To say that certain form values are expressive of mood and feeling, for example, is equivalent to saying that the natural power of optics does not exist in a visual vacuum but is quintessentially related to touch and movement, to a felt life; similarly, just as to ignore the living body of the Other as an organic totality is to ignore "reciprocal incarnation," so to ignore form values is to ignore "inwardness." To ignore form values is moreover to be blind to others in the Sartrean sense of indifference: one's gaze spreads over the surface of the body of the Other in the way it spreads over the surface of a thing. Such a gaze is indifferent to those expressive values that are natural semantic dimensions of intercorporeal life.[59] In sum, what Portmann's work emphasizes and challenges us to notice are the multiple ways in which visible form is part of the very fabric of animate social relations with respect to *meaning,* and how in each case intensification of meaning is an intensification of form.[60]

IV. Implications of an Evolutionary Perspective

The phylogeny of the power of optics is transparent in such acts as brandishing, throwing, and stamping, in the spatial values of size—natural, augmented, or diminished—and of front/back and high/low spatial positionings, and in the optical gestures of staring and averting the eyes. The bodily acts, like the spatial values and optical gestures, are species-specific and species-overlapping invariant expressions of power relations. That these bodily phenomena are indicative of power relations is only to be expected given the fact that *social relations are always intercorporeal relations,* and that intercorporeal social relations among members of the same species necessarily involve corporeal invariants. The discussion of Portmann's work conclusively attests to this fundamental relationship and its foundation in animate form. Species-specific behaviors and correlative interanimate under-standings would hardly be possible short of archetypal intercorporeal postures, gestures, and so on. Where accounts of social relations, and by extension, accounts of power relations, can go astray or reach dead ends is where, in lieu of a central concern with animate form, relations with others are conceived in terms of an abstract *intersubjectivity*.[61] Bodies seldom come into the picture in such accounts, or if they do, it is certainly not with the precision demanded by the corporeal facts of the matter. Indeed, explanations of intersubjectivity sometimes verge on explanations of *clairvoyance;* seeing into other minds, reading other's intentions, feelings, and so on are in the end preternatural accomplishments. In contrast, one has only to look out at the actual world of living creatures to see that intersubjectivity is first and foremost an *intercorporeality;* it has to do with meanings engendered and/or articulated by living bodies. Understanding how these meanings originate and how they are mediated, understanding how invariants are built-in dimensions of intercorporeal life and how they are differentially expressed across species and across cultures, such understandings lead ultimately not to behaviorism and certainly not to an eliminative materialism, but to a semantics far less wedded to the semiotics of language and objectified behavior than to a semasiology of the body. Indeed, it is to these deep and complex understandings of the body that one must turn in order to shed light on the genesis of power relations and the very concept of power.

The work of many semioticians, ethologists, and anthropologists already bears out the importance and richness of this corporeal approach, its profound evolutionary significance, and its promise of both opening up and enriching understandings of power. Brief examples will exemplify the value of their work.

For ethologists and anthropologists, an evolutionary perspective has been especially significant in understanding the socio-political nature of bodily space and spatial behaviors—territoriality, overcrowding, and social distance, for example. Anthropologist Edward T. Hall, for instance, who initiated *proxemics*—study of the ways in which humans use space and structure it—staunchly affirms that "The more we learn about animals . . . the more relevant these studies become to the solution of some of the more baffling human problems."[62] Ethologist and former Director of the Zurich Zoological Gardens Heini Hediger was obviously equally convinced of the importance of an evolutionary perspective when he wrote that "human behavior can never be understood as something isolated, but only in its phylogeny as revealed by comparative studies."[63] Hediger's observation that flight distance must be modified in adult wild animals introduced into a zoo environment is of great interest in this regard. He noted that such animals' "flight distance"—the point at which animals react to the degree of closeness of others—became proportionally smaller than their flight distance in the wild. The captive animals would otherwise be virtually in a constant state of panic due to the proximity of humans and other animals.[64]

Hediger's observation underscores the fact that the way in which animals, including human ones, structure space is expressive of certain power relations. Studies of overcrowding testify in a similar way to the relationship between power and space. In human populations these studies have shown that lack of sufficient personal space increases aggressive and violent behavior; overcrowding in rat populations has shown the same kind of behavior to result.[65] In fact, as "proxemic" anthropologist O. Michael Watson states, "It has been known for some time that crowding is almost always harmful to animals"; population density increases stress.[66]

Though emphatically focused on a corporeal approach to social behavior, the above examples remain at a fairly general level with respect to actual intercorporeal practices, whether of human or nonhuman animals. Where the finer interplay of socio-political relations becomes transparent is at the level of the body itself. Here the filigreed infrastructure of power relations is apparent in the actual realities of an intercorporeal semantics, and this because at the level of living bodies themselves, basic facts emerge. A group of semioticians call attention, for example, to the fact that "tongue-showing" is not a human invention; other primate species display their tongues in a manner similar to human tongue-showing and in situations similar to human tongue-showing. The display thus has an evolutionary history.[67] Another semiotician, David Givens, in his study "Greeting a Stranger: Some Commonly Used Nonverbal Signals of Aversiveness," affirms anthropologist Gregory Bateson's contention that "human nonverbal communication appears to express information pertaining to the most fundamental contingencies of social relationships, to dominance, submission, affiliation, and aversion. It predates spoken language and is more closely related to other mammalian signaling systems than to linguistic behavior."[68] Givens stops short of affirming species-specific/species-overlapping invariants, however. He states that "It would be tempting, but premature, to state that all of the aversive behaviors observed in this study are unlearned or innately predisposed. The temptation exists because so many of them are performed at an early age (e.g., gaze avoidance from 14 days, automanipulations and aversive mouth behaviors from 1-to-2 years) in many cultural settings, are found in the Hominoidea generally, and are exhibited by congenitally deprived youngsters."[69] His hesitation is difficult to understand given the fact that human or nonhuman, we are all bodies, and that in virtue of being the bodies we are, we share certain common structures. It is, in other words, not surprising that facial displays of aversion have an evolutionary history. Being a body means discovering all of its particular possibilities, calling into play all of its movement potentials, maximizing to the full its tactile-kinesthetic capacities. In the economy of nature, there is virtually no corporeal stone unturned. *Everything* works, *everything* goes into making a living. For creatures belonging to the same (biological) family, this economy of nature readily spills over into shared expressions, gestures, actions, values, and meanings—into corporeal archetypes.

It might finally be noted that where intersubjectivity rather than intercorporeality constitutes the conceptual point of departure for understanding socio-political relations, there is a risk of losing oneself in words, and this because *explanations* or *definitions* tend to take the place of *descriptions* and in so doing can deflect attention away from and, indeed, get in the way of describing what is actually there, namely, corporeal matters of fact. To become involved in intentional systems analyses,[70] for example, on the order of "I believe that X is feeling angry and is going to attack me physically because he thinks that I am going to report him to the authorities," is to ignore the intercorporeal sense-making that disposes one to the explanatory account in the first place. In such accounts words hide bodies just as the Look hides the eyes. Indeed, Sartre's description of the Look as "going in front of the eyes" offers a quite apposite analogy. With the Look, the Other's eyes "remain at a precise distance . . . whereas the look is upon me without distance."[71] Just so when words go in front of the living body articulating them: the living body "remains at a precise distance, whereas its words are upon me without distance." Close understandings of power and power relations demand, on the contrary, close understandings of the intercorporeal semantics that are their foundation.

3

Corporeal Archetypes and Power: Preliminary Clarifications and Considerations of Sex

The natural body is . . . a discursive phenomenon.

Ladelle McWhorter[1]

All significant differences between men and women are thoroughly historical, social and cultural. *. . . I am indeed denying that the difference is fundamentally or significantly biological.*

Carol C. Gould[2]

Although it is probably true that the physiological disturbances characterizing emotions . . . are continuous with the instinctive responses of our prehuman ancestors and also that the ontogeny of emotions to some extent recapitulates their phylogeny, mature human emotions can be seen as neither instinctive nor biologically determined.

Alison M. Jaggar[3]

There is nothing about being "female" that naturally binds women. There is not even such a state as "being" female. Biology is historical discourse, not the body itself.

Donna Haraway[4]

Not only is there no causal relation between sex and gender, but more recent readers of Simone de Beauvoir have suggested that "sex" itself is a misnomer, and that the ostensibly biological reality that we designate as sex is itself an historical construct and, indeed, a political category.

Judith Butler[5]

An earlier version of this chapter was published in *Hypatia* 7/3 (Summer 1992).

I. Introduction

It is not uncommon to find feminist disavowals of biology. The spectre of essentialism is a perennial threat to many feminists. Words uttered in a variety of politically reclamative and transformative contexts have gone some way toward mitigating the threat, but they have failed to cause biology to fade compliantly into the sunset. This is because however powerful the arguments they present, they cannot make evolutionary facts of corporeal life disappear. The purpose of this chapter is to demonstrate the significance of such facts, and in the process show the oppressive results when evolutionary scientists themselves fail to bring these facts to the fore. Insofar as the chapter goes against the current constructionist grain of much feminist thought, an introduction is called for to situate the presentation within the framework of current constructionist feminist theories and to indicate how wholesale disavowals of biology are precipitous. A brief critical look at the claims and writings of the above-cited authors will anchor the introductory remarks.

The best place to begin is with the naive, simplistic view of the natural as a pure, unadulterated realm untainted by culture, a realm that is "clean" and even good in contrast to the realm in which repressive forces enslave us and are, one might say, bad for our health. This view of the natural is sentimentalized—or can be—in various ways, so that it appears a haven for righting all that is wrong in twentieth-century Western society. But this construal of the natural is a *mis*construal. It is well exemplified in Ladelle McWhorter's attempt to correct certain misinterpreters of Foucault (un-named in her article) who, according to McWhorter, think Foucault in his writings is urging us to turn away from oppressions of the body and reclaim that natural, "clean" body which underlies all the inscriptions that subdue it.[6] While McWhorter is wholly right in thinking this clean body does not exist, she is wholly wrong in thinking that this body is *the natural* body. A natural body is the product of a natural history. It is in this sense a Darwinian body, a body not just shaped in morphological ways by evolution but shaped semantically—which means kinetically, gesturally, spatially, behaviorally. Because we are all natural bodies in this sense, we ourselves have a history. Our fundamental human habits and beliefs have an evolutionary past: burying our dead has an evolutionary past; so also does drawing, counting, and language. And so also does our intercorporeal semantics—as this chapter will go on to show specifically in terms of sexual signalling behavior. When McWhorter approvingly cites Nietzsche's remark about gruesome beasts grinning at us knowingly,[7] she—like Nietzsche—

fails to realize that in a substantive evolutionary sense those beasts are us. Fundamental aspects of our humanness cannot be written off as mere cultural inscriptions. They have to do with a history more ancient than we, a history in which the body is precisely not a surface on which any culture can leave its marks—arbitrarily and willy-nilly—but a three-dimensional living natural form that itself is the source of inscriptions—meanings. As I have elsewhere shown, this natural form is in fact our original semantic template; it cannot be "discoursed" out of existence.[8] In a concrete phylogenetic sense, it is a carrier of meanings, an emitter of signs, but the richness and complexity of its intrinsic, ancient semantics cannot be acknowledged until the conception of the body *exclusively* as a surface is recognized as the myopic cultural conception it is, that is, until the surface is seen to be the literal outer skin of a far deeper and denser body, and indeed, in a fundamental sense, to depend upon the inscriptions of that deeper, denser body.

Carol Gould's dismissal of biology, while not resting on a postmodern foundation, can be questioned on similar grounds. In seeking to uproot Essentialism (her capitalization), she distinguishes between an abstract and a concrete universality, the latter categorization allowing, in contrast to the former, the inclusion of what distinguishes individuals within a class as well as what is common to them. On the way to setting forth arguments in support of her thesis, she notes that while "there *are* biological differences between men and women . . . there are infinitely many differences among individuals, any of which is logically an equal candidate for making group distinctions among humans," and adds that "in denying that the biological difference is an essential or fundamental difference, I am asserting that it becomes one only through its historical and cultural development."[9] A good part of the problem is that what is being denied remains amorphous. While thinly suggested (in support of her claim of "infinitely many differences," Gould notes that a body has a certain hair color and is right-handed or left-handed), what is being denied is virtually nowhere to be found; whatever the biological difference(s) between men and women that are acknowledged, they are never specified. What is meant by "biological difference(s)" is thus by easy implication reduced to anatomy and written off before it is even identified, much less examined. Such slighting and inattentive treatments of the body demonstrate a lapse in understanding. They strongly suggest that an overzealous preoccupation with Essentialism can quickly lead one to forget evolution, which is not only the thread connecting the whole of biology from the molecular to the macroscopic, and on that account requiring careful attention, but also the thread connecting we humans to our own history—a history that must figure centrally in any just and rigorous appraisal of what it is to be human, in particular, a human body.

A similar write-off of the natural body is again apparent in Alison Jaggar's remarks on emotions. Jaggar writes that there is nothing biological in mature human emotions. While there are biological determinants and instinctive emotional behaviors in infants and developing children, mature human emotions are thoroughly social constructions. Apart from the implausibility of the view—that is, the unreasonableness of the claim that instincts and biology disappear, that what is in the nature of things in the beginning is *completely effaced* by culture—how in fact does one get from the one side to the other, from an instinctive, biological emotional life to one that is entirely cultural? What is the nature of *development*? Does one grow up in such a way that at a certain age, and regardless of culture, instincts are swept away? To claim that there is nothing biological in the emotions of mature human individuals can only mean that the human living body undergoes some radical transformation in the course of developing; indeed, that its affective insides molt.

One might begin weighing the soundness of this claim by considering smiling. Infants smile. In fact, Darwin jotted down in one of his Notebooks, "Seeing a Baby . . . smile & frown, who can doubt these are instinctive—child does not sneer."[10] But of course young children and adults also smile. Moreover when an infant, child, or adult smiles, the tendency of others is to smile back. If these simple observations appear to lack authority for want of experimental evidence, consider the cross-cultural studies of Irenaus Eibl-Eibesfeldt, which show, for example, that surprise is consistently expressed by a widening of the eyes and a raising of the eyebrows.[11] Consider further the cross-cultural studies of Paul Ekman and others which show, for example, that disgust is consistently expressed by a wrinkling upward of the nose and a consequent pursing of the upper lip.[12] From a Darwinian point of view, the findings of both Eibl-Eibesfeldt and Ekman—ethologist and psychologist, respectively—which show basic facial expressions of emotion to be universal are not surprising. Darwin long ago identified and analyzed basic emotional human expressions and their links with nonhuman animal life.[13] Although he did not examine situational-cultural shadings or reworkings in the manner of Eibl-Eibesfeldt and Ekman, he showed that basic human emotions exist in virtue of a common heritage, an evolutionary lineage which we hominids share with each other, and which we share in major ways with our primate relatives.

Evolutionary continuities aside, one might equally question the rectitude of Jagger's claim on the grounds that in the throes of actual life, feelings are lived through, not just written or spoken about. Indeed, taking a cue from psychotherapist Jane Flax, one might note that there are literally de-centered individuals ("borderline" people), not just academically de-centered ones.[14] The latter's stance from a Flaxian viewpoint might well be termed a pseudo-posture, *pseudo* because what is de-centered is not a living body reverberating with feelings that keep it in a state of perpetual imbalance, but words which may be made to appear with the scratch of a pen or the wag of a tongue. Moreover, the feelings of a literally de-centered individual, while perhaps associated with present-day Western styles of living, are themselves through and through pan-human ones. Anxiety, fear, self-doubt—such feelings are not peculiar to any particular culture, any more than the human body itself is. In this context, we might ponder Susan Bordo's remark that "Whatever the effective role played by biology in human life it never exists or presents itself in 'pure' form, untouched by culture."[15] If this statement is true, then rather than a disavowal of biology, an intensive, sustained, and unabridged examination of the body is in order precisely to elucidate those biologically invariant structures, emotional and otherwise, that vary thematically from culture to culture. We need precisely to understand "the effective role played by biology," which means we need to turn toward the body. As I have noted elsewhere, "the linguistic turn produced extraordinary insights . . . a corporeal turn would assuredly do no less." This is because "the corporeal turn, like the linguistic turn, requires paying attention to something long taken for granted."[16]

The idea that we might successfully combine the best of two possible epistemological worlds—that of the relativist and that of the foundationalist—is put forth by Donna Haraway as the problem of "how to have *simultaneously* an account of radical historical contingency . . . *and* a nononsense commitment to faithful accounts of a 'real' world."[17] In urging a solution to the problem through "situated knowledges," Haraway begins with concerns about objectivity, which quickly become attached to concerns about an embodied or disembodied objectivity. Her point is that "we need the power of modern critical theories of how meanings and bodies get made . . . in order to build meanings and bodies that have a chance for life."[18] But the program as envisioned elides the body in a fundamental sense; it shifts to linguistic and theoretical considerations. That this is so is clear from statements such as, "We need to learn in our bodies . . . how to attach the objective to our theoretical and political scanners in order to *name* where we are and are not, in dimensions of mental and physical space we hardly know how to name."[19] Clearly the purpose is not to understand the body but to catapult it into language. Furthermore, the term "embodiment" often covers over a schizoid metaphysics, a metaphysics that has not in fact resolved Cartesian dualism because it has not in fact taken the body and bodily experience into close and full account.[20] The term is a lexical band-aid put on a three-century old metaphysical wound. It is not surprising, then, that the program of "embodied knowledges"[21] actually takes the body itself for granted in its entire epistemological enterprise; functioning as an indexical, the body is simply the place one puts one's epistemology. While it is true that "we must be hostile to easy relativisms and holisms built out of summing and subsuming parts,"[22] we should also be wary of an "embodied objectivity" that, amid "ethnophilosophies," "heteroglossia," "deconstruction," "oppositional positioning," "local knowledges," and "webbed accounts,"[23] in truth thoroughly distances itself from the body except as an epistemological receptacle. Unless we are wary, we will easily find ourselves distanced from the real, living body that is the very ground of our knowledge for it lies buried at the bottom of the barrel.

The quest for transformative knowledge must of necessity bring that body to the surface, both because it leads to an understanding of the fundamental semantics of intercorporeal life that informs our lives and because power relations are interwoven in complex ways in those semantics. Certainly we may acknowledge as threatening the idea that the body is intrinsically tied to knowledge: immediately one thinks of biological sex differences, and biological sex differences lead straight to essentialism. But we must also acknowledge the possibility of the threat's blindering us, thus keeping us from examining what is actually there. Clearly, we might discover something other than what we are *assuming* is there. These unexplored assumptions might help explain why Haraway, particularly with her impressive biological background, skirts the real, living body: "Feminist embodiment . . . is not about fixed location in a reified body, female or otherwise, but about nodes in fields, inflections in orientations, and responsibility for difference in material-semiotic fields of meaning." What do these nodes and inflections, and this responsibility have to do with breathing, pulsing, locomoting, sensing creatures? By "reified body," Haraway presumably means mere matter: the biologically given anatomical specimen splayed out like a cadaver on a laboratory table. No wonder then that she continues her description of feminist embodiment by saying that "embodiment is significant prosthesis."[24] What Haraway urges us to do is extend our bodies before we even understand them. A concern with objectivity and bodies should, on the contrary, translate from the very start into a concern with animate form, particularly so if the ultimate concern is with understanding and reclaiming power. As this chapter will show in further detail, the animate form that is a living human body is not simply inscribed with power from without; it carries its own inscriptions.

In the context of the above critique of Haraway's program of embodied knowledges and her neglect of the actual living body in the construction of meaning, a few words should be added about Helen Longino's recent book in philosophy of science and Lorraine Code's recent book in epistemology, for each book, in a quite different way, suggests that something more is possible in the way of a body than simply "embodied" knowledges. Longino's thesis that scientific knowledge is social knowledge and her emphasis on the experiential nature of knowledge do not lead to an "unbridled relativism."[25] On the contrary, a socially constructed science that ties knowing to experience relies on there being a basic commonality to which all appeal and may appeal when accounts of the world differ: "There is always some minimal level of description of the common world to which we can retreat when our initial descriptions of what is the same state of affairs differs."[26] This "minimalist form of realism," as Longino terms it,[27] is not that distant from the practice of phenomenology, in which one verifies by one's own experience what is described, and, if verification is not forthcoming, one goes back again (and again) "to the things themselves"[28] to see if one has missed something, if one has been sufficiently attentive to what is actually there, and so on. The procedure—to go back to the things themselves—is, of course, not unlike the way in which present-day science normally proceeds when disputes arise. But the notion that there is something basic "out there" is an epistemological claim as well as the basis of an epistemological procedure. When applied to the living body, it means not only that there is the possibility of reconciling differing viewpoints on corporeal matters of fact, but also that there are basic corporeal matters of fact to be described in the first place.[29] Interestingly enough, Code's recognition and support of phenomenologically oriented studies which link knowing and the body—"all knowing is permeated with mood, feeling, sensibility, affectivity"—and which "ground their analyses in experience"[30] dovetail with Longino's espousal of minimal realism and its implicit vindication of the possibility of fundamental corporeal descriptions. While Code points out that "phenomenological discussions . . . [are different from] the mainstream discourse in whose terms Anglo-American epistemology and philosophy of science are predominantly discussed," and adds that "purists might find it illegitimate to introduce [phenomenological discussions] into a book that engages primarily with mainstream epistemology," she goes on to affirm that "part of the project of feminist critique is to uncover the suppressions and exclusions that received ways of thinking have effected and to challenge disciplinary, methodological, and ideological boundaries."[31] Although Code's own explorative remapping of the epistemic terrain does not incorporate the suppressed and excluded body that so often lies buried under what now appear to be traditional feminist theoretical concerns (including theory itself),[32] instances of where this excluded body could gain entrance are readily apparent for this body hovers at the very threshold of her provocative discussions. When she singles out the problem of a Foucaultian subject who is never there, for example, and points out the dangers of "discursive determinism,"[33] rather than limit herself to a defense of Foucault on theoretical grounds, she could well recall in addition her own earlier listing of convergences between her own project and that of phenomenology. The latter studies, she writes, "ground their analyses [not only] in experience" but also "in praxis and embodied existence; . . . [they] concentrate on *particular* experiences, specific modes of existence, from the conviction that it is in particularity and concreteness that generality—essence—can be known. . . . [They] do not privilege vision: perception engages all of the senses; objects are known by touch, holding, sensing their whole presence. Indeed, perception also engages a 'sixth sense' . . . 'proprioception': the sense of themselves through which people 'feel [their] bodies as proper to [them] . . . as [their] own' and position themselves in the world through an awareness of the shape, [and] capacities . . . of their bodies."[34] Clearly the Foucault-induced problem of subjectivity and agency, which as Code points out remains contested among Foucault readers, can find resolution in deeper studies of the body, which is to say in descriptive analyses of what is actually there, both corporeally and intercorporeally.

In a recent book, Judith Butler attempts to spell out how "language itself produce[s] the fictive construction of 'sex'" and how, by adopting "a performative theory of gender acts" as strategy, bodily categories—of sex, gender, sexuality, and the body itself—can be disrupted.[35] In her zeal to validate the constructionist claim that with respect to sex and gender, there is "no foundation all the way down the line,"[36] she attempts to show that the body is a set of boundaries, "a surface whose permeability is politically regulated."[37] In contrast to a Foucaultian reading of the body that requires something there—a materiality—prior to inscription, Butler wants a body with no past, with no prediscursive significance or even ontological status: "the body," she writes, "is not a 'being'."[38] On these spirited-away grounds, Butler can justify a performative over expressive body. What she sacrifices, however, is a history replete with corporeal matters of fact, including expressive matters of fact.[39] Unlike a body that can be read as mere synecdoche for a social system,[40] the evolutionary body cannot be so reduced: it does not stand for, refer, or function as a trope in any way.

Moreover, unlike a body that has no stable identity and whose aim is to disrupt and destabilize,[41] the evolutionary body has an established identity that, however flukey its existence, circumstantial its form, or minuscule its lifetime, abides over time and is part of an unbroken, continuous historical process. Constructionist theories that fail to take the evolutionary body into account not only ignore the relational ties that that historical process describes and that bind us to certain corporeal acts, dispositions, and possibilities, and to a certain related intercorporeal semantics; they also put us on the edge of an unnatural history. It is as if we humans descended *deus ex machina* not just into the world but into a ready-made culture, a culture that, whatever its nature, can only be the product of an immaculate linguistic conception. Indeed, short of an accounting of the evolutionary body, we are, unlike all other living creatures, products of grammatological creationism. It is significant that, with respect to the body, Darwinian evolutionary theory began not in speculative explanations or thoughts about the body nor with programmatic goals but with descriptive accounts of what is actually there: living, moving bodies prior to discourse.

The preceding critical remarks show that evolutionary facts of corporeal life need to be recognized and attentively examined. This chapter urges just such an acknowledgment and closer examination of biology. In fact it urges a re-thinking of biology—in evolutionary terms and the intricate ways in which cultures rework and translate those terms. Such a re-thinking constitutes a beginning step toward showing that there is more to biology than anatomical parts; in other words, more to the body than meets an anatomical eye—or I. There is animate form. In developing an evolutionary perspective and in considering its cultural reworkings and translations, this chapter focuses on out-there-in-the-world-anyone-can-observe-them corporeal matters of fact: for example, nonhuman female primates have estrous cycles and many of them present their swollen hindquarters to males either as a sexual invitation or as a response to male sexual solicitations; human female primates do not have estrous cycles and are pan-culturally enjoined from showing their genitals. Corporeal facts such as these are points of departure for assessing our biological heritage in the form of corporeal archetypes. Persistent examination of these archetypes—as engendering certain power relations and as constituting invitational sexual signalling behaviors—will culminate in an elucidation of how corporeal archetypes undergird our Western female-denigrating "biology of human sexuality."

The program of re-thinking is written in the form of a peripatetic investigation; that is, rather than writing up results, I am inviting readers to go through the process of re-thinking, to make the meditative journey themselves. In this way, the full import of an evolutionary beginning will become apparent. The chapter begins with a further elaboration of the relationship between animate form and power and goes on to examine corporeal archetypes as sex-specific and sex-neutral acts. By juxtaposing nonhuman and human animate form and sexual archetypes in the light of sexual signalling behavior, and by reckoning with the reigning Western biological paradigm of human sexuality, the chapter brings to light corporeal matters of fact that subtend female oppression and the elevated sense of power that establishes and sustains that oppression.

II.   Animate Form: The Conceptual Basis of Power

The concept of power and concepts of how power can or might be wielded arise on the basis of *animate form*. One has only to imagine what it would be like to be another body to realize the truth of this statement. That one's concept of power derives from being the particular body one is means first of all that originally, in abstraction from any cultural overlays or constructions, whatever they might be, there is the body *simpliciter*. At the most basic level this body *simpliciter* is a human body, a crow body, an ant body, a dog body, or an orangutan body. Whatever its specification—whatever its *species*-fication—this body clearly has certain distinctive behavioral possibilities and not others: certain sensory-kinetic powers are vouchsafed to it in virtue of the animate form it is. As an *individual* instance of the animate form it is, this body may from the very beginning be blind, have a cleft palate that impedes normal articulation, have a heart defect that prevents normal exertion, have a stunted arm or one leg longer than the other such that normal range of motion or normal locomotor rhythms are precluded, have a spinal abnormality that prohibits twisting, and so on. Moreover, through some gross trick of biological fate, this body might have no ears, a big "little" toe instead of a big "big" toe, or a second thumb on one hand. Furthermore, in the course of growing up, it may develop illnesses or abnormalities such as asthma, undescended testicles, deafness, scoliosis, breasts though it has a penis, or through injury, it may lose a leg. Moreover in the process of aging, its powers will necessarily change: it cannot run as far, see or hear as acutely, lift as much, dodge things as agilely as before; it may develop emphysema, be spatially amnesic, suffer Parkinson's or Alzheimer's disease. Clearly, one's concept of power and one's concepts of possible deployments of power arise on the ground of the body one is.

Translated into terms of the body *simpliciter*, the concept of power is initially generated on the basis of species-specific 'I can's'.[42] In the case of a human body, this means specific capabilities and possibilities such as I can stand, I can run, I can throw, I can speak, I can oppose thumb and fingers, I can climb. It naturally means at the same time certain incapabilities such as I cannot bark, I cannot brachiate through a forest canopy, I cannot fly, I cannot live either underground like a mole or in water like a fish. Many species-specific human 'I can's' are shared with nonhuman primates—with chimpanzees, for example, who can also stand, throw, run, oppose thumb and fingers, and climb. Shared actions will in each case be performed in a species-specific corporeally idiosyncratic manner, but whatever the variations in performance, they do not diminish or in any way nullify the basic primate commonality. Ample and consistent evidence for this claim is found in the primatological literature. We have in fact earlier seen how a basic primate commonality is apparent in the acts of staring and of averting the eyes; in the spatial valencies that exist with reference to size, and to high and low, and front and back intercorporeal positionings; and in the performance of exaggerated body spectacles.

As suggested above, the body *simpliciter* is the source of individual 'I can's' that qualify its species-specific 'I can's'. On the positive side, this might mean that a particular individual cannot only threaten, but given its size, out-threaten others; that another individual cannot only run, but given its superior endurance, out-run others, and furthermore not only throw, but given its superior strength, out-throw others; that still another individual in virtue of superior coordination cannot only move but move more agilely than others; and so on. In this sense, individual 'I can's' qualify the basic distinctions of animate form that characterize the individual as human (chimpanzee, shark, or raven). The concept of power and concepts of possible deployments of power are thus subsumed in a repertoire of 'I can's' that derive from the animate form one is in both a species-specific and individual sense.

Sexed bodies are one aspect of animate form. The opening task, then, is to delineate in a beginning but precise way how sexed primate bodies are linked to animate form, and to determine, again in a beginning way, whether and how, in virtue of that linkage, they are linked inescapably to certain power relations. Though a possible topic given the task, we will not be concerned with the ways in which an individual's repertoire of 'I can's' might qualify its individual attractivity from the viewpoint of sexual selection theory.[43] In other words, that an individual has a longer and thicker penis, for example, and that those particular biological characteristics may be of service in procuring a mate by enhancing display or by enhancing "internal courtship" powers,[44] will not figure in the analysis that follows.[45] The concern is rather to identify and elucidate fundamental behavioral possibilities that by their very nature, their spatio-kinetic dynamics, instantiate certain power relations. In effect, by unearthing the linkage between fundamental behavioral possibilities and certain power relations, we will unearth corporeal archetypes of power.

The term *archetype* will become progressively elaborated in the course of completing the task, and indeed, in the course of the book as a whole. To be emphasized in these preliminary clarifications, however, are the common present-day practices of regarding sex and gender as exclusive though interactive aspects of anatomy (or biology) and culture, respectively,[46] or of making no distinction between sex and gender on the grounds that sex is as culturally determined as gender.[47] The practice fails to recognize animate form and the concrete, complex corporeal relationships that exist between animate form and power. As described above, animate form is not *mere anatomy*. Animate form is indeed *animate;* it is not the simple having of certain bodily parts and not others but more importantly, the having of certain sensory-kinetic possibilities and not others. The idea that gender is thoroughly derivative of culture and the idea that even sex is a thoroughly historical construct fail to recognize how basic features of animate form resonate experientially and behaviorally in sex- and ultimately gender-specific ways. Indeed, to appreciate how fine the relationship is, we must interrogate male and female bodies as specific forms of livability in the world. We must specify concrete realities of bodily life, in particular, those mundane, archetypal realities of the primate body *simpliciter*. Sexed, and ultimately gendered, human bodies can in this way be ultimately seen as fundamentally tied to certain concepts of power and to concepts of deployments of power irrespective of cultural overlays, constructions, diversifications, and so on. As suggested above, staring, averting the eyes, being higher than/lower than, facing front-end toward/facing back-end toward, and making a visual spectacle of oneself are each fundamental primate behaviors that instantiate just such archetypal deployments of power; each behavior is either part of the normal repertoire of 'I can's' peculiar to primates in general, or peculiar to certain primates in particular. We will approach the task of delineating corporeal archetypes of power by considering *presenting,* a complex but excellent example of a certain form of livability in the world precisely because it is both a sex-specific and sex-neutral act.

III.   Presenting as Corporeal Archetype

Insofar as corporeal archetypes are behavioral forms natural to being the body one is, they are species-specific—in the case of presenting, the form is even quasi-order-specific.[48] Moreover, like the psychic archetypes which

Jung describes as structurally homologous to instincts,[49] corporeal archetypes too may be regarded homologous to instincts: they are not learned as such but are intuitively enacted and understood. Presenting is just such a form in this further sense. The basic difference between a Jungian and a corporeal archetype centers on meaning: corporeal archetypal meanings are not expressed in symbolic motifs but are themselves directly evident in the spatio-kinetic dynamics of everyday bodily behaviors. The meanings "submission" and "vulnerability" with respect to *presenting* are apt examples. The meanings are a built-in of primate bodily life. Why? Because primates generally have face-on, front-end defense systems and aggressive displays. In presenting, they face their hind-ends to a conspecific. They thus place themselves in an inferior position insofar as they cannot easily see the presented-to animal nor monitor its behavior, nor can they easily defend themselves in such a position. In addition, in presenting they frequently lower themselves toward the ground, thus giving the presented-to animal an advantage with respect to the all-pervasive biological value, *size*. The back-end presenting posture necessarily adopted by a female nonhuman primate in dorsoventral copulation thus raises a fundamental question. Viewed as a corporeal archetype, is the position one of submission for the female? Does a sexually presenting female feel vulnerable? In finer terms, do the same power relations obtain in a socio-sexual context as obtain in a socio-aggressive context?

Field observations and descriptions in the literature might readily incline us toward an affirmative answer.[50] In her presenting copulatory posture, the female appears in fact dominated by the male. She is not only vulnerable—as occasional neck bites by the male might indicate—but she cannot move freely on her own. Indeed, depending on where the male rests his feet (i.e., on the ground or on the female), she may assume his entire weight. *Being on the bottom or underneath the male, and being frontally turned away from him as well,* the female clearly appears to be in an inferior position—on the nether end of power relations. In these senses, at least, she appears to submit to the male and to be vulnerable to him.

All the same, the question is difficult to answer. It may actually be the wrong one to ask, and for the following reason. While presenting behavior does not differ essentially in its two contexts—copulation on the one hand, and submission, conciliation, or deference on the other—its meaning does. Our third-person human observer classification of the behavior reflects this difference: we classify it on the one hand as a socio-sexual act, and on the other as a socio-aggressive act. The two distinct meanings, however, derive fundamentally from the nonhuman participants themselves, the creatures actually involved in the two "contexts of utterance." Hence, we could only ask the female participants themselves for a decisive answer to the question, for only female chimpanzees themselves could testify to *a necessary affective linkage between the two contexts.* In other words, only female chimpanzees themselves could tell us whether in copulatory contexts presenting, simply in virtue of the intercorporeal postural relations it instantiates, is necessarily felt as a submissive and vulnerable posture. Given this fact, the critical and more properly informative question to ask would be whether *any* archetypal intercorporeal postures invariably carry with them certain feeling tones which, though stronger or weaker in some instances, are never muted whatever the context of utterance. Only in the answering of this broader question could we humans be relatively certain that presenting in its copulatory context is a sex-neutral and not a sex-specific archetype. Only then would we have supportive evidence for extrapolating affective meanings from one context of utterance to another.

It is worthwhile noting that by putting the question of affective linkage in the broader perspective, both the assumption (tacit or otherwise) in our very posing of the initial question and our immediate understanding of what is at stake in presenting are thrown in relief. Our immediate understanding of presenting indicates an unacknowledged common heritage of bodily understandings, that is, a heritage of understandings anchored in primate animate form. Our intuitive comprehension of certain aspects of chimpanzee presenting behavior—their "nervous and even fearful" feelings in socio-aggressive presenting, for example[51]—rests on our own personally experienced understandings of the archetypal intercorporeal meanings of front and back facings, and higher and lower positionings. With respect to the assumption that female chimpanzees feel themselves in a vulnerable or submissive position relative to males in dorsoventral mating, we extrapolate, wittingly or not, from our own sentient understandings of the archetypal intercorporeal meanings of front and back, and above and below, to the sex-specific act of another primate. Assumption and understanding alike indicate an unexamined disposition to read into sex, meanings of animate form. It should be emphasized that, from a methodological point of view, our comprehension and assumption are both rooted in *introspective* evidence: it is by consulting our own experience, whether in so many reflective acts, or in self-intuitive ways that are sedimented into our very being the body we are, that we readily come to the experiential estimations we do of "what it is like to present," that we in turn label the behavior an act of submission or an invitation to copulation, and that we readily transfer (or entertain the possibility of transferring) meaning from sex-neutral to sex-specific context of utterance. The methodological point underscores the earlier epistemological one, that an answer to the question of whether the position assumed in presenting always carries with it overtones of submission and/or vulnerability irrespective of context such that female chimpanzees consistently feel a dorsoventral mating position to be one of risk and inferiority with respect to power relations could only be had by canvassing female chimpanzees themselves.

These preliminary considerations of presenting lead to the conclusion that power relations engendered in the archetypal act of presenting are not necessarily sex-specific; they are rather more likely a function of fundamental intercorporeal postural relationships. This possibility—that submissive behavior is not a female behavior but the enactment of a certain intercorporeal spatio-kinetics (in effect, that to be a female primate is not necessarily to be submissive)—does not surface in descriptive analyses of presenting.[52] For example, ethologist Konrad Lorenz categorically states that submissive behavior is in essence female behavior, that is, it derives from, and is equated to, *female copulatory* posture: "Expression movements of social submissiveness, evolved from the female invitation to mate, are found in monkeys, particularly baboons. . . . 'I am your woman' and 'I am your slave', are more or less synonymous."[53] The same view, muted but still apparent, is found in assessments by primatologists: "Ritualized gestures [presenting and mounting], which reduce tension and prevent fighting, are similar to and possibly derive from the mating postures of male and female."[54] In brief, it is not the spatio-kinetic intercorporeal relationships that are noticed but the sex of the bodies concerned. To *demonstrate* sex-specificity, however, the broader question must be addressed because *that* question is the only one humans can reasonably answer. A brief homespun example that phrases the question in terms of human intercorporeal positional valencies immediately brings this fact into close focus: in human copulation, is that body which is below another body "doomed" to an inferior status? Not only are we clearly in a position to answer that question, but the question brings to the fore a basic similarity between traditional (or traditionally pictured) human female copulatory positioning and primate female copulatory positioning in general. In fact it suggests interesting analogies, all of which would need examination.

The above preliminary considerations show that the relationship of sex to animate form raises complex issues as a result of an evolutionary viewpoint and of deepened corporeal analyses. They do not, however, tell the whole story. Indeed, we must consider that in the economy of nature generally, and in the economy of primate nature in particular, just as the same behavior may have more than one meaning, so the same behavior *may be performed by male and female alike*. Put in this perspective, the posture of female chimpanzees relative to male chimpanzees in copulation clearly does not have a basically or exclusively "female" meaning. The posture's fundamental archetypal meaning is sex-neutral: facing backward to, being lower than, and not being in eye contact with are of the essence of a sex-neutral intercorporeality in which one individual—male or female—is in an inferior position relative to another.

This priority of a sex-neutral archetypal meaning could nevertheless be questioned. It could be claimed that by presenting in socio-aggressive encounters, the inferior individual is reduced to "behaving like a female." Rather than being seen as adopting a posture that has inherently inferior or submissive status, the individual is viewed as engaging in female behavior. But precisely insofar as *both* sexes present in nonsexual primate encounters, and to the same as well as opposite sex—just as *both sexes mount* in nonsexual primate encounters, and mount creatures of the same as well as opposite sex—the claim of an exclusive "female" meaning is not evidentially supportable. Presenting and mounting are socio-aggressive acts performed by females and males alike. It is because both sexes engage in both behaviors, and intrasexually so, that the fundamental, sex-neutral archetypal meaning of presenting is substantiated. Facing backward to, being lower than, being out of direct eye contact with—each is itself an archetypal primate behavior within the global primate archetypal behavior, presenting. The re-enacted dominance behavior of two female gelada baboons demonstrates this fact unequivocally. In a newly established gelada colony, one of the females successfully paired with the single male: she presented to him, he mounted her, and she then groomed his cape. When a second female was introduced, "[she] was eventually accepted, but only after the first, dominant female had gone through the pair-forming process with female number two, with herself in the role of the male."[55] (The phrase "in the role of the male," unless a simple if misleading way of identifying a different *position,* obviously assumes sex-specific archetypes—females are submissive, males are dominant—and reflects precisely the kind of biases that human observers may bring with them and that precipitously cloud over the complexity otherwise apparent in examining the relationship between animate form and sexed bodies.)

The question of whether presenting is an essentially female behavior, *and then* a socio-aggressive behavior common to all members of the species (in which case the behavior in socio-aggressive contexts has the "female" meaning it has because the behavior is proper to females) or whether presenting has an essentially sex-neutral archetypal meaning (the position being an inherently vulnerable one with no primary sexual significance) is thus at this point answerable in a straightforward way. So also is the question of whether mounting is an essentially male behavior, *and then* a socio-aggressive behavior common to all members of the species, or whether mounting has an essentially sex-neutral archetypal meaning, the position being an inherently powerful one with no primary sexual significance. In neither case does the answer controvert the possibility that power relations are a built-in of sexed bodies. The answer means only that power relations are not reducible to sexed bodies *in the case examined*. With respect to the latter, power relations are fundamentally the expression not of a sex-specific invariant but of an invariant in the spatio-kinetic semantics of animate form.

IV. From Nonhuman Primate Presenting to Human Sitting Postures

To elaborate the point, let us consider a contrapuntal, preeminently human example: sitting postures. Anthropologist Gordon Hewes's study of human postures worldwide documents male/female differences both descriptively and pictorially.[56] His study strongly suggests sex-specific postural archetypes. What Hewes found in surveying 480 different cultures or cultural subgroups was that certain sitting positions (or variations thereof, i.e., actual sitting, kneeling, crouching, or squatting) are typically female and certain others are typically male, irrespective of culture. For example, females seldom sit in positions in which one leg is drawn up and the other is to some degree lower and flexed. Neither do they typically stand in what is called "the Nilotic one-legged resting stance" in which an individual, holding a pole that rests against the ground and stabilizing himself with it, stands on one leg, the other leg being raised and the foot placed in some position against the standing leg. (Cattle herders in the Nilotic Sudan regularly assume this posture, hence the name.) These two postures are typically male ones. In addition to these two, males also typically sit in a position similar to a squat, but with the knees only partially flexed. Hewes notes that in our own culture this sitting or resting posture is "frequently assumed by males (and by trousered females)."[57] Though he does not explicitly draw attention to the fact, it is thus apparent that clothing may influence what one does with one's legs, and that rather than being a veritable sex archetype, a presumed sex-specific posture may in fact be taken up by the opposite sex and thereby become sex-neutral. But this interpretation misses an important point.

As might be indicated by his reference to "trousered females," Hewes suggests at several points that what a female does with her legs has, or may have, sexual significance. He states, for example, that "the role of taboos against female genital exposure in determination of acceptable or nonacceptable feminine postures is presumably important, but the evidence is slight."[58] On the other hand, he associates "one of the best cases of a feminine postural habit"—"sitting on the ground or floor with legs stretched out in the midline, sometimes with the ankles or knees crossed" —with tasks commonly performed by females: weaving, for instance, or nursing.[59] Labor rather than genital exposure is in effect suggested in explanation of the female posture. Yet seemingly quite similar sitting postures, in which rather than both legs being outstretched together or crossed at some point, one leg is stretched out and the other is variously flexed at the knee "so that the foot lies above the opposite knee, beneath it, or is sat upon," are *not* common to females.[60] In such postures, the legs are sufficiently spread so as partially to expose genitals. It is thus possible that avoidance of genital exposure rather than labor (or other factors) is the motivation for the posture. In an instance where Hewes comments that "exposure avoidance" might explain the posture—a typically female posture in which the legs are folded to the side—he refrains from endorsing exposure-avoidance straightaway as an explanation because the posture may be the result of there being no chairs or benches to sit on, or the result of clothing restrictions. In sum, while an answer to the question of genital exposure-avoidance appears mooted for lack of clear-cut evidence, the evidence does support, and in a strong sense, sex-specific postural archetypes.

It is curious of course that no mention is made of possible *male* genital exposure or of *male* exposure-avoidance. Indeed, in another study of sitting postures, this one focused on typical Western ones, graphics and accompanying descriptions show that males typically sit with their legs spread, females with them closed.[61] Taken together, the graphics presented in the two studies indicate that there is indeed a difference between male and female sitting postures with respect to the spatial relationship between the legs and the degree of triangulation the relationship establishes vis-à-vis the genitals. Clellan Ford, an anthropologist, and Frank Beach, a psychologist whose work on sexuality is informed by an evolutionary perspective, studied "190 different societies . . . scattered around the world from the edge of the Arctic Circle to the southernmost tip of Australia" and observed that

> the provocative gesture of exposing the genitals has become the subject of widespread social control in every human society. There are no peoples in

our sample who generally allow women to expose their genitals under any but the most restricted circumstances.[62]

The question is why males are not likewise restricted. Indeed, since their genitals are far more anteriorly situated than those of females, why is it that males are not equally if not more assiduously controlled? When males sit with legs spread, their genitals are in full view. Whether they are literally or figuratively in view is in the present context of no consequence; in other words, clothes or a lack thereof is beside the point. There is clearly a double standard. But let us put the question of male genital exposure temporarily on hold and pursue the question of female genital exposure. Ford and Beach state that "although there are a few societies in which both sexes are usually nude, there are no peoples who insist upon the man covering his genitals and at the same time permit the woman to expose her genital region." They go on immediately to say that "exposure of the genitals by the receptive female seems to be an almost universal form of sexual invitation throughout the mammalian scale. Descriptions of the mating patterns characteristic of various subhuman primates . . . [show] the ubiquitousness of feminine exposure. . . . The female ape or monkey characteristically invites inter- course by turning her back to the male and bending sharply forward at the hips, thus calling attention to her sexual parts."[63]

The proscription against human females adopting "legs-spread" sitting postures, and the correlatively unconstrained adoption of "legs-spread" sitting postures by human males, are thus linked to a sex-specific, i.e., female, evolutionary behavioral pattern. We are, in effect, led back to the act of presenting and in turn to the question, is the apparent pan-cultural proscription against human female genital exposure really a safeguard against presenting? If so, presenting might appear to be a sex-specific archetype after all in the sense of its *invitational* meaning. The earlier analysis and discussion of presenting showed only how submission and vulnerability are sex-neutral meanings of presenting; how, in other words, animate form does *not* constrain sexed bodies to certain power valencies to the exclusion of others; how, in nonhuman primate societies, both males and females present (and mount), intersexually and intrasexually. What was not considered earlier is how animate form *does* constrain sexed bodies. Though not analyzed in such terms, animate form in Ford and Beach's evolutionary account of *sexual invitational behavior* constrains human females to certain postural possibilities to the exclusion of others, and, from a proscriptive cultural point of view, to certain sitting postures (as documented by Hewes's studies) to the exclusion of others. This evolution- ary perspective leads us necessarily to a deeper consideration of presenting as socio-sexual act, specifically, as sexual invitation and sign of "the receptive female." It leads us also, however, to consider whether animate form, with respect to sexual *invitational* behavior, does not constrain males in any analogous way to certain possibilities to the exclusion of others. That males are posturally unconstrained is in Ford and Beach's account not considered in need of explanation. Yet it should be because primate males, at least chimpanzee males, invite copulation by penile display. Indeed, Jane van Lawick-Goodall, in her original report on Gombe Stream Reserve chimpanzees, notes that males took the initiative in 176 out of 213 witnessed matings,[64] a proportion that, as I have noted elsewhere, is "already indicative of the attention properly due male sexual display."[65] A consideration of male sexual invitational behavior will bring to the fore the question of human male genital exposure. Oddly enough, initial consideration of this topic will end by shedding considerable light on presenting (leg-spreading) as sexual invitation and sign of the "receptive" female. It will in other words lead us to a first appreciation of a corporeal archetype and its cultural reworkings.

V. The Question of Human Male Genital Exposure

"Soliciting by the normal male [chimpanzee] is highly stylized and involves squatting *with knees spread wide* to display an erect penis" (italics added).[66] Primatologist C. M. Rogers thus initially describes male chimpanzee sexual invitation. He goes on to note that "most wildborn males accompany [penile display] by slapping the ground with open palms. If a female does not present to him, he may after several seconds rise to an erect posture and execute a short dance in some respects similar to a threat display. He will then frequently alternate from one pattern to the other if not interrupted by a sexually presenting female."[67]

Primatologist Frans de Waal writes that "courtship between adult chimpanzees is almost exclusively on the initiative of the male. He places himself at a little distance . . . from the oestrus female. He sits down with his back straight and *his legs wide apart* so that his erection is clearly visible" (italics added). De Waal goes on to note that the chimpanzee "sometimes . . . flicks his penis quickly up and down, a movement which makes it all the more obvious. During this show of his manhood the male supports himself with his hands behind him on the ground and thrusts his pelvis forward."[68]

Both accounts call attention to the position of the male's legs. Both substantiate the claim that leg-spreading focuses attention on the male's penis. Both furthermore indicate that leg-spreading is a voluntary act, at least in the sense that it is a deviation from usual positioning, that is, from the natural alignment of the legs in sitting, crouching, standing, and so on. Where purposefully enacted for sexual ends, male chimpanzee leg-spreading clearly invites copulation. In these very same ways, *bipedal* male leg-spreading—human or nonhuman—can be an invitational act. Where purposefully enacted for sexual ends, it too invites copulation. Hence, regarded simply from the viewpoint of animate form irrespective of species, leg-spreading for male and female primate alike is a way of inviting copulation. Though posturally distinctive with respect to sex—female leg-spreading is typically non-bipedal, male leg-spreading may be bipedal or not—in each case the act focuses attention on the individual's genitals and constitutes an invitational display. Where purposefully enacted for sexual ends, genital exposure equals sexual invitation.

Given the placement of primate male and human female genitalia, it is not surprising that there should be a sex difference between the two overall body postures in which leg-spreading occurs and/or is perceived as sexual invitation. A bipedal stance for a human female accomplishes nothing in the way of genital exposure; it accomplishes even less for a quadrupedal nonhuman primate female whose genitalia face posteriorly. When a human or nonhuman female primate stands up, her genitals face downward, to varying degrees depending on the species. When a human female primate is in a sitting posture, however, and specifically as her legs are spread, her genitals become quasi-visible and accessible. It is of interest in this context to point out that the clitoris of some New World monkeys such as the spider monkey is quite large and pendulous and that, as sexual invitation, females "abduct the thigh," i.e., *spread their legs,* to display an erect clitoris.[69] Animate form in this sense is clearly the constraining force that pan-culturally restricts the human female to certain postural possibilities as a safeguard against sexual invitation. Similarly—though inversely constraining and quite otherwise posturally—for a primate male. As suggested in the earlier reference to van Lawick-Goodall's field studies, and as explicitly shown by the above citations, penile display is a regular behavior in the repertoire of male chimpanzees (as it is also a regular behavior in the repertoire of other male primates).[70] Insofar as his penis is on the anterior surface of his body, and insofar as customary bodily facing is forward in all animals with a head-end—all such animals move forward *toward* something—a primate male can easily call attention to his penis, especially and most specifically his *erect* penis, by spreading his legs. Through the latter act, what in squatting or sitting male primates—human or nonhuman— would be partially obscured by, or figure less prominently by being visually behind, knees or legs, is put up front and made the focal point of attention; similarly, what in bipedal male primates—human or nonhuman—would normally be visible but not necessarily seen is made the focal point of attention: leg-spreading accentuates the transformation and translocation of the penis in erection. Its sex-distinct postural expressions notwithstanding, leg-spreading is thus clearly an invitational sexual act for both primate males and human primate females. Enacted for the purpose of sexual invitation, it is clearly a corporeal archetype, one whose evolutionary roots run deep, as deep as the acts of staring and averting the eyes, of making oneself higher than or lower than, of frontally facing or backing toward another, and of making an exaggerated spectacle of oneself. (The latter archetype is incidentally exemplified in Rogers's description: the ground-slapping and "the short dance"—more properly called *the bipedal swagger*[71]—of the male chimpanzee are both spectacle behaviors, ones enacted for the purpose of gaining attention.) *All* of these acts are primate corporeal archetypes.

There may perhaps be a sensed unevenness about the above conclusion. After all, we are speaking of leg-spreading as a human corporeal archetype when human males, unlike their biological primate cousins, do not ordinarily spread their legs to invite copulation, and when human females do not, in any pan-cultural sense, invite copulation by spreading their legs. If we extend our identification of corporeal archetypes by following Ford and Beach's evolutionary lead and acknowledge female genital exposure to be the rule in mammalian sexual invitation, however, and at the same time follow the clue given by primatologists and recognize male genital exposure to be the equal rule in primate sexual invitation, we see clearly that *genital exposure is a biological archetype: leg-spreading is an invitation to copulate.* We see this all the more clearly when we consider that *exposure-avoidance is not* a biological rule. Exposure-avoidance indeed has no place in nonhuman primate sexual relations. If we ask how there can be a corporeal archetype when its regular enactment is not in evidence, we find the rather obvious answer that cultural overlays mask the archetype; exposure-avoidance is precisely the cultural mask. Whether through the wearing of clothes, sheaths, or other ornamentation, or whether through tacit or explicit codes of conduct, genitals are consistently hidden from view. The point, of course, is that consistent hiding of the genitals not only consistently hides the genitals, but consistently hides the corporeal archetype as well. In so doing, it keeps at bay the possibility of what culture-laden humans likely envision an untempered sexuality, i.e., wild orgies typical of "the beasts."

The sexual sobriety of nudist colonists aside, it should be parenthetically noted with respect to the presumption of an untempered sexuality that if we take our cue from chimpanzees on the grounds that we are genetically closer to them than other primates, then we must literally reckon with *their* animate form and its relationship to *their* sexual practices when it comes to using what paleoanthropologists term "the comparative method," i.e., using extant chimpanzee behaviors as a standard for fleshing out ancestral hominid behaviors, and for enhancing our own self-understandings.[72] We cannot, in other words, simply assume that short of culture, our sexual behavior would mimic chimpanzee sexual behavior, nor can we with reason simply assume categorically that *their* behavior is *untempered.* On the contrary, precisely because there are uniform sexual signalling behaviors in chimpanzee societies, there are, by precisely human standards, *uniform* rather than untempered sexual behaviors. These basic considerations aside, there is furthermore no reason to assume without question that over millions of years what was practiced in our hominid beginnings did not change over time, particularly with the emergence and development of cultures. Indeed, perhaps early hominid sexual practices were more like those of extant gorillas, offspring appearing regularly but copulation being infrequent by present-day human standards. The priority of "sex," especially as it is known today in the West, was not necessarily the priority of yesteryear, especially a yesteryear reaching into the millions. It is of interest to note in this context the observation of one primatologist vis-à-vis chimpanzee, hamadryas baboon, and Guinea baboon male behavior: "The maturing male in all these groups seems to be concerned more with status, control, and courtship than with copulation, an activity he would have perfected in early adolescence."[73] It is well to keep in mind too that where primates do not have to forage for themselves—that is, where food procurement is not an everyday, time- and effort-consuming activity—sexual behavior and preoccupations are heightened. The priority of "sex" in the United States today appears related to just such leisure. The relationship is born out by those "kept" situations where chimpanzees, as in de Waal's study,[74] live in a natural environment but feed wholly unnaturally, i.e., on food apportioned by human caretakers: their sexual— and political—interactions are intensified. With their food needs cared for, the chimpanzees, after all, have not much else to do but interact with each other. Can something similar be said of present-day Western humans?

In sum, while humans, like other creatures, are constrained by animate form to certain sexual invitational possibilities to the exclusion of others— possibilities that originally, no doubt, served in positive ways to perpetuate the species—they are also reversely constrained by culture. Their repertoire of invitational 'I can's' is in consequence a complex conjunction of evolutionary and cultural factors. The complexity of the conjunction is substantively apparent in the act of leg-spreading—its archetypal primate heritage, and its cultural restrictions and freedoms. By describing in finer terms how, with respect to animate form, the advent of consistent bipedality radically changed the nature of primate sexual invitational behavior, we can come to a deeper understanding of primate sexual signalling behavior and the present-day pan-cultural proscription against human female leg-spreading. We can, in short, come to a deeper understanding of a radical shift in sexual signalling powers.[75] We can in turn begin to elucidate the formidable socio-political implications of present-day biological explanations of human sexuality by way of "the receptive female."

VI. Sexual Signalling Behavior and the Reigning Western Biological Explanation of Human Sexuality

With the advent of consistent bipedality, female genitalia were no longer permanently on display; male genitalia were. Moreover while extant nonhuman primate penes are normally hidden and extruded only upon arousal— and presumably, ancestral nonhominid primate penes were similarly hidden and extruded only upon arousal—a human's penis is permanently exposed —and presumably an ancestral hominid's penis was similarly permanently exposed.[76] Furthermore, the anatomical placement of female genitalia changed; the placement of male genitalia did not change. While penes were still anteriorly located on the primate body, vaginas became more anteriorly located. In brief, the standard primate morphological/visual relationship that heretofore obtained with respect to male genitalia and to female genitalia was radically altered by consistent bipedality: what was normally hidden in quadrupedal male primates became exposed in bipedal male primates; what was normally exposed in quadrupedal female primates became hidden in bipedal female primates.[77] This radically changed relationship necessarily meant a radical change in sexual signalling behavior, behavior that is an essential part of normal mating for any species whose members must come in direct contact with one another in order to mate. From the perspective of animate form, what was before accomplished through both penile display and presenting—what might basically if figuratively be described as invitation by mutual "leg-spreading" (taking *presenting* in Ford and Beach's sense of "genital exposure"), and basically if idealistically described as a situation of mutually recognized readiness and potency—was now accomplished only through penile display. Exclusive reliance on penile display is axiomatic given hominid animate form: only what is exposed, visibly apparent, can minimally count as a sexual signal, and then, through enhancement of some kind, become a display.[78]

Adolf Portmann calls attention precisely to such basic visual signalling when he shows in detail how visual patternings on animate forms signal "genital-end" and "head-end," for example, and in turn form the basis of behavioral displays.[79] Evidence of this same basic visibility and subsequent enhancement through display is apparent in the present context: a human male does not first have to spread his legs to call attention to his penis; being an upright creature, his penis is already there, exposed, in full visible sight.[80] What leg-spreading does is enhance or intensify exposure—as Western guitar-playing, rock-singing males might readily testify. In biological terms, they are enacting a human male sexual display, a sexual *display* in precisely the sense defined: by their actions they accentuate *an already present signal*. Indeed, in conjunction with the advent of consistent bipedality one might speak with reason both of *an intensification of male sexuality with respect to signalling readiness and potency through penile display, and correlatively, of a loss of natural sexual signalling in females.* With the advent of consistent bipedality and the anterior shift in female genitalia, not only was presenting per se no longer an anatomically possible way of inviting sexual relations or of demonstrating sexual readiness and potency, but a female who no longer had estrous cycles would no longer have anything to display to begin with in the way of periodic swellings and coloration changes. The question is, what did hominid females do to invite copulation in the absence of presenting?

The only minimally sure way of answering the question is by appeal to leg-spreading: primate sexual invitation is by genital exposure. The question has never been answered on these grounds because the question has never been asked to begin with; and it has never been asked to begin with because the terms of the discourse have been changed. The question has been swallowed up in the fundamental Western biological characterization of human, that is, hominid, sexuality as "year-round receptivity." With the central concentration on females being receptive "year-round"—hominid females are not constrained to estrous cycles like other primates—a familiar biological picture of human sexuality emerges: hominid males are conceived as being always ready for any sexual adventure and hominid females are conceived as being always available—even if not always willing. There are multiple issues involved in this viewpoint.[81] Our exclusive concern here is to show how the shift in discourse—from sexual invitational signalling behavior to sexual "receptivity"—by trading basically in physiology rather than in corporeal archetypes and their cultural proscriptions, undercuts female sexuality at the same time that it propels it to evolutionary prominence.

Conceived simply as being receptive "year-round," human female primates are conceived as lacking any natural sign of sexual readiness and potency, certainly not any natural sign on the order of vulval swellings and changes in coloration, or on the order of penile erection with its visible upward movement and its changes in size, texture, and shape. Conceived as lacking any natural sign of readiness and potency, human female primates are in turn conceived as lacking natural powers of sexual expression. This means that on the one hand, there is no bodily organ to be consulted to determine a female's true sexual disposition, and on the other, that there is no bodily organ by which a female can express her sexual yearnings. She is simply open and available "year-round." From this current Western, essentially male point of view, a human female invites copulation simply by being there. She herself is an invitation. Hence, wherever human sexual readiness and potency exist, there is a human male—and a female to be found to satisfy him. Sociobiological anthropologist Donald Symons affirms this human state of affairs unequivocally in his text *The Evolution of Human Sexuality* when he states that "women inspire male sexual desire simply by existing."[82] Insofar as his book is used as a textbook and is considered a major formulation of human sexuality, it will warrant consistent attention in the continuing critical analysis of female "year-round receptivity." His account of female sexuality may indeed be treated as a paradigm of the prevailing Western biological view and as such is an excellent prism through which to show the intricate cultural translations and reworkings of leg-spreading as corporeal archetype of primate sexual invitational behavior.

Not unexpectedly, "year-round receptivity" reduces human female sexuality not only to male readiness and potency but to male fantasies. Symons affirms this reduction as well when he writes that "pornotopia is and always has been a male fantasy realm; easy, anonymous, impersonal, unencumbered sex with an endless succession of lustful, beautiful, orgasmic women reflects basic male wishes."[83] The basic theorem underlying "basic male wishes" is unmistakable: human males are sexually insatiable; females are receptive "year-round." By neglecting serious study of the body itself,[84] and instead fastening on pornography and on cultural studies of sex practices to locate "human universals" (Symons's term), such putatively biological accounts cannot hope to come to grips with the origin and evolution of human sexuality. Not only do they mistake current practices for originary/evolutionary ones,[85] but they ignore from the very beginning both the radical shift in primate animate form that defines consistent bipedality and the related questions that radical shift raises about hominid sexual signalling behavior.[86] It is quite easy to see how, through such formulations of the origin and evolution of human sexuality, the notion of females as sex objects is strengthened at the same time that it is steadfastly maintained. Not only this, but it is easy to see how females themselves become transformed into two opposing cultural psychic archetypes: the witch/temptress and the goddess/mother, a man's she-devil or the figure who gives him birth. As Jung points out, the real woman dissolves in such primitive images. She dissolves because her individuality gives way to "infantile dominants," that is, to archaic visions tied on the one hand to fear and on the other to worship.[87]

A natural sign is unequivocal—"where there's smoke, there's fire." A female conceived simply as receptive "year-round" can give no such definitive invitational signal. Her true sexual dispositions of the moment are never visibly apparent. Certainly she can spread her legs, but she cannot spread them without going against cultural proscriptions, thus going against what is deemed "moral." To reclaim her natural sexuality would mean in effect to be wanton, corrupt, degenerate, even wicked. More than this, it would put her at severe risk. By adding leg-spreading to simply being there, she doubles sexual invitation and runs a double risk. As if this were not enough, in taking the double risk she not only emphatically shows herself open and available to the male to whom she addresses her display, but she may well be considered showing herself open and available to any and all males who might witness or hear of her leg-spreading. Exposure-avoidance in this sense guarantees both her reputation and her safety. At the same time, however, her "year-round receptivity" puts and keeps her in a perpetual double bind: from a male point of view, she is "damned if she does and damned if she doesn't." The double bind is reflected in the notion that females are capricious; they do not know what they want. They will flirt, they will hold your hand, they will kiss you, but they will not "go all the way"; or, they will "go all the way," but they are either loose women to begin with or sorry and unsatisfied women in the end. Once more it is easy to see how females are transformed into a cultural psychic archetype of destruction. Capricious temptresses, they are incarnations of the devil.

As briefly indicated earlier, there is another remarkable side to the implications of "year-round receptivity." Human females are regarded as being without desire. Since they no longer signal naturally like most other female primates—since they have no organ proper to sexual desire—they cannot possibly have sexual desires. Since they are without sexual desire, they spread their legs not as a genuinely felt invitation or response to an invitation but as an allowance. Symons's characterization of copulation as a "female service" indicates precisely this conception of females.[88] Females can *always* spread their legs; it is merely a matter of their choosing to do so. The principle of female choice—the principle first described at length by Darwin whereby females select certain males over others in light of their superior qualities, e.g., more ornamented, better armed, more vigorous, and so on[89]—is egregiously weakened, even negated, by the conception. Conceived as rendering a service to males, females do not actually choose males in a Darwinian sense; they choose only to accommodate a male. The warpedness of this emaciated version of female choice is particularly well highlighted by recent, highly detailed accounts of female choice that focus on female discrimination of male genitalia on the basis of their "internal courtship" powers.[90] The significance of this latter research is best appreciated in the context of internal fertilization since much of Darwin's original evidence for female choice came from studies of birds, creatures who do not actually copulate. Accordingly, the recent studies of female choice alluded to are concerned with features of male genitalia precisely because male genitalia contact female genitalia directly in sexual reproduction by internal fertilization, because male genitalia vary more than female genitalia, because males compete for females, because specialized structures in male genitalia tend to evolve rapidly through female choice, and because male genitalia in turn are taxonomically significant with respect to speciation. It is understandable why in *this* context, female choice—in the strong, original Darwinian sense—is of central concern. In those creatures who reproduce by internal fertilization, females choose one male over another in terms of his ability to stimulate her more fully "by squeezing her harder, touching her over a wider area, rubbing her more often, and so on."[91] Reduced simply to "year-round receptivity," a human female is far removed from such discriminations. She is simply open and available "year-round"; even her capacity for orgasm—if she has orgasms or had them—is, in Symons's sociobiological account, "a byproduct of selection for male orgasm."[92]

Now if a human female has no sexual desire, she can hardly be motivated even for the sheer pleasure of "internal courtship" to choose one male over another for his greater stimulatory powers. She can merely *allow herself to be copulated.* The locution is the natural extension of the phrase some scientists have used to describe female sexuality: "continuously copulable."[93] The phrase gives ample testimony to the conception of female sexuality as a sexuality both without desire and without choice in any discriminating and autonomous sense of the word. In fact, female choice in Symons's work cryptically passes over into male hands; it is actually defined in terms of *male power*. Symons writes that "Darwin identified two types of sexual selection: intersexual selection, based on female choice of males (*'the power to charm the females'*), and intrasexual selection, based on male-male competition ('the power to conquer other males in battle')" (italics added).[94] As is apparent, in Symons's account sexual selection is synonymous with male power in *both* intersexual and intrasexual selection. His use of the term "*intersexual selection*" notwithstanding, there is not the slightest hint of *female choice*. But then the parenthetical definition that Symons gives of intersexual selection is actually one of his own making; though attributed to Darwin, the definition is not Darwin's. Both of Symons's definitional quotations are in fact taken from a passage in which Darwin is describing not sexual selection per se, but advantageous modifications in male animate form that result from sexual selection. The modifications are advantageous, Darwin explains, insofar as they "make one male victorious over another, either in fighting or in charming the female."[95] In the specifically relevant passage, Darwin, after commenting on how sexual selection can confer advantages beyond those of natural selection, goes on to say that "we shall further see . . . that the power to charm the female has been in some few instances more important than the power to conquer other males in battle."[96] In this context of *advantageous modifications,* Darwin is indeed speaking of, and in fact contrasting, two male powers; *he is not in this passage defining the two types of sexual selection*—intersexual and intrasexual. In Darwin's account, as in the accounts of evolutionary biologists generally, there is no question but that the former type of selection proceeds by female choice, the latter by male-male competition. The former is hence not a power of males but of females. To deny that the power exists is one thing. To transform it surreptitiously into a male power is quite another.

As if the usurpation were not sufficient, Symons in fact completely and explicitly devalues female choice as a factor in the evolution of humans, affirming in its place *intrasexual* selection: "Although copulation is, and presumably always has been, in some sense a female service or favor, hominid females evolved in a milieu in which physical and political power was wielded by adult males."[97] His judgment is peculiar given both the substantiated evolutionary importance of female choice (it is the major factor creating "runaway selection" of certain male traits, for example)[98] and the quintessential linkage of female choice to male-male competition: males compete with one another in a variety of ways to win females, but they do not win females automatically with the winning of the competition. As Darwin originally noted, "males which conquer other males . . . do not obtain possession of the females . . . independently of choice on the part of the latter."[99] The above-mentioned studies of male genitalia are of considerable interest in this context. Were we to take seriously Symons's definition of intersexual selection, we would have to ask in what "the power to charm the females" consists. Since he nowhere spells it out, we could genuinely wonder in what specific capability this male power would have rested in terms of the evolution of human sexuality—and in what it might rest today. As noted above, for creatures who reproduce by internal fertilization, internal courtship powers can be of singular importance in the successful pursuit of a mate. The genitalia of hominid males would in this fundamental sense be no different from the genitalia of males of other species who reproduce by internal fertilization. The stimulatory powers bequeathed by animate form in both a species-specific and individual sense would have to figure in any account of "the power [of hominid males] to charm the females." Indeed, in a properly exemplified view of hominid intersexual selection, hominid penile display—particularly the display of an erect penis—would have to be taken into account since it would have visually indicated to females the internal courtship powers of the displaying male.[100] On *this* view of the evolution of human sexuality, the power of male genitalia is not simply the power to deliver sperm. It is the power to give pleasure, and it would thereby understandably have been a power subject to sexual selection by females.

Spelled out succinctly in the terms of the critical analysis thus far, "year-round receptivity" reduces female choice to two logically related facts: a human or ancestral hominid female has or had no sexual desire, and therefore she cannot have or have had any motivation for choosing one male rather than another or even any motivation for choosing sheer sexual pleasure; lacking any motivation for choosing one male rather than another or for choosing sheer sexual pleasure, she is or was simply choosing to spread her legs or not. From this summary perspective, it is once more readily clear how the received Western biological account of human sexuality gives way on the one hand to the notion of females as sex objects—founts of orgasmic pleasure for men—and on the other hand to the transformation of the female herself into a cultural psychic archetype. On the former reading, female invitation—leg-spreading—is a mere allowance; on the latter reading, it is a trap: the female body is an insidious lure that tempts but does not allow, or if it allows, is evil. Klaus Theweleit, in his thoroughgoing analysis of Nazi sexual politics, practices, propaganda, dictums, images, metaphors, and more, gives perhaps the ultimate picture of the female body as the archetype of evil.[101] Whether allowance or trap, sexual invitation has no felt significance for the female. What is spreading or not spreading its legs is not a subject but an object.

Clearly the negation of choice is the negation of power. Autonomy disappears. To say as above that a female is *"choosing* to spread her legs or not" is true only in the abstract. That this is so becomes immediately apparent when we consider *allowance* concretely in the given paradigmatic context of "continuous copulability" or "year-round receptivity." When a female *allows* herself to be copulated, she is *allowing* at some point along the continuum, or at the extreme, of two possible senses of allowing, neither of which engages her as an authentic subject. She can allow copulation in the sense of letting it happen as a "service or favor," in which case it is difficult to imagine her in any vigorous or involved way caught up in the act, or she can allow it in the sense of submitting physically to male sexual power. In essence, these two senses reduce to the possibility that a female can spread her legs voluntarily or they can be spread for her by force. Given the two possible senses, it is not surprising that with the conception of females as receptive "year-round," and with the denial of female choice, rape becomes a distinctive and most obvious male possibility. Neither is it surprising that rape figures as a prime factor in Symons's explanation of "copulation as a female service." And neither, finally, is it surprising that in that explanation, Symons finds rapists' actions justly explicable. "Many feminists," he writes, "call attention to male anger as a motive in some rapes but fail to note what is obvious in many interviews with rapists, that the anger is partly sexual, aimed at women because women incite ungratifiable sexual desire."[102] It is in this context—a single sentence later—that he remarks "women inspire male sexual desire simply by existing." We are, in effect, back to what is perhaps the first cultural reworking of the primate corporeal archetype of exposure as sexual invitation, the notion that females invite copulation simply by being there. Appended to this notion, however, is the seemingly added consequence that if their existence puts them under perpetual risk of sexual attack, that is *their* problem. Though not a "facultative adaptation," rape is an act natural to males: "the evidence [from rapists] does appear to support the views . . . that human males tend to desire no-cost, impersonal copulations . . . hence that there is a possibility of rape wherever rape entails little or no risk."[103] Interestingly enough, the section of the book in which Symons discusses rape is titled "Forcible rape." Is there actually another kind? Is a person less a person in an ethical sense if mentally retarded, asleep, or unconscious?

One last aspect of the reigning Western biological explanation of human sexuality vis-à-vis sexual signalling behavior warrants special comment. As discussed earlier, when human sexuality is characterized as "loss of estrus" —i.e., a human female is receptive "year-round"—it is virtually a foregone conclusion that a human female cannot possibly have any sexual desire. She has nothing to have it *with*. There is a further scientific dimension to this deficiency that should be pointed out. All functions need an *organ*. In Western thinking as evidenced by Western medicine, there is no function without a structure. A vagina is a mere passageway, an opening.[104] A clitoris is obviously a structure, but it is not an *organ of desire*, the stuff of which sexual yearnings are made visibly present in one's own experience and in the experience of others. As a belatedly (in the West) discovered feature of female anatomy, a clitoris is regarded only as an organ of sexual pleasure. The contrast with a penis is manifestly remarkable: a penis is *both* an organ of desire and an organ of pleasure. There is no mistaking male sexual desire and pleasure—no equivocation, no ambiguity. Male sexual desire and pleasure are indeed commonly conceived as wrapped up in one. The visible disequilibrium in animate form between human male and human female in effect reinforces the Western biological explanation of human sexuality as "year-round receptivity." It reinforces the notion of women as *always* capable of spreading their legs, of *allowing* something sexual to transpire, of being in Freudian terms "the passive sex," so passive in fact that, being under no periodic pressure to copulate, being simply receptive "year-round," the human female is under no pressure to copulate at all. Only her desire to have a child—presuming, of course, she knows of the connection between copulation and pregnancy—impels her "naturally" to copulate. Sexual desire is thus conclusively for the female a desire *manqué;* what she experiences is nothing more than reproductive desire.

VII.   Year-Round Receptivity and Penile Display: Uncovering the Hidden Relationship

The foregoing critical analysis of "year-round receptivity" can be extended to reveal a further socio-political dimension of the cultural reworking and translation of genital exposure as corporeal archetype. That evolutionary scientists concerned with the evolution of human sexuality have failed to remark on the obvious—on penile display—and that some of them have either reinterpreted Darwin's fundamental principle of female choice in sexual selection or forgotten it altogether in invoking the principle of "year-round female receptivity" show just how clearly the question of animate form and of sexual signalling behavior has been overlooked. How an ancestral male or female hominid would have invited sexual union is a question nowhere raised; in turn, the question of genital exposure within the evolution of human sexuality—from ancestral to present-day times—is nowhere raised. Oversight of the obvious is particularly queer in light of the enormous literature on, and concern with, sexual signalling behavior in nonhuman primates. It is all the more queer in light of the fact that, as indicated earlier, nonhuman primate behavior is regularly used as an analogical standard to infer ancestral hominid behaviors. How could the sexual impact of consistent bipedality be missed? How in turn could penile display be missed? It is important to emphasize the oversight, the focus instead on "year-round receptivity," and the enormous primatological literature because present-day Western cultural attitudes are so heavily influenced by science. We are indeed the inheritors *and practitioners* of precisely the *scientia sexualis* Foucault describes. By failing to see the obvious and reckon with its archetypal significance, and by using data on nonhuman primate sexual behavior only to contrast estrous cycles with "year-round receptivity," evolutionary scientists, perhaps justifying their own personal predilections, have skewed the picture of human sexuality and sizably influenced present-day Western cultural attitudes in the following way.

The characterization of hominid sexuality as "year-round receptivity" can itself be characterized as the replacement of periodic leg-spreading (presenting as per estrous cycles) by continual leg-spreading. It is difficult not to judge this view of hominid sexuality as a shallow male account, male because it focuses *all* attention on the female and fails to investigate male genitalia and sexual signalling behavior, shallow because it similarly fails to investigate consistent bipedality and animate form, and gives no reference to penile display. It avoids the question of male genital exposure entirely. In a broad sense, the view coincides with Western cultural practice generally; that is, a male's body is not anatomized nor is it ever made an object of study in the same way as female bodies. The net result is that the penis is never made public, never put on the measuring line in the same way that female sexual body parts are put on the measuring line. On the contrary, a penis remains shrouded in mystery. It is protected, hidden from sight. What is normally no more than a swag of flesh in this way gains unassailable stature and power. To call attention to it by actually exposing it, let alone actually displaying it by leg-spreading, is in consequence readily regarded not as an invitation, but as a threat.[105] As an object perpetually protected from public view and popular scientific investigation, it is conceived not as the swag of flesh it normally is in all the humdrum acts and routines of everyday life but as a Phallus, an organ of unconditioned power. Indeed, a new corporeal archetype is born, the archetype of power par excellence. This archetype is a cultural translation of penile display. The translation deviates considerably from the evolutionary original in that the archetype is not *of* the body itself but a *symbol* of the body, a corporeal archetype once removed, so to speak. We know this because however much and in whatever ways we are seduced or intimidated by the symbol and myth of the Phallus, we know that the real everyday thing pales in comparison; it is no match for perpetual engorgement. Nonetheless, the archetype prevails: it is as if Western human males were in a steady state of tumescence. Ironically, of course, male exposure-avoidance protects this image by not giving it the lie. But the cultural masking is not what basically sustains the archetype. What basically sustains the archetype is "year-round receptivity." *The male correlate of "year-round receptivity" is perpetual erection.* The two go hand in hand. Our eyes focused wholly on "year-round receptivity," the myth of a perpetually erect penis is kept alive: males are *always* ready for any sexual adventure. Our attention caught up wholly in "loss of estrus," penile display, its male signalling correlate, remains invisible and thereby forever present as possibility if not actuality: females can always provide a service; they can always be threatened. In these circumstances, it is no wonder that the Phallus remains indomitable, retaining its sovereign hold on the Western sexual psyche, and that sexual power is conceived to be a solely male prerogative. The overarching power of the Phallus derives from an imaginary repertoire of 'I can's' peculiar to human males—males who pin their research and their dreams on "year-round receptivity," and who are thereby always ready and potent, never uncontrollably flaccid or sexually spent for all that "easy, anonymous, impersonal, unencumbered sex with an endless succession of lustful, beautiful, orgasmic women."

4
Corporeal Archetypes and Postmodern Theory

A cancelation (sic) can't go wrong. . . . Absence is perfection. "Thou ceaseth to be something thou hadst done better never to become." . . . [D]emonstration of the void in the thing and the thing in the void. Also a demonstration of subjective continuity. . . . [A] 500-foot paper man on Fifty Third Street between Fifth and Sixth Avenues. . . . It took 100 hours to make and 10 minutes to destroy. . . . Gilreth International Company . . . makes this dissolvable paper. . . . It's a sterile edible material. Spies can now eat their information. . . ."This paper has undone me."

Jill Johnston[1]

I. Introduction

The previous chapter exposed in detail the underpinnings of a biology geared to the maintenance of a certain concept of power and blind to the significance of animate form. The purpose of this chapter is to expose in equal fashion the underpinnings of postmodernism that are similarly geared to a certain concept of power and blind to the significance of animate form. Through successive deconstructive *explications de textes,* the chapter offers radically new insights into how Derrida's "epoch of writing" is kept alive by a denial of the living body and an affirmation of a scripturally copulating genital, or in other terms, by a replacement of animate form with grammatological form.

Gayatri Spivak in her justly lauded preface to Derrida's *Of Grammatology* speaks praisingly of Derrida's brilliance in finding the trace, the contradiction which is more than a simple conceptual lapse, the contradiction which in any given work remains unresolved, which haunts any reputed finalities of meaning and in consequence de-centers knowledge. She speaks laudingly of

An earlier and shorter version of this chapter was presented at the March 1992 American Philosophical Association Pacific Division meeting in Portland, Oregon, at a symposium titled "Philosophy of Bodymind."

Derrida's "spectacular readings"—those "acts of controlled acrobatics" in which he fastens upon seemingly innocent moments of a text. She writes, "The strategy of deconstruction . . . often fastens upon such a small but tell-tale moment."[2] Later she describes this moment as "the moment that is undecidable in terms of the text's apparent system of meaning, the moment in the text that seems to transgress its own system of values."[3] She gives examples of this moment in Derrida's reading of Rousseau, of Plato, and of Aristotle.[4] Later still, she writes that "Perhaps all texts are at least double, containing within themselves the seeds of their own deconstruction."[5]

There is a moment in Derrida's *Of Grammatology* in which just such seeds present themselves. The seeds are found in the form of a footnote, an appropriate Derridean niche. In the footnote, the entire program of deconstruction as subsumed in the name of *writing* affirms a denial of the living body, that is, a denial of animate form, the fundamental corporeality of human life. This buried-at-the-back-of-the-text denial that effectively puts the living body out of play is in truth a pseudo-phenomenological backstage manoeuvre that thinly hides its Husserlian borrowings. What the *explications* will show is how these borrowings keep the deconstructive enterprise alive, and how Derrida, in resolutely suspending the life of the living body altogether, goes much further than his psychological mentor—Freud—who merely left the body behind and unattended. What the *explications* will in turn serve to exemplify is how postmodern theory, in effacing the living body, blinders us to our heritage of corporeal archetypes. These archetypes, primordial forms of behavior deriving from our primate past, cannot be deconstructed by the stroke of a pen. However culturally reworked, elaborated, distorted, or even suppressed, they inform our lives. They come with our being the bodies we are. Only a metaphysics of absence could make us think otherwise. As we shall see, such a metaphysics is in fact precisely what sustains Derrida's relentless critique of a metaphysics of presence.

II. On the Way to Forging a History of Writing

In his book *Of Grammatology,* specifically in his chapter titled, "Of Grammatology as a Positive Science," Derrida's aim is to construct a history of writing consonant with the semantic and metaphysical claims of deconstruction. To achieve his aim, Derrida must in fact construct an evolutionary history different from the Darwinian one we know. Such a history is essential if the tenets of deconstruction are to be secured. Derrida begins by invoking the paleoanthropological research analyses of French archaeologist André Leroi-Gourhan, in particular those found in Leroi-Gourhan's book

Le geste et la parole, a book which addresses the question of evolutionary relationships among several specific hominid accomplishments: early hominid stone tool-making, paleolithic cave art, verbal speech, and writing. By taking from Leroi-Gourhan only that which serves grammatological purposes, Derrida infers to the reader that Leroi-Gourhan's work is a testimonial to the tenets of grammatology. For example, he appropriates Leroi-Gourhan's term "mythogram." Mythogram is the name given by Leroi-Gourhan to *superimpositions,* paleolithic cave drawings in which animals and lines are painted over one another. Derrida appropriates the term without relating it to paleolithic cave art at all. He invokes the term first in the context of conjuring "a past of nonlinear writing," and then of conjuring immediately an abrupt assault and annihilation of this past: "Writing in the narrow sense" he says, "and phonetic writing above all—is rooted in a past of nonlinear writing. It had to be defeated. . . . A war was declared, and a suppression of all that resisted linearization was installed. And first of what Leroi-Gourhan calls the 'mythogram' [was suppressed], a writing that spells its symbols pluri-dimensionally; there the meaning is not subjected to successivity, to the order of a logical time, or to the irreversible temporality of sound. This pluri-dimensionality does not paralyze history within simultaneity, it corresponds to another level of historical experience. . . ."[6]

One can see straightaway why Leroi-Gourhan's characterization of "superimpositions" as multidimensional mythograms (*le graphisme multidimensionnel*)[7] appeals to Derrida: pluri-dimensional mythograms are coincident with grammatology. Not only this but, so described, they are concrete forerunners if not exact ancestral equivalents of Freud's mystic writing pad, a pad that, at the same time it offers a fresh surface, it preserves a trace. The mystic writing pad was Freud's model of the psyche. Derrida in later writings appropriates the Pad as the model of grammatology, of writing "in the broad sense." More will be said of the mystic writing pad presently. What is of moment here is that Derrida wants to situate grammatology within an evolutionary history without getting caught in the latter's actual complexities, relationships, or scope. He can do this by giving it a metaphysical history, and that is his actual achievement, for the evolutionary history of grammatology is forged at the expense of the living body which necessarily undergoes both a radical physical and metaphysical transfiguration.

Derrida begins by telling us that we must disturb the familiar historical knowledge that Leroi-Gourhan gives us. In less jostling terms, he says, we must *"awaken* a meaning of hand and face in terms of that [familiar historical knowledge]" (italics added), and this because the history of writing is not erected on the basis of bodily changes, changes, we should note, that are actually changes in animate form and that in a strongly Husserlian sense opened up new bodily possibilities that in part expanded, in part fundamentally altered, a repertoire of 'I can's'.[8] Quite otherwise. "The history of writing," Derrida tells us, "is erected on the base of the history of the *gramme* as an adventure of relationships between the face and the hand."[9] Without having even mentioned upright posture previously, he proceeds to proclaim that "this representation of the *anthropos* is then granted: a precarious balance linked to manual-visual script."[10] ("Manual-visual script" is Derrida's cryptic and assumptive way of referring to mythograms.) Derrida of course grants no more because his deconstructionist history of writing will go beyond what he has previously characterized as "mechanistic, technical, teleological" analyses of the relationships between "gesture and speech, body and language, tool and thought."[11] The historical task, as Derrida envisions it, is to recapture the unity of body and language, gesture and speech, tool and thought *"before* (italics added) the originality of the one and the other is articulated. . . . These original significations," he says, "must not be confused *within the orbit* of the system where they are opposed." In other words, the original, *Ur,* relations between face and hand that characterize grammatology are not to be confused with those relationships in which body and language, gesture and speech, or tool and thought are, *according to Derrida,* oppositionally dichotomized. Thus, to recapture the original significations we must utilize Leroi-Gourhan's science but, as Derrida forewarns, we must at the same time go *vastly* beyond its limits. "To think the history of the system," Derrida writes, "its meaning and value must, in an *exorbitant* way, be somewhere exceeded."[12] Accordingly, grammatology must make its own history, a history in which the body must undergo radical transformation to find its "profound unity"[13] with writing. In true deconstructive fashion, the history of grammatology will thus both do away with familiar relationships between gesture and speech, body and language, tool and thought, and simultaneously carry them forward to their new meanings.

III. The Moment and Its Analysis

Because Derrida never refers specifically to the ancestral hominid tools or drawings that are the subjects of much of Leroi-Gourhan's attention, "manual-visual script[s]" are easily enfolded into the new hominid history that Derrida is writing and, as noted above, actually provide its point of departure through their linkage with a delicately maintained upright posture—"a precarious balance."[14] But if "the history of writing is *to be erected* on the base of the history of the gramme" (italics added), then of course the precarious balance must be *upended*. Our feet must in other words be exed out; in Derrida's favored term, they must be put *"sous rature"*—under erasure. However much Derrida has emphasized an *Ur* unity of body and writing, it is clear that he does not mean body in the sense of animate form. On the contrary, that familiar, living human body must, in a radical and deeply fractionated sense, be physically and metaphysically transfigured. Feet will henceforth no longer be privileged as our base of support but will be replaced by the gramme. To write, and especially to write the history of writing, it becomes in effect understandable why Derrida finds it necessary at the very start to give the body a *precarious* balance; he can take advantage of its instability, knock it off its feet, and replace its ordinary, everyday base of support with a trace, with *différance*. It is in turn not surprising that in continuing to forge his history of writing, Derrida's next move is to affirm magisterially that "This [precarious] balance is slowly threatened." To support this menacing and causally unexplained assertion, Derrida gives snatch quotes from Leroi-Gourhan. He writes, "It is at least known that 'no major change' giving birth to 'a man of the future' who will no longer be a 'man', 'can be easily produced without the loss of the hand, the teeth, and therefore of the upright position. A toothless humanity that would exist in a prone position using what limbs it had left to push buttons with, is not completely inconceivable.'" We begin to realize just what Derrida has in mind when he writes that the meaning and value of the history of the system must be exceeded "in an *exorbitant* way." An exorbitant history is indeed being written: what must be exceeded is the living human body—animate form. Enter evolution which slowly takes back upright posture, and with it, hands and teeth. Enter evolution which will lay the body low—in fact, flatten it. But wait! If this knocked-down button-pushing specimen is the issue of what is *outside* the orbit of the familiar system, then it must be the product of another process. Derrida's covert use of Leroi-Gourhan's actual writings is at this point ingenious. What he wants to inject is a historical—a veritable evolutionary—idea of a human who is not upright, and he does this by means of a button-pushing specimen which, he intimates, Leroi-Gourhan thinks conceivable.[15] In the specific paragraph from which Derrida has selected his quotes, Leroi-Gourhan is in actuality first briefly describing cranial enlargements in answer to the basic question with which he is concerned: whether humans could evolve still further. He states that further evolution would be a matter of redoing the cranium, but goes on to affirm that "in order to remain human such as we conceive humanness physiologically and mentally, one can hardly admit another sizable gamble"—in other words, another enlargement the size of previous hominid cranial enlargements. He then speaks of people at the end of the nineteenth century who, inspired by the anatomy of the foetus, were led to imagine certain changes by the end of the twentieth century: humans would have "a large cranium, a minuscule face, and a scanty body"—"*un corps étriqué.*" He notes that what these nineteenth-century prophets did not foresee was that any major change could hardly be produced without the loss of the hand, the teeth, and in consequence, loss of upright posture. *It is in this context* that he says that it is not inconceivable to end up with a button-pushing specimen, adding immediately as part of the same sentence that certain science fiction writers, hatching all possible scenarios, have created Martians and Venusians that approach this evolutionary ideal.[16] What we thus discover is that in his covert appropriation of Leroi-Gourhan's train of thought, Derrida is writing one body out of existence and another body—its replacement—into existence. It is all done with the neat stroke of the pen, a knock-out performance. In place of the familiar body with its feet on the ground is the ideal specimen, a grammatological rather than animate form. The appropriation is clearly the work of an unerring grammatological surgeon.

What concerns us now is the specific statement that encapsulates this button-pushing evolutionary ideal and that Derrida makes in a note at the foot-end of his text in conjunction with his snatch quotes from Leroi-Gourhan. Following his page reference to Leroi-Gourhan, Derrida writes, "In a totally different context, we have elsewhere specified the *epoch* of writing as the suspension of *being-upright.*" (Note Derrida's italics.) Derrida then refers the reader to two essays in *Writing and Difference.*[17] I will begin with the footnote itself and subsequently examine the relevant passages in the two essays.

For anyone familiar with phenomenology, Derrida's note rings with an unmistakable vocabulary—it is thick with phenomenological significance. Derrida's epoch of writing specifies not only the time of writing as an historical phenomenon. It specifies a suspension with respect to that history, a graphic epoché akin to philosopher Edmund Husserl's phenomenological epoché. Like the phenomenological epoché, the graphic epoché refers both to a field of experience and to a suspension of the natural attitude with respect to that field of experience. In the case of the graphic epoché, however, the suspension is not a suspension of the natural doxastic attitude toward the "fact-world," but a suspension of the natural *bodily* attitude, the attitude of "being-upright" toward the world. Derrida's graphic epoché thus specifies a postural shift in fact—one of course *sits* to write—*and* in natural bodily stance—in animate form itself. Because the suspension of upright posture is not just a suspension in fact but in attitude as well, it is necessarily a suspension of being bodily in an everyday, living sense—a suspension of the body in its natural bearing toward the world. And indeed, with Derrida's declarative founding of the epoch of writing as a suspension of being-upright, the body is put out of gear, bracketed, parenthesized. It is still there, of course, but it has no *actuality;* its everyday existence is put out of play, as are its living resonance and possibilities. It is, in short, a body whose 'I can's' are reduced to a single possibility: writing. The postural shift in attitude in consequence effectively cancels out the living body of the subject who writes—in effect, the subject herself/himself—at the same time that it magnifies the script that writing leaves behind—leaves behind in a scatological sense etched out in one of the essays Derrida refers us to and to which we will presently turn. For the present, it is clear that writing, as Derrida conceives it, threatens and finally violates animate form, the natural upright attitude of the human body that is the product of a natural history, and that is the familiar, everyday center of our being. What the epoch of writing initiates and makes possible is grammatology: words that strut their stuff across the page and whose meanings can be endlessly played on.

Now when an end is put to my being upright, clearly my feet no longer support me; I am no longer grounded on this earth. As Derrida himself has told us, evolution can conceivably lay the body low. The second meaning of *upright* in the OED is in fact "to lie prone," and in truth, when an end is put to my being upright, I might well be laid out horizontally, flattened—even dead. If not dead, I might be lying down, perhaps on a couch, a possibility of enormous psychological significance, as we shall see. Alternatively, *being suspended,* I could well be strung up overhead; I might be hanging, deposed and depotentiated, a non-animate dangling form devoid of effective presence. In this mode of groundless posture, a suspension of my being upright has just such a doubly emphatic significance. Dangling from overhead, I am both spatially out of play—I am no longer on the scene—and kinetically out of play—I can do naught. Strung up in this way, I am altogether powerless, a once-living form whose animation has been put *sous rature.* Moreover, dangling in this way, I am dangling in a sense akin to the sense in which a certain bodily organ dangles when it is powerless or impotent. The relationship between full body flaccidity and a flaccid penis is topical not only because Derrida links textuality and the endless deferral of meaning with the opposite of flaccidity, with *dissemination,* an act which, however intertwined by Derrida with hymeneal structures, is nonetheless itself linked with erection, and in actuality does not take place without erection, and not only because in both cases, it is a question of a body incapable of signifying in peculiarly, though not exclusively, human ways,[18] but because the relationship leads us directly to passages in one of the essays

Derrida singles out as relevant to an understanding of the *epoch* of writing. The essay "La parole soufflée" is on Antonin Artaud, revolutionary theater artist of the 1960s.

Derrida writes at length of Artaud's view of the work of the artist as something that does not stand up by itself, that collapses; something that is akin to excrement. Derrida states that, "Like excrement, like the turd, which is, as is also well known, a metaphor for the penis, the work *should* stand upright. But the work, as excrement, is but matter without life, without force or form. It always falls and collapses as soon as it is outside me."[19] The lifeless work of the artist thus clearly has both scatological and sexual associations. A few paragraphs later, in commenting laudingly on Artaud's visionary theater, Derrida states that "The uprightness of the work . . . is the reign of literality over breath."[20] In other words, the work poses as something lifelike when it is not anything but marks made on a page, marks that hide traces and that await interpretation. This literal uprightness that Derrida links to the "uprightness of the letter or the tip of the pen" is to be denounced. Though not described as such, the uprightness of the work is clearly an *im-posture*. In this same context, Derrida connects honorific notions of standing upright and dancing with Nietzsche and with poets such as Hölderlin and affirms that these positive, animate acts contravene Artaud's "essential decision,"[21] that is, Artaud's decree that the artist's work is, well, crap. In consequence, these positive, animate acts must in some way be put *sous rature;* they must be scratched out and over. Derrida's move here is to suggest that perhaps they are merely metaphorical acts since "erection . . . is not obliged to exile itself into the . . . literal uprightness of the letter or the tip of the pen." But then he also affirms that "It is metaphor that Artaud wants to destroy. He wishes to have done with standing upright as metaphorical erection within the written work."[22] So neither a literal uprightness nor a metaphorical uprightness are to be countenanced. Nullification of the literal is of course accomplished by a suspension in which the body sits or is made to sit, and which, in sitting, can be neither upright nor dancing. Nullification of the metaphorical is, in Derrida's words, accomplished "by summarily reducing the organ." This precipitate surgery warrants examination.

To begin with, we may ask how the surgery is carried out. It is carried out by the body itself in speaking, in excreting, in ejaculating. The scatological and sexual associations mentioned earlier come to the fore in support of the reduction: a full organ empties itself. It is a *self*-reducing machine. Within the particular context of Artaud's theatrical program, summarily reducing the organ means first summarily reducing the power of the literal text, and thus the power of the author, the director, and the actors, and their undeviating iterations of the given, primal words before them.[23] To summarily reduce the organ is, in effect, to take the literal words literally out of their literal mouths. But furthermore, organs, Derrida tells us in elaborating on various texts of Artaud, are just such things that can be emptied—organs like the penis that in ejaculating dis-seminates, or the body that in excreting leaves something behind. An organ, Derrida tells us, is a "place of loss because its center always has the form of an orifice. The organ always functions as an embouchure"—a mouthpiece. The organ is thus the text, the penis, the mouth, the anus, and no doubt more. In fact although Derrida does not say so, we can say that it is the body itself insofar as the body is the Organ of organs. Let the air out of the body and it expires. Emptied of life, it collapses. Hence, whether it is a question of summarily reducing the organ that is the text or summarily reducing the organ that is the penis, or summarily reducing the organ that is the living body, the organ in each case becomes flaccid and impotent; whatever its original nature, it is now silenced, spent, flattened, dead. In short, "the work [is] destroyed."[24]

The suspension of *being-upright* has an oblique interpretation in this organ-reductive context. The suspension of *being-upright* may be understood to raise questions of moral rectitude. Can one, with moral impunity, knock the body out or string it up? In more graphic terms, does the characterization of artistic works as excrement and their resurrection through deconstruction raise questions of foul play? In whichever guise the questions might surface, it is clear that the suspension of *being-upright* has an ethical twist. Moreover taking place as it does within a graphic epoché—it is all done with words—the suspension appears the work of a roguish lexical alchemist. The alchemy performed is reversible—baser words are transmuted into golden ones and golden ones into baser ones—but dancing feet, as we shall see, are always converted to lead. Because it is all done with words, the result is ever the same: author and work are by one transformative stroke after the next summarily reduced. Carte-blanche powers are legitimated such that whatever is written or produced becomes fair game for all comers—for all interested in disseminating the otherwise lifeless work. In effect, the *epoch* of writing—the graphic epoché—raises the question of foul play not only by authorizing free play in the dirt, but because in summarily reducing the organ, it summarily suspends integrity.

Derrida's second essay, "Force and Signification," is precisely relevant at this juncture. In the course of his disquisition on the relationship between force and form, Derrida has occasion to invoke Nietzsche's affirmation that one must dance "with the feet, with ideas, with words, and need I add that one must also be able to dance with the pen—that one must learn how to write?"[25] Derrida immediately adds, "But Nietzsche was certain that the writer would never be upright; that writing is first and always something over which one bends." Even before this, immediately prior to quoting Nietzsche as above, Derrida, in agreeing with Nietzsche's estimation of Flaubert's estimation of himself—that the author is nothing, the work is everything—states that if this is so, "We would have to choose . . . between writing and dance."[26] In other words, unless the work is everything only in the sense that it is not something created but is merely a *pre*-text for interpretation, we would have to deal with the dancing body producing the work. Clearly, in the name of writing, the dancing body must be put out of play and in more than one sense. Insofar as the *epoch* of writing is the suspension of *being-upright,* then, to begin with, that is, in the literal sense of writing, the literal dancing body must be made to sit. Here the meaning of a suspension of upright posture—the postural shift in fact—is straight-forward enough. One bends over one's writing. One gives up uprightness— dancing—in order to write. This literal meaning comes close to capturing the *labor* of writing, a *felt bodily labor,* not merely in the sense of actually moving pen or fingers, but of a bodily engaged thinking in which one is bent-over in thought, concentrated, intent. To be concentrated in this creative bodily way over one's writing is actually to be extraordinarily potent. Powers of thought are scrutinously keen. Furthermore, to be concentrated in this creative bodily way is to be wholly geared into the task at hand; attentiveness is undistracted and unswerving. In brief, there is a total bodily immersion in the labor of creating. Short of this intent, centripetally focused posture, there would be no writing, writing to be sure in what Derrida at times derisively calls "the narrow sense," but writing that is nonetheless in a prior way *de rigueur* to writing in the broad sense. But there is in fact no such intensively concentrated, writing body within the graphic epoché. This concentrated body, this *corps engagé,* is precisely the one that has been suspended, that has been summarily reduced, emptied; though essential to the task at hand, it has been declared an unauthorized body. It is for this reason a body whose absence can be read the result of foul play, a body that, purely in the name of grammatology, has been stealthily removed. Indeed, the dancing upright body is on this oblique reading a purloined body, a body carted off and put *sous rature.*

But this is not all. Another body from time to time is surreptitiously made to stand in place of the purloined body. This other so-called body, this counterfeit form—a corps not *engagé* but *étriqué*—parades as a synecdochic specimen that would *stand* for the whole body. It is patterned on that familiar Freudian symbolism in which one piece of anatomy purportedly signifies the whole. In truth, however, the specimen is a synecdoche manqué. Being a thoroughly fragmentary and detached form, it is, metonymically speaking, impotent, a *corps dégagé* that cannot possibly bear the weight of the whole. We meet this disengaged pseudo-body, this synecdochic pretender, along with the purloined dancing body in the same second essay to which Derrida has referred us.

Derrida writes of writing at the close of that essay that "It will be necessary to descend, to work, to bend in order to engrave and carry the new Tables to the valleys." He has noted in an epigraph at the beginning of the last section of the essay and with due reference to Freud that "Valley is a common female dream symbol." At the end of the section and essay, he writes of writing that it is "the outlet as the descent of meaning outside itself within itself"; it is "excavation within the other toward the other"; it is "submission in which the same can always lose (itself)"; it is "two affirmations *espous[ing] each other*"; it is the other being "called up in the night by the excavating work of interrogation." After affirming all this of writing, he finally concludes the essay by affirming that "Writing is the moment of this original Valley of the other within Being. The moment of depth as decay. Incidence and insistence of inscription."[27] In short, writing is equivalent to copulation, that is, to male copulation—intromission and ejaculation—and male copulation defined as both a grammatological happening—incidence of inscription—and a grammatological compulsion —insistence of inscription. Males apparently do not just circumstantially leave marks; they insist on leaving marks. In his use of the writing-copulation analogy, Derrida is actually repeating Freud, who already made the connection between writing and copulating in *Inhibitions, Symptoms, and Anxiety*. For both gentlemen, there is no doubt but that the body of concern is not a living body but a genital organ of one kind or another. A genital of course does not dance; it copulates—or writes. But a scripturally copulating genital is a failed synecdoche because its writing, no matter how fluent, is always and only writing. A copulating genital in other words lacks corporeal possibilities because it lacks corporeal stature. The difference between a living body and a copulating genital is the difference between animate form and grammatological form.[28] Once the latter form is positioned on stage and writing begins, there is no way a dancing body can intervene and make its presence felt. It has been removed from the scene and has not a leg to stand on; as we saw earlier, it has been put spatially and kinetically out of play. Moreover as Derrida himself has said, we must *choose* between writing and dance: if we choose writing, then we cannot possibly choose a spatially and kinetically alive body. On the contrary, we have already exiled that body, suspended it, whether by stringing it up from the rafters or by laying it to an early rest.

IV.   A Brief but Telling Note on the Scene of Writing or Exposing the Metaphysical Underpinnings of Grammatology

Derrida's conceptual linkage to Freud is borne out by more than the writing–copulation analogy. It is borne out by his explicit analogical patterning of grammatology on Freud's psychoanalytic of consciousness, specifically as that psychoanalytic was itself conceived by Freud to be analogically patterned on "the mystic writing pad."[29] Brief examination of the writing pad analogy will show that just as Freud ignores his original psychoanalytic insight—that, in his own words, "The ego is first and foremost a bodily ego"[30]—so Derrida, following but at the same time vastly improving upon his psychological master, resolutely absences the body from the scene of writing. The philosophical consequences that follow are heavy indeed.

In "A Note Upon the 'Mystic Writing-Pad'," Freud likens perceptual consciousness and the unconscious to a writing apparatus children play with, an apparatus that allows one to write on a surface, then wipe off the writing by lifting the surface sheet, thus leaving in effect, a clean slate on which to write again. The writing, however, leaves an impression on the wax slab beneath the writing surface. Thus, one has at one's disposal a perpetually fresh surface on which to write and at the same time a permanent record of all writings. In the same way, one has, psychoanalytically speaking, perpetually fresh perceptions and at the same time an indelible record of each of them.

In appropriating Freud's analogy to grammatology, Derrida states specifically that "the deconstruction of logocentrism is not a psychoanalysis of philosophy,"[31] but it is clear from remarks at the end of the essay which aim toward "a new *psychoanalytic graphology*"[32] that the program of deconstruction is equivalent to a psychoanalysis of language, that in effect and in the name of grammatology, Derrida puts language on the couch. The import of this move is momentous. Language *is* the psyche: its surface and its indelible underside. No wonder then that the living body is resolutely removed from the scene of writing, that in consequence, there is no subject but only a disengaged body part that writes.[33] A living body could only distract the psyche just as, platonically speaking, it could only distract the soul. Humans, in effect, are grammatological forms—all psyche and no soma.

Now however covertly related the suspension of the living body to the placement of language on the couch, both moves are of a metaphysical piece. Moreover, they are not only of a metaphysical piece with respect to each other but with the synecdochal specimen discussed earlier, and this because however failed, the genitalic synecdoche too is in the service of the same metaphysics, a metaphysics of absence. In particular, it pretends to call forth an absent body but gives us only itself, a fragment, a singularly specialized token member of the whole. The synecdoche is thus thoroughly successful precisely in virtue of its failure. It is the means whereby the living body, while putatively invoked, is kept at a remove, resolutely distanced from the scene of writing. Possibilities of corporeal signification are thus also hidden away, possibilities which are not even considered integral to the domain of meaning, in all probability because such meanings are not even recognized or conceived in the first place. The living body is not perceived as having any inherent signifying powers. Only genitalia have such powers, and, with particular respect to writing, only male genitalia at that. Why, then, bother to suspend the living body if it cannot signify? Because, as Derrida himself has intimated in the beginning, the living body *threatens* writing. It threatens it in the form of a subject, a subject who, as we have seen, not only creates meaning, but a subject whose fundamental concepts emanate in both a phylogenetic and ontogenetic sense from its living body.[34] Hence, it becomes clear why the living body's balance is by fiat deemed "precarious" to begin with, and why the "precarious balance" is threatened. Being-upright *must* be threatened and defeated if the epoch of writing is to be installed. Of course Derrida does not tell us what threatens the body, but surely we know that it is grammatology. When language is put on the couch, it displaces the living body, the subject. Language does away with the living body in one fell swoop of the grammatological logos. Any vestige of a subject is in effect erased. With respect to the body, only genitalia, putative representatives of the living body, can enter the scene of writing. Cast now in leading roles, appearing now as trace figures, they find regular work because they are cut out for it. One can play endlessly and at a respectable cognitive distance with genitalia without ever acknowledging a subject, a full-bodied presence. Through such play, one might dupe oneself into thinking that the body is substantively entwined with grammatology, but in fact it is graphic fun and games that one has with genital cut-outs at the expense of understanding the body as a whole—understanding the body as animate form—including genitalia.

In sum, as an historian of writing, Derrida works in devious ways his grammatological powers to unfold, but with respect to the living body, those powers reduce to an act of grammatological creationism.

V. Corporeal Archetypes and the Metaphysics of Absence

The metaphysical absence created and sustained by failed synecdochic practice reinforces the metaphysics of absence created by placing language on the couch and by the body's original suspension to begin with. In such circumstances, not only can one not recant that original suspension without recanting writing—without doing away with the *epoch* of writing—so one cannot successfully call forth the living body synecdochally to the scene of writing without recanting writing. The absence of the living body is thus not an absence that can be attenuated in any way. The absence of the living body is integral to writing. The whole graphic manoeuvre—the suspension of being upright in all its guises, from button-pushing specimen to summarily reduced organ to synecdoche to grammatological form—is in the service of a metaphysics of absence, an absence designed to leave no trace, and one that, but for Derrida's attempt to forge a history of writing, would have left no trace. The absence would thus have been *final,* one which the living body could never have erased or overcome in any way, for if it could, it would appear, and in appearing, it would immediately instantiate a subject, and thereby immediately signal the death of grammatology. Clearly, the epoch of writing is kept alive by a denial of the living body.

The metaphysical history rendered by absencing the living body is actually a metaphysical historicism in which there is but one deconstruction after the next.[35] The mystic writing pad, after all, offers on the one hand a perpetually clean slate, and on the other hand traces that are layered but that bear no internal relationship to one another. Just so with postmodern theorists who, in speaking of "situated" knowledge,[36] accord a spatio-temporality to something that is not of itself a spatio-temporal thing at all, and moreover give a historical position to something which itself has no history. In being accorded a location in space and time and a historical position, situated knowledges pretend to a life, thereby usurping what properly belongs to living bodies. But in fact quite unlike living bodies and the subjecthood that goes with them, they are still-life drawings on a mystic writing pad. Their pointillist rendition of history is in fact a *trompe l'oeil.* Lacking the concrete presence of animate form, "situated" knowledges, like

Derrida's own history of writing, lack any genuine connections, internal progressions, or formal continuities. No matter that Derrida is not concerned with "familiar knowledge," and must exceed familiar forms to forge his history of grammatology, the absence of animate form means the absence of a veritable history. What Derrida renders in place of the latter is an unnatural history, and not just an unnatural history of writing, but necessarily in conjunction with his grammatological creationism, an unnatural history of the body itself. That history too is a pointillist history held together by words. It is exemplified by the postmodernist's credo of the body which, conceiving the body as a thoroughly social construct, conceives it from the beginning as a somatological tabula rasa. But such a body is the product of an immaculate linguistic conception. Indeed, it has no form other than a grammatological one.

The metaphysics of absence that undergirds the tenets of deconstruction and its postmodern expressions is a metaphysics that has in fact haunted Western philosophy from its very inception. With few exceptions, animate form has been consistently and rigorously suspended. No wonder the mirror of nature is cracked. What distinguishes Derrida is his formal— albeit covert—installation of this metaphysics of absence. The metaphysics is integral to his critique of the metaphysics of presence. His revealing footnote, in exposing the annulment of the living body, exposes the metaphysical foundation and infrastructure of deconstruction. Derrida has in this sense done an inestimable service to philosophy. By resolutely suspending the body, by carrying language to a reified extreme in the inauguration of grammatology, by building deconstruction on the analogy of the mystic writing pad and thereby putting the body to its purportedly final rest—with words—he gives us the possibility of illuminating from at least one perspective the absence at the heart of Western philosophy, an absence held in place by a long and deep mesmerization with language and mind that has its own dogmatic history.

In his Introduction to Husserl's *The Origin of Geometry,* Derrida purportedly shows how cultural traditions are language-dependent. There are ways, however, in which cultural traditions are *not* language-dependent, ways that derive from an evolutionary history and that do not fit into the grammatological writing mold Derrida fashions for all meaning. In this overlooked dimension of history, there are living subjects in a living present; there is an evolutionary heritage; there are animate forms whose fundamental powers of signification are impervious to deconstruction because in these instances what is signified is what it is and is not another thing—and another thing, and another thing. Moreover it is precisely because it is what it is and invariantly so that archetypal primate behaviors can be used to deceptive ends—for self-protection or to gain advantage, for example.[37] These fundamental powers of signification reach far back into a history that is through and through a human history. In this history, inscriptions are written in an *intercorporeal semantics*. In an intercorporeal semantics, the body is neither a text nor an organ; neither something to be read nor summarily reduced. Neither is it a fragment which pretends to stand for a whole. It is something to be fathomed as the animate form it is. Indeed, Derrida does not need to *forge* a history. Grammatological creationism is unnecessary. The evolutionary history that can be read, should one stop long enough to perceive the living body and give it its due, provides insights enough into fundamental human meanings.[38] As suggested in chapter 2, these fundamental meanings are easily missed in part at least because common linguistic and conceptual focus is wrongly placed; an inter*subjectivity* is more properly conceived (and labelled) an inter*corporeality*. As suggested too in chapter 2, when understanding other minds and reading another's intentions, feelings, and so on are treated as if preternatural events, explanations of intersubjectivity verge on explanations of clairvoyance. On the contrary, creatures are there for each other first of all in the flesh. An understanding of intercorporeal archetypal meanings demands that we first of all recognize this fact. It demands secondly that we recognize the seenness of each other and the communal somatic verities that go with that seenness. Third, it demands that we recognize not merely what we *do* as forms of life, but recognize ourselves as a form of life.

A caveat should perhaps be inserted at this point regarding the significance of examples of nonhuman primate behaviors given in earlier chapters to exemplify primate corporeal archetypes. In no case are the examples to be construed merely as giving support for a claim. They are part of the claim itself. Otherwise stated, the examples have *historical* significance; they are not offered as evidence for a comparison between *us* and *them, here* and *now*. This simple comparison misses the point completely. It regards the examples as *filler* to support a view rather than as statements of *paleoanthropological* import. As I have elsewhere pointed out, all humans are hominids, but not all hominids are human.[39] There is, in effect, no great gap in kind between "us and them," for we have ancestors, ancestors who were conceptually astute, articulate, and inventive. Furthermore, the comparative method in evolutionary biology is utilized as a means of gathering evidence about hominids as a historically developing genus. The noting of similarities between human and nonhuman primates is accordingly in the service of recognizing ourselves as products of a natural history. Not only are our pentadactyl limbs similar to those of other primates (and other vertebrates), for example, but certain fundamental behaviors are also similar. Such behaviors originated and were carried forward on the basis of what is common to primate animate form. Similar anatomies (not to mention physiologies) are the basis of similarities in behavioral repertoires, repertoires that include, at bottom, fundamental patterns, archetypal forms of behavior, which are themselves the foundation of an intercorporeal semantics.

It follows that an absence of the living body is unnatural. The absence deanimates us and in turn de-historicizes us, separating us from the natural world in which we have our origin. Grammatology wrenches us from our place in the living world leaving a metaphysical void. What it inserts in place of the void is a metaphysical apparatus that writes, that emits signs, that leaves marks behind, but that has none of the archetypal powers of signification common to animate forms. These fundamental forms of signification are not equivalent to hearing oneself speak nor to writing. They do not yield a third thing in the sense of reified meanings divorced from subjects who create them. Hence they are not detached or detachable idealities nor are they susceptible to deferrals. Where corporeally enacted, they are unqualifiedly and unequivocally present. They are rooted in one's being the body one is. This means that however culturally elaborated or reworked, archetypal human significations, like the human animate form from which they derive, are pan-culturally invariant. We can begin to get a sense of why this is so by recalling Husserl's original description of the body as the zero-point of orientation, a corporeal center on the basis of which "[we intuit] space and the whole world of the senses."[40] Nonlinguistic meanings and the intercorporeal archetypes that instantiate them can be understood as an elaboration of Husserl's basic somatological insights.

In sum, corporeal archetypes are non-deferrable truths of the body because they are *intercorporeal meanings*. They are not the enterprise of an isolate psyche; they are not *writings,* an individual act in any sense. They are corporeal matters of fact needing recognition, analysis, and painstaking investigation. They are part of an intercorporeal semantics more ancient than we, which is to say that we come into the world as animate forms which are already inscribed in originary ways. By absencing ourselves as bodies, we erase ourselves as subjects, thus opening the way to grammatology, to putting language on the couch and making it speak of, by, and for itself in endless, and imperishable ways. Our intercorporeal semantics and its rich and complex archetypal forms in turn disappear without a trace; language is designated the only site of meaning, and meaning is in turn reified. Were we to turn as intensively toward the body as we have turned toward language, we would find that that corporeal turn would yield insights no less extraordinary than those of any kind of linguistic turn. Those corporeal insights would moreover not only have broad interdisciplinary impact; they would lead to irrefutable foundational understandings, and this because from beginning to end, ontogenetically and phylogenetically, we are all bodies.

VI. A Brief Afterword

Consultation of Derrida's two *Geschlecht* essays, essays explicating Heideggerian texts and considered "immediately related to the topic [of the body],"[41] strengthens rather than contravenes the claim that grammatology selectively dis-members the living body and with that dismemberment, elevates certain bodily parts and accords them privileged status. On the surface, the explicative essays appear to have nothing to do with Derrida's grammatological creationism and his suspension of the living body, but in fact they validate both, and this even though the concern of the first essay is sexual difference and ontological difference, and a major concern of the second is the hand and its distinctively human nature. The essays validate the charge of grammatological creationism and a suspension of the living body because what they offer are explications of the abstract by the more abstract. The living body is, in other words, present in grammatological dress only. In the first essay, titled *"Geschlecht:* Sexual Difference, Ontological Difference,*"* it is evanescently clad, its synecdochic form discretely veiled in lexical insinuations on the theme of dissemination. That Derrida's vocabulary and descriptive narrative have their ultimate if unnamed source in an ejaculating phallus is clear from at least two perspectives, each of which we will briefly specify.

To begin with, *dissemination* is insistently there from the moment of its introduction when Derrida attempts to tie Heidegger's "carefully ob-serve[d]" *"'metaphysical isolation* of man'"* with sexual difference and "the dual partition within *Geschlectlichkeit."*[42] The word's formidable Derridean import is admirably underscored in Heidegger's phrase "transcendental dispersion." "'Transcendental dispersion',"* Derrida tells us, at the point he is glossing on Heideggerian texts, "belongs to the essence of *Dasein."*[43] But earlier he has told us that dispersion rests upon dissemination, that is, dissemination is foundational to dispersion; it belongs to *Dasein before* dispersion: "An 'originary dissemination' already belongs to the Being of *Dasein. . . .* This originary dissemination (*Streuung*) is from a fully deter-mined point of view *dispersion (Zerstreuung). . . .* "[44] In short, there is no dispersion without the prior act of dissemination. More specifically, Derrida insists that with respect to *Dasein,* dissemination is the "originary possibility," and that dispersion marks the possibility of its factual manifestation. Tellingly and fittingly enough on behalf of his claim, Derrida muses that in the specific Heideggerian text he is explicating (what he refers to as the "Course" Heidegger gave at the University of Marburg/Lahn in the summer semester of 1928), "the word *Streuung* appears but once, it seems, to designate that originary possibility, that disseminality (if this be allowed). Afterward, it is always *Zerstreuung.* . . ."[45] The idea of a foundational dissemination is thus suggestively grounded in Heideggerian textual fact. The message of Heidegger's text as of Derrida's own claim can in consequence become emphatic within an unmistakable reading: the "originary possibility" of "disseminality" makes transcendental dispersion into factual existence possible. Translated into non-abstract, concrete bodily terms, the message and claim can hardly mean anything other than that the spritzing of sperm is foundational to what Derrida terms "the dispersion of *Dasein.*"[46] Indeed, and with all due acknowledgment of Derrida's ardent desire "[to] protect the analytic of *Dasein* from the risks of anthropology, psychoanalysis, even of biology,"[47] the sentence "Transcendental dispersion is the possibility of every dissociation and parcelling out (*Zersplitterung, Zerspaltung*) into factual existence"[48] reads like a metaphysical translation of a typical sociobiological pronouncement on how genes get represented in the next generation and a species number is thereby increased: through male promiscuity and insemination.

The second perspective has to do with deconstruction itself and the preservation of its principle of multiplicity over fixity. Thanks to an irreducible, ever-voluble Organ, Derrida can productively read off multiplicity from an original Oneness, the factual existence of individuals from an *Ur* disseminality, as when he states, for example, that *"Dasein* . . . shelters in itself the internal possibility of a factual dispersion,"[49] or more pointedly, "It is the originary disseminal structure . . . that makes possible this [factual] multiplicity."[50] No matter that Derrida attempts to underscore the sexual *neutrality* of Heidegger's singular-appearing *Streuung,* the "internal possibility" of *Dasein* is cast from a decisively—if always veiled—male bodily mold, thus strongly suggesting that females "are mere egg repositories waiting for something to happen."[51] (Its veiling also recalls Lacan's well-known claim that the phallus can only play its role when veiled.) Precisely by keeping the bodily mold—animate form—veiled, by assuring that it is nowhere explicit or even remotely outlined in his narrative, Derrida can hope successfully to defend his thesis of sexual neutrality. The result, however, is "sexual difference, ontological difference" and ultimately philosophy—by innuendo. Indeed, Derrida's reading of Heidegger's

Course as offering an *"order of implications"* (Derrida's italics) with respect to a non-dual sexuality, to "a sexual difference . . . not sealed by a two,"[52] is soft-shoe dancing without shoes, a *baseless* dancing in fact, for a foot has never so much as left a mark on these pages. Nimbleness on behalf of multiplicity is all in the head.

A grammatological history and a suspension of the living body are secured in Derrida's second *Geschlecht* essay, *"Geschlecht* II," by explicit synecdoche and by the dichotomization "humanity versus animality."[53] It is in this double context that "the hand" comes to its unique human stature. Although Derrida declares that "Thinking is not cerebral or disincarnate; the relation to the essence of being is a certain *manner* of *Dasein* as *Leib,*"[54] he explicitly states, following Heidegger, that the hand "is *not* part of the organic body."[55] Granted that Derrida, following Heidegger, wants to dissociate *the hand* as a material object from *the hand* in its essential human aspect, even to the point of dissociating that essential aspect from "conceptual grasping" as well as object-grasping,[56] the result is "a thing apart"[57] that becomes the fleshless emblem of grammatology. In brief, the hand no longer has reference to a living body; it has reference only to thought and language (and of course thought and language are not conceived as having any substantive reference to living bodies). Although Derrida criticizes Heidegger for taking no account of "zoological knowledge," and for raising his "nonknowing" with respect to "animality" to a dogmatic "tranquil knowing," he goes on to insist that Heidegger's pronouncements on animality nevertheless trace out "a system of limits within which everything he [Heidegger] says of man's hand takes on sense and value."[58] What Derrida will go on to claim in delineating this sense and value is that "man's hand" is not a separable organ but "a thing apart" in the sense of being "different, dissimilar from all prehensile organs (paws, claws, talons); man's hand is far from these in an infinite way through the abyss of its being. This abyss is speech and thought."[59]

What Derrida seems to take away with one hand, however, he gives back with the other, utterly transformed, of course, but handy all the same. The absolute separation of humanity from animality is attained for Derrida as for Heidegger by a synecdochic reading of the hand. The hand stands for thinking, for crafting in a way no mere animality can craft. The watershed is speech and thought. Derrida quotes Heidegger: "Only a being who can speak, that is, think, can have the hand and can be handy. . . ."[60] In the name of grammatology, Derrida adds that "Man's hand is thought ever since thought, but thought is thought ever since speaking or language. That is the order Heidegger opposes to metaphysics: 'Only when man speaks, does he think—not the other way around, as metaphysics still believes'"[61]—and, one might add, that virtually all paleoanthropologists and primatologists likely believe on the basis of multiple facts and artifacts: for example, extant nonhuman primate deceptive behaviors and stone tools made by early hominids.

Indeed, the "system of limits" and grammatological history that Derrida invokes on the basis of Heidegger's meditations on the hand fly in the face of paleoanthropological and primatological evidence. To begin with, stone tool-making is craft, handiwork, thought. Stone tool-making dates back two and a half million years to a period of time that virtually all paleoanthropologists agree preceded the invention of language.[62] Further-more, when Derrida invokes the hand's two vocations, that of showing and that of giving itself, he first quotes Heidegger, who in part writes that the distinctive human hand "receives its own welcome in the hand of the other," and then goes on to insist that animality does not *give;* it takes.[63] What are we to make, then, of a description which reads: "There are many postures and gestures in the repertoire of [chimpanzee greeting] behavior. They consist of bobbing, bowing, and crouching, touching, kissing, embracing, grooming, presenting, mounting, inspecting of the genital area, and, occasionally, hand-holding."[64] The Heideggerian–Derridean claim that is being made, "that a hand can never upsurge out of a paw or claws, but only from speech,"[65] falsifies evidentially verifiable evolutionary continui-ties, not only with respect to one hand "receiv[ing] its own welcome in the hand of the other," but in hands that fashion sticks for the purpose of termite-fishing.[66] Surely these hands are crafting hands. Surely these hands are akin to the hands of those early hominids who crafted stone tools. Surely too these hands are akin to those later but still nonhuman hominids who buried their dead and laid flowers around them. Surely we must recognize that we humans have an evolutionary history.

In light of that history, comment is also due on Derrida's claim regarding the hand's specific linkage to the origin of language. Derrida writes that "the most immediate, the most primordial manifestation of this origin will be the hand's gesture for making the word manifest, to wit, handwriting, manuscripture, that shows and inscribes the word for the gaze."[67] Two points in particular warrant mention here, the first quite briefly in the form of a simple lead question: what is one to make of humans who do not write? More specifically, what is one to make of people whose culture does not include writing? Are these people without thought? Are they something other than human, other than "man," as Heidegger and Derrida consistently identify humankind? If we accept grammatological history, we can only conclude that these humans are *not* human (or in a lighter vein, not *man,* as well they might not be). They are on the animality side of the divide.

But we must ask ourselves, what do they know that we do not know? How do they come to have the stories they have? How do they come to *wonder* about the cosmos and their relationship to it? How do they come to *describe* both the cosmos and their relationship to it? That we must also ask ourselves about how "the word" was ever invented such that it could be shown and inscribed "for the gaze," and about arrogance and a certain academic *préciosité* is beyond question.

The second point regarding Derrida's invocation of, a "system of limits" and a grammatological history concerns a seemingly banal topic by comparison, the typewriter, but Derrida's positive estimation of it, *contra* Heidegger, ties in specifically with the idea of a button-pushing specimen and its conceivability. Derrida writes at length of Heidegger's indictment of typewriters as machines that "degrade" the word. For example, quoting Heidegger, he writes that "In typewriting, all men resemble one another."[68] While applauding Heidegger's privileging of manuscripture, however, Derrida clearly does not consider the typewriter a degradation of thought. He underscores in a footnote the fact that "one can write on the typewriter, as I have done, with three hands among three tongues," meaning that typewriting certainly did not impede his manuo-lingual dexterity in French, German, and English. Now if we take Derrida at his word to the effect that key-pushing does not de-privilege the hand and degrade thought, that is, if manuscripture is in no way debased *or made impossible* by a writing machine, then surely it is in no way debased *or made impossible* by button-pushing. By the same token, if a key-pusher is grammatologically conceivable, so also is a button-pusher. In short, insofar as buttons can replace keys, and button-pushers key-pushers, it is conclusively evident that the living body is not essential to the task of writing conceived as grammatology. Whatever accomplishes manuscripture—manifesting the word to the gaze—is grammatologically correct. Multiplicity reigns here too, then. The techno-logical possibilities of manuscripture are endless as are the linguistically engineered possibilities of manuscripters themselves, for unlike animate form, a grammatological body can be whatever one wants it to be—even though it might be forced to eat its own words and come undone.

5
Corporeal Archetypes: Sex and Aggression

When it comes to the signification of power, dominance, aggression, control, and related notions in human culture, the penis is, as Lacan would say, "le signifiant sans pair."

Daniel Rancour-Laferrière[1]

Rape is a horrifying crime, but what is the meaning of this horrifying act of rape? . . . In order to explain my position, I analyse a rape attack. In this instance, I want to use the words and actions of the rapist who attacked me along with the effects of rape trauma that I experienced in order to demonstrate the meaning of rape. The use of this attack for data enables me to have an investigator-victim perspective not otherwise conceivable.

I argue that the rapist tried to completely define himself into my existence. . . . The attack began when the rapist awakened me. . . . When I initially failed to respond to his orders, he beat me numerous times. His fists to my face made me realize that his attack of rape was my escape and only hope from physical death. By his actions, the rapist had given me a choice: to suffer his attack or to face my physical death. Of the two, I chose to live through and survive . . .

Cathy Winkler[2]

I. Introduction

A critical examination and elucidation of year-round receptivity as cultural archetype and a beginning appreciation of its covert male counterpart, perpetual erection, led us to a beginning appreciation of that preeminent cultural archetype: the Phallus. More recently we have uncovered a mythico-symbolic scion of that cultural archetype in the guise of original dissemina-tor of the word. In grammatological hands, the Phallus becomes the fount of language and is duly invested with all the powers proper to that orifice. We need now to continue the elucidation by interrogating male and female bodies further. In particular, this chapter will begin by examining the basis on which the cultural equation "male threat/female vulnerability" is structured and played out. In the course of doing so, it will necessarily introduce archetypes of animate form that clearly distinguish female and male bodies. These archetypes will be fully examined in the chapter that follows. The central purpose here is to elucidate the cultural equation on its own grounds, so to speak, showing how a domain within the human repertoire of 'I can's'—a domain in which a sense of agency and autonomy is paramount—is intensified, negatively on the one hand, positively on the other, by rape. The chapter will then go on to examine major scientific claims about the relationship between sex and aggression. The purpose of these later sections is to bring into critical focus both the data and implications of leading studies that explain the linkage between sex and aggression in general and the act of rape in particular.

II. Invitation and Threat

Whether a natural and readily visible signal or whether behaviorally intensified in display, human genital exposure is possible in only two ways: penile display and leg-spreading. The former possibility, understood as the mere possibility of revealing the penis—whether from inside clothing, a sheath, or an ornament—is a sex-specific act; the latter possibility is a sex-neutral act. As described earlier, primate males, by spreading their legs, *display* their (erect) penis. For human males the position heightens a focus on an already visible penis by calling attention to its transformation and translocation in erection. Human females, by spreading their legs, similarly *display* their genitals. Their act, however, is basically quite different. In spreading her legs, a female not only makes particular parts apparent that are hidden from sight; she exposes them to begin with in a way they are not normally exposed. In other words, and as has been emphasized from various perspectives, in terms of animate form, human females (and presumably all hominid females) have no *sexual signalling organ.* When they display, they are not heightening attention either to something readily accessible to vision—something that a shift in observer position, for example, would immediately bring into view—or to something already straightforwardly visible, e.g., a swollen pudenda or an erect clitoris. They are exposing themselves and calling attention to their genitals in one and the same act. *The act of exposing themselves is hence not separable from display, that is, from invitation.* The act is precisely for this reason a vulnerable act. Gynecological examinations and typical Western positioning for childbirth consistently testify to this fact. Not that females feel they are inviting sexual union by their exposure on the examining or delivery table, but that being positioned in such a way, they can easily feel vulnerable. The relationship between a supine/knees-flexed/legs-spread position and the feeling of vulnerability merits examination. With respect to a standing observer, for example, the position is obviously not unconnected to the intercorporeal high/low spatial archetypes described earlier. The more immediate point here, however, is that exposure may be taken as an invitation whether it is just so purposefully enacted by a female or not. It is thus clear why female genital exposure-avoidance is pan-culturally proscribed. There being no separation of signal from display, behavior carries the full burden. Human females do indeed have to be circumspect about what they do with their legs. In effect, it is not surprising that the first behavior biologist Ruth Hubbard mentions to exemplify "the interplay between biological and cultural factors" (or how "we cannot sort out our biology from our social being") focuses on what little girls learn that they must not do with their legs: "move [them] in ways that reveal their underpants."[3] Human females are clearly vulnerable.

What we need to understand are the ways in which male animate form is different, that is, how and why male animate form has the inherent potential to be threatening: what is it about an erect penis that allows it to be regarded a threat? Indeed, *how does what is invitational sexual display become aggressive display?* There is a further question, in part adumbrated by science writer Beryl Benderly. "[In] a world infinitely full of facts and phenomena," she writes, "here's one with far-reaching consequences: male genitals permit intercourse with an unwilling partner, but female genitals do not."[4] Clearly, the act of penetration can carry with it a sex-specific positive power valence, not only with respect to rape, as Benderly suggests, but with respect to male homosexual practices as well. Foucault has spoken—with repugnance—of just this power valence in his study of ancient Greek sexuality. In answer to the question, "Why does sex have to be virile?" he responded that "The Greek ethics of pleasure is linked to a virile society, to . . . an obsession with penetration. . . . All that is quite disgusting!"[5] How the act of penetration comes to be conceived not only life-threatening but also an act of unconditioned male power thus demands investigation as well. It will be helpful before pursuing these several topics and questions directly, however, to examine first the female side of the equation. By analyzing female vulnerability more closely, we will be able to appreciate more fully the intricate ways in which the sexual equation male threat/female vulnerability—essentially a finely tuned "equation in intensities"—ultimately sheds light on, and in fact defines from yet another perspective, "year-round receptivity" and perpetual erection.

III.   Female Vulnerability

A.   *Existential Vulnerability*

It is not a great conceptual leap from the fact that *mere genital exposure is invitation* to the notion that females invite sexual relations simply by being there, in Symons's words, to the notion that "women inspire male sexual desire simply by existing." Indeed, in patriarchal cultures the two ideas easily compress into one: for females to expose themselves at all in a male world is to be sexually vulnerable. Simply to be there, literally present in the flesh, in an environment with males is to be at risk, and this whether they are sitting with legs spread or not. In this compressed view of female sexual invitation, females are equivalent to their genitals. The person and the part are one and the same. Cartoons and cultural lore in the West testify to this equivalence, as when males are pictured as undressing women in their mind's eye in the very act of actually perceiving them. That the equivalence exists was implicitly suggested in chapter 3 in the context of showing how females, by expressing their "natural" sexuality, put themselves at double risk: by spreading their legs—by "doing what comes naturally"—they at the same time brand themselves as loose (or evil), and this because in addition simply to being there, an invitation in themselves, they are intensifying invitation by displaying their genitals. But their vulnerability too is decidedly heightened by the display. The reason it is hinges precisely on the fact that in the compressed patriarchal view, *females are themselves sexual signals.* The act of exposing themselves is thus indeed not separable from display, but not simply in the former sense that human females lack a signalling organ on the order of a pudenda or penis so that exposing, and behaviorally calling attention to, their genitals constitute one and the same act. Rather, exposure is here not separable from display in an existential sense: they—females—exist; they—females—are seen. As "natural" sexual signals themselves, their genital exposure intensifies a signal already present. What may be designated their inherent existential vulnerability is heightened by a perceptual transfiguration in which the whole is the part (the whole is the hole), and in which, conversely, the part is the whole (the hole is the whole) in the sense that it is *all* that exists. Reduced to their genitals in this synecdochic double sense, females are indeed conceived as nothing but objects of sex for males and as such are perpetually vulnerable. Foucault's fecund concept of docile bodies is more than topical here; it is in many ways definitive of the view of females as just such sex objects. Examination of the infrastructure of these sex objects will provide insight into the intercorporeal power relations that typically constitute and haunt human sexuality as twentieth-century Westerners have come to know it.

B. *Females as Docile Bodies*

Females are turned into docile bodies by keeping them at risk of exposure, that is, by keeping them under the threat—actual or potential—of male eyes. In such a situation, females do not themselves have to *do* anything to be at risk; they simply have to be there and know themselves to be there, visibly present to males in consistent, simple, everyday ways, whether walking down the street, attending an office meeting, traveling, attending a party. Reinforcing and capitalizing on this visual liability of females, the advertising industry—in the United States if not elsewhere—echoes and re-echoes the theme of ever-present male eyes. Where females respond submissively to the threat, where they accede to being docile bodies, they are subservient to the idea of themselves as "being in the eyes of males." They comport themselves in accordant ways. Being both actually looked at, and under perpetual threat of being looked at, they keep themselves at all times *comme il faut*—in the gender-specific sense of that phrase. Whether a matter of clothing, makeup, behavior, occupation—whatever—they comport themselves as *he* requires. Being submissive to the demands of male eyes, docile females live in a world in which what matters most is their visual body and how it measures up.

One might say that such docile bodies have, by their own choice, compromised their autonomy. In face of their existential vulnerability, they have chosen to become docile. But one might equally say that they have been forced to compromise their autonomy by the intensity of the threat. The threat is not some fitful menacing gaze, breaking off here, being interposed there. It is continuous intimidation. If "women inspire male sexual desire simply by existing," then of necessity, male sexual desire is not something that comes and goes but is something that is there unremittingly. There are no reductions, interruptions, remissions, or gaps in the intensity of sexually desiring eyes. Equally, there are no reductions, interruptions, remissions, or gaps in their focus, any more than there are reductions, interruptions, remissions, or gaps in female existence. "Being in the eyes of males" is thus understandably an intercorporeal structure held in place as much by desiring eyes as by the docile bodies to which the eyes attach like leeches. To compromise one's autonomy is one way of living with incessantly agglutinous eyes.

Foucault has of course shown, through an examination of madness, punishment, sexuality, and other Western social institutions, how what I have termed "an equation in intensities" obtains in power relations and how it holds those relations in place. The above-described intercorporeal situation is no different. The female visual body that is being constantly surveyed heightens itself in just those ways that conform to the sexually desiring eyes that see it and that perpetually fix and hold it in their visual field. This equation in intensities—male sexual threat and female docility—is readily contrasted with what Sartre describes as "reciprocal incarnation." Rather than two subjects coming together and "incarnating" each other in desire, there is on the one side an aggressively self-indulgent presence and on the other an accommodating but empty presence. There is, in short, no reciprocal incarnation at all but a reciprocal instantiation of power. To uncover deeper understandings of the two sides of this instantiation of power requires finer examination of the question of autonomy and the complex relations that generate and perpetuate the perception of females as docile bodies.

We might begin by considering the fact that where females can be said to accede to the threat of male eyes, one can accuse them of living in bad faith. They are allowing male-defined cultural standards and practices to dictate to them not only what they should look like, how they should behave, what they are good for, but even more deeply, "their place in nature." They allow not mere cosmetics but the whole of their very existence to be defined by an Other. There is in consequence no possibility of a docile, acquiescent female separating being from docility; they are one and the same, in the same way that her being and her genitalia are one and the same. Indeed, her genitalia are *docile* in the very sense Freud describes when he equates femaleness with objective passivity.[6] Living in bad faith, a docile female defines herself by the "he (or they) for whom I am." This "he (or they) for whom I am" is subsumed in the Look, or rather, in unceasing Looks, since it is a question of eyes that are always there, actually or potentially present at all times, eyes that are moreover impersonal in the sense of belonging now to this male, now that one. Whoever the male might be at any particular moment, "I" am always in his eyes. Clearly, to live one's life in this way—as no more than a visual reflection—is to turn away from autonomous being and to accede to being an object for Others. Clearly, the charge of bad faith is justifiable.

Viewed in this perspective, "being in the eyes of males" is as much an instance of irresponsibility as an instance of Foucaultian docility. To compromise one's autonomy in this way is, after all, not to compromise an intermittent moment in one's life. It is to choose a life itself: the life of a docile body in the form of a sex object. Through such a choice, the docile body becomes the fulfillment of a corporeal archetype. The archetype is culturally spawned; it is, one might say, not a natural but an acquired taste. As suggested by its generation in bad faith, it is an archetype into which one steps, a cultural ready-made as it were. It is, all the same, held in place by nature and could not exist short of those natural dynamics that fundamentally structure intercorporeal life. In other words, a radically accentuated power of vision subtends the socio-political structure of male threat/female docility, just as it does the socio-political structures of all those docile bodies that Foucault has described. Equally, the intercorporeally transforming power of the Look that subtends the possibility of a docile body is the same power of the Look that subtends the possibility of a shameful body discovered at a keyhole. In brief, the same finely tuned equation in intensities is apparent at both the socio-political and personal levels: the same power of optics that generates and perpetuates social institutions on the order of the Panopticon and everyday interpersonal situations on the order of shame, embarrassment, and humiliation generates and perpetuates both the cultural institution and interpersonal relationship of males as sexual subjects and females as sex objects. By drawing on this archetypal power of optics, males instantiate females as bodies that are in themselves powerless, bodies that are there to be subservient and useful to males; and females, acquiescing to this archetypal power of optics, instantiate males as bodies that are in themselves powerful, bodies that are there to hold sway and dominate.

Pauline Réage's *The Story of O* testifies in chilling terms to the corporeal archetype of female docile bodies. Females in this story are construed not as witches or as goddesses but as thoroughly inferior creatures whose sole *raison d'être* is to serve their male masters in whatever sexual ways the latter demand. Females who construe themselves in this way no longer construe themselves as autonomous subjects. To paraphrase a remark by Merleau-Ponty, they 'manipulate their bodies and give up living in them'. That the story is an immediately understood tale, however outrageous or painful one judges it to be, is evidence of its archetypal roots.

Still other dimensions of docility (and inferiority) surface in the context of autonomy. The power of optics aside, docile bodies are an essentially male-spawned corporeal archetype of human female sexuality. The archetype is formulated on the basis of "animate form," but animate form reduced to the mere having—or rather, not having—of a certain anatomical part. By this reductive act, of course, animate form is in fact no longer animate. Shorn of its livingness, reduced to mere materiality, and indeed, to nothing more than a particular material *part,* this "animate form" has none of the 'I can's' proper to a bona fide subject. The particular part singled out is itself of moment to consider here in some detail, and in two senses: on the one hand, the part gives rise to the idea of females being "in the form of a hole"—a Sartrean characterization[7]—and on the other hand, to the idea of female sexual being as a lack. We will consider both, though we will defer a complete discussion of the former until the next chapter.

A hole, unless it is already filled, is always open to being filled. The idea of females being "in the form of a hole" is thus clearly connected with the idea of females being receptive "year-round." In both cases one can presume human males having only to wait their turn at the watering hole, so to speak. A ready way of spelling out the connection between females as holes and as receptive "year-round" is to enumerate the several ways in which females are considered sexual receptacles. I itemized these ways on the basis of a review of the literature in *The Roots of Thinking* and will quote from that earlier work.

> Receptivity is used as a synonym for *responsiveness* (e.g., "a circumscribed period of time in mammals during which the female is responsive to courtship and mating");[8] as a sign of *willingness* (e.g., "oestrus should be a strictly behavioral term signifying willingness to mate");[9] as closely synonymous with the *openness* or *closedness* of a receptacle for males (e.g., an estrus female is one who is not "continuously copulable," hence who cannot provide continuous service, hence whose receptacle is not perpetually open hormonally and theoretically for male use);[10] and as synonymous with being on the *receiving* end of copulatory transactions (e.g., "In most [mammalian species,] these simple activities ["assumption of the mating posture which facilitates the male's achievement of intromission, plus maintenance of this position until intravaginal ejaculation has occurred"] comprise all of the female's receptive behavior").[11]

As I pointed out in the earlier book, insofar as females as well as males invite copulation, receptivity is a behavior that may be predicated of males as well as females. In primate societies, for example, when a female invites copulation by presenting, and when the male accepts the invitation by mounting, the male may be said to be *responsive* and *willing*. Not only this but he may be said to be *providing a service* to the female insofar as (*a*) she is soliciting something from him, namely an erect, sexually potent penis, and (*b*) he is acceding to her solicitations. Only in the literal sense that the female is receiving something from the male—a penis and sperm—is receptivity a concept applying exclusively to females. A concern with sexual being in terms of a psychological ontology must then take *penetration* into equal consideration. If female sexual being is conceived as being "in the form of a hole"—a receptacle for something—then the "strange flesh," as Sartre terms it, which fills the hole and which is equivalent to male sexual being, must be spelled out along with the hole. In short, a psychological ontology of "that which penetrates holes" warrants correlative analysis. Such an analysis constitutes a large and complex task, and this because "that which penetrates holes" may be viewed from a number of different perspectives. For example, that which has the power to penetrate is also "that which dangles,"[12] hence the necessity of distinguishing between flaccid and erect penis. The task in this chapter is limited to delineating penetration as male threat and act of unconditioned power, at the same time specifying its relationship to a construal of females as inferior to males—in the immediate context, to a construal of females as equivalent to their genitals, i.e., to a hole or lack.

The idea that females lack a body part which, if they had it, would give them wholly other life options than those entailed by being in the form of a hole—being "year-round" receptive bodies—needs examination precisely in terms of the archetypal power relations it upholds, relations in which males are dominant—in virtue of penes—and females are submissive—in virtue of a lack thereof. Of interest in this context is Sartre's remark that for an autonomous subject in the midst of its projects, there is no *lack* which can be predicated of it, a lack which, if somehow satisfied, would have changed the nature or outcome of its projects. The lack, in Sartre's words, *"can in no way limit"* an autonomous subject.[13] Sartre is actually speaking of inventions and techniques which exist today and which one cannot say were *lacking* to certain individuals in earlier times. He says, for example, that "It would be absurd to declare that the Albigenses lacked heavy artillery to use in resisting Simon de Montfort; for the Seigneur de Trencavel or the Comte de Toulouse chose themselves such as they were in a world in which artillery had no place; they viewed politics in that world; they made plans for military resistance in that world; they chose themselves as sympathizers with the Cathari *in that world;* and as they were only what they chose to be, they were *absolutely* in a world as absolutely full as that of the Panzer-divisionen or of the R. A. F."[14] What Sartre is insisting upon here is that no autonomous subject would have turned out differently if such and such objects had been available to him at the time he lived. He goes on to state quite specifically that "For him, who has no relation of any kind with these objects and the techniques that refer to them, there is a kind of absolute, unthinkable, and undecipherable nothingness. Such a nothingness *can in no way limit* the For-itself which is choosing itself; it cannot be apprehended as a lack, no matter how we consider it."[15]

What Sartre says of objects can in an attenuated way be said of sex: the possibilities one is are not a matter of what one has at one's disposal.[16] There is in other words no sexual lack that existentially conditions a female (or a male for that matter), no sexual parts that limit her choices such that, *had they been different,* they would have given her—*a female*—a different life. To think that there is, is to think that by having had different parts, as by introducing artillery, that *the subject herself* would have been different. In effect, it is to think that the choices of an autonomous subject—female or male—are genital-contingent, that by adding such and such a genital and, correlatively, by subtracting such and such a genital, *a given life* would have turned out differently. To reduce a person to her genitals in this way is to privilege—and *de*-privilege—pieces of the body, to valorize them, to make *them* rather than the person her/himself the source of meaning. In such a conception, a person's possibilities in the world reduce to her or his bodily possessions. An individual's life in consequence perpetually hangs in the balance, and understandably so: genitalia have been made the source of one's personal autonomy and the channel through which that autonomy is expressed. Should one have "the right stuff," there is of course no problem; should one not have the right stuff, the *lack* irrevocably limits and alters one's life. But in fact, for a female, the idea that a penis "would have changed her life" is "a kind of absolute, unthinkable, and undecipherable nothingness."

Sex operations that turn males into females or females into males support this very point. The operations are performed not out of a desire for *parts* but out of deference to the felt sense one has of oneself, to what might be called "life meanings." They do not change a female into a male or a male into a female so much as—in the usual phrasing—free the female trapped in the male body or the male trapped in the female body.[17] One might justifiably speak of an innately conditioned sexuality in this bodily felt sense (though not as if it were completely impervious to cultural influences). An individually felt, and not part-reducible, sexuality would in this view be an aspect of one's facticity, a dimension of one's individual given nature such as having a disposition to move in certain ways, to desire certain kinds of relationships, to enjoy certain affections, and so on, a nature readily exemplified by the sensed discomfort a trapped male or trapped female feels prior to surgical changes. In such instances, having this or that anatomical part available and at one's service is not the point; being able to live with others in accordance with one's feelings, motivations, and desires is. In just this sense, the trapped person lives in the middle of an unresolved life. It is the turmoil of such an existence that prompts consideration of an operation. The operation is a beginning step toward resolving the turmoil. *Parts* thereby undeniably change a person's life, but not by satisfying a *lack*. While on the surface one may objectively point to changed parts, it is fundamentally the person her/himself who is changed—into the person s/he feels her/himself to be and to have been all along.

The idea that innate sexual dispositions are central to one's felt bodily sense of oneself merits further consideration. As suggested above, one is by nature motivated to act in this way rather than that, disposed to express desire in this way rather than that, and to find pleasure in this way rather than that. The seemingly separate, partitioned-off realm of being called one's "sex life" is thus not simply a function of culture. Indeed, individual predispositions toward certain acts, desires, and pleasures anchor one's sexuality no less than what Foucault calls "cultural formulae" that specify certain acts, desires, and pleasures as proper.[18] Desire, act, and pleasure, he says, are "the poles" of sexual behavior, each particular culture having its own particular formula, that is, each particular culture delineating how the three poles are to be valued with respect to each other, and consequently specifying the way in which sexual behavior is properly conceived and practiced. That cultural and personal "formulae" are in actuality overlaid, that a particular formula constitutes the cultural cadre in which individual sexual predilections emerge, is evident in the fact that individual "sex lives" are either compatible with the particular formula of the reigning regime or they are at odds with it. In the latter case, the individuals are cast as sexual misfits. This evaluation is not made on the grounds of a cultural inspection of their parts. Cultural practice, especially in the form of disciplinary technologies, is not so naive as to think that merely having a part in any way guarantees any particular behavior[19]—one need only think of nuns and other celibates. Those individual sexualities at odds with the reigning regime are cast as misfits because of a propensity toward acts, desires, and/or pleasures contrary to those deemed proper. Misfits thus exist not because a part is lacking to them, but because a certain felt life is culturally lived at risk, is unlived, or unlivable.

Whether a question of females, transsexuals, or homosexuals, the practice of reducing sex to a body part perpetuates the practice of conceiving certain others as something less than human. The concept of females as incomplete humans has had a particularly long and varied history. It can be traced back on the one hand to Aristotelian biology and on the other to Freudian psychology. Like the concept of females as receptive "year-round," the concept solidifies just those power relations that maintain male dominance and female docility. The two concepts are in fact distinctly related: the person who penetrates has power; the person who simply receives is powerless. Foucault's account of ancient Greek males' "obsession with penetration" puts this conception of the powerful and the powerless in broader perspective in that the conception, or "obsession," clearly legislates not only female/male relationships but male homosexual relationships as well. According to Foucault, young Greek boys were not the ones to penetrate in homosexual relations; penetration was the privilege of the older dominant males. The idea of penetration as an act of unconditioned power and the idea of the person penetrated as being always open and at the continual service of the one who holds power thus appear to have as lengthy a Western history as the concept of females as incomplete. The dual conceptions are not of course limited to the West. Preeminent power is given to and enjoyed by the penetrator in the homosexual practices of a present-day New Guinea culture, for example. Young boys, seven to ten years of age, in a series of traumatizing, continuously terrifying initiation rites, are removed from the care of their mothers and made to cohabit with older boys with whom they must perform fellatio. Not only this, but semen, they are told, is like mother's milk; it will make them strong; it will make them grow; it will make them *men*. The young boys are in the beginning ashamed, frightened, and/or repelled by the idea of sucking penes and swallowing semen. They are, however, forced into accepting their homosexual role: "The initiation experience involves considerable coercion on the part of the initiators and death threats are necessary for men to accomplish their task."[20] Another anthropologist writes that once "brainwashed," the young boys eventually come to "enjoy" their role "and seek to multiply their sexual contacts so that they will rapidly grow to be manly warriors."[21] Thus, at the same time that they become habituated to their role as passive consumers, the initiates come to look forward to the time when they will be in the dominant position, that of the penetrator, though "they are anxious about losing their own semen."[22]

In sum, sexual penetration and docility are potentially power-weighted cultural themes that extend beyond the mere dual categorization of male and female bodies. All the same, the pervasive view of females as docile sex receptacles for penetrating males is far more negative and threatening than is the view that young boys are docile sex receptacles for penetrating older males, and this precisely because the status of females is irremediable. Not only is the situation such that, over the years, nothing will change in virtue of age, but given the positive valorization of the act of penetration, females cannot appeal their case. Penetration is a strictly male prerogative. It could remain a male prerogative and *not* entail female docility only if males were not set up culturally over against females, that is, only if 'he who penetrates' were not positively valorized at the expense of females. Since he who penetrates *is* given privileged status in Western culture and of course many others besides, the strictly male prerogative rules. It in fact literally dominates. Penetration, after all, demands that something be there to be penetrated; it demands docile bodies. Female bodies provide a natural and ready receptacle; hence docility is proper to females. In this way, the concept of females as being "in the form of a hole" and the concept of females as being incomplete can again be seen to solidify just those power relations that cast the female in the role of docile object. What the biological concept of "year-round receptivity" does is give scientific testimony at one stroke to both their hole and their deficiency. In Western and other patriarchal cultures, female vulnerability is, in effect, compounded daily.

IV. Sex and Aggression

The question to be posed now concerns the relationship of females being "in the form of a hole" to various formulations of the link between sex and aggression. The link has been explained all the way from evolutionary antecedents, male hormones, cerebral representation, chromosomal make-up, and other related neurophysiological-biological structures and functions, to cultural concepts of gender inequality. A central issue in light of corporeal archetypes is whether there are evolutionary antecedents not only for the equation in intensities—male threat/female vulnerability—but for human male assault/human female fear. As might be anticipated, the act of rape will come to figure centrally in the sections that follow.

A. Female Fear/Male Assault

To make fundamental sense of females as being in the form of a hole means returning once again to the tandem "biological" principles that females are receptive "year-round" and that they incite male sexual desire simply by existing. The purpose here is not to examine the principles further, but rather to call attention to their corollaries; namely, *that males are perpetually tumescent and hence a perpetual threat, thus that they instill fear—simply by existing.* These corollaries are nowhere stated in the literature, socio-biological or otherwise. Yet they are clearly entailed on adaptive if not logical grounds by the "biological" principles. If it is an adaptive truth that females are receptive "year-round," then "year-round" erection necessarily follows as an adaptive truth. If it is axiomatic that females incite male sexual desire simply by existing, then female vulnerability to male sexual desire is also axiomatic. The corollaries are in fact experientially born out by "the female fear"—rape.[23]

"Forcible sexual access . . . [is] a significant social problem worldwide,"[24] and in present-day American culture at least, the fear of rape is pervasive.[25] Several aspects of female fear warrant mention straight off, the first one being the not uncommon male notion that females naturally resist sexual advances but "really want it." As psychologists Margaret Gordon and Stephanie Riger state, "The underlying assumption [is] that it is the natural masculine role to proceed aggressively toward the stated goal, while the natural feminine role is to 'resist' or 'submit'."[26] "Fear" in this male-spawned cultural scenario is not fear at all but a natural reticence; in resisting, a female is just "doing what comes naturally." A second aspect rests not on male interpretation of female behavior at all but on an actual fact met with earlier in perhaps its initial form: little girls are taught early on to be careful of "what they do with their legs." As they grow older, they in fact learn they must do more than be careful about what they do with their legs; *they must watch their step in the quite critical sense of being careful where they go.* As one woman described it, and in a way emphatically documenting both the principle that females are at risk simply by existing and the correlate principle of a built-in, "year-round" receptivity: "[Rape] could happen anywhere on the street, anywhere that you are walking *that you are exposed* . . ." (italics added).[27] Not only this but if a woman happens to encounter an aggressive man at a particular place, then she may well feel impelled to avoid going to that place in the future. For example, one woman, after encountering a man in a store who displayed his penis to her and said, "Here, look at this!," exclaimed when interviewed later about the experience, "I avoided that store after that!"[28]

What is there in a male's presence or genital exposure that is frightening to a female such that she has to be careful where she goes and will choose to avoid certain places if she has been intimidated or assaulted there? Rape has been an intensively researched act. Both men and women have written about it and from various points of view. There is a considerable literature on its evolutionary ties, on its incidence in large American cities, on its central relationship to understanding the link between sex and aggression, on its being a form of psychological perversion, and so on.[29] Even so, few of the studies focus in depth and directly on what is so very frightening about being raped. Subtitles on the order of "Management and Intervention"—the genuine concern of the researchers notwithstanding—make the act of rape something like a bodily malfunction of some kind, an occurrence that society in general and raped females in particular can learn to *handle* in the same way that they *handle* stress, or post-stroke problems.[30] The fear of being killed aside, the question, Why do females fear rape? is of singular importance because its answer ultimately illuminates the existential plight of those creatures who are perceived "in the form of a hole," those creatures who are receptive "year-round" and at risk simply by existing.

The most dramatic way of initially setting forth the fear of rape is by considering its psychological if not physical equivalent: fear of castration. What if males incited female aggression simply by existing? What if males were susceptible "year-round" to violation by females? Would they not in their everyday lives be fearful of females for reasons similar to those females presently have for being fearful of males? After all, just as rape is believed by some to be a desire of all males, so castration might be conceived to be a desire of all females. Indeed, suppose that our culture were such that females were dominant and males were submissive. Given this gender inequality, females could at any time assault males. No matter that they have no *organ* of assault, they might without difficulty appropriate a tool or weapon by which they could violate males. Males would thus undoubtedly find themselves living in fear. That fear would limit their freedom; their ostensible 'I can's' would become self-imposed '*I cannot's*.' Thus, they might not venture out alone at night; they might be careful about what they said and what clothes they wore; they might think twice about working alone in the office after dark; and so on. Should they actually be violated, their lives would change in drastic, long-range ways, and this because fundamentally, their sense of themselves and of their freedom to do—their 'I can's'—would be irrevocably altered. Many if not all would find the crime something more than a mere "sexual offense"; many if not all would likely blame themselves for not behaving more judiciously if they were so violated; many if not all might be judged by others as having "brought it on themselves."

The scenario is not completely unfounded. Castration figures centrally in psychoanalytic theory, its central role being presumably based in part at least on male child-to-adolescent concerns and reports of dreams. All the same, fear of castration is not a fear that is prominent in our culture. Castration is in fact in the Western world a topic virtually exclusive to the psychology of sex, most especially to the (Freudian) psychology of infant and child sexuality. It does not commonly figure in discussions of adult human sexuality or human aggression—except in one instance. It is figuratively invoked as a reason for rape. Psychoanalytically oriented treatment programs give central stage to "castration anxiety." A castrating female—mother, girlfriend, or wife—is the *raison d'être* of a rapist's actions. This explanatory mechanism—a female in the wings—of course leaves something to be desired in the way of responsibility. But furthermore, as dissenting psychiatrists Gene Abel, Edward Blanchard, and Judith Becker point out, "No measures concomitant with analytic treatment have been obtained to substantiate that resolution of castration fears leads to increased arousal to adult females and to reduction of arousal to rape cues."[31] In addition, were female fear of rape to give way in like manner to actual female aggression toward, and assault of, males, thus to a prominence of male fear of castration, a female could similarly explain her castration of males on the basis of her "rape anxiety." Even a *figuratively* castrating female could explain her behavior on the basis of "rape anxiety."

Clearly, just such corollary relationships must be taken into account. If it is true that females incite male sexual desire simply by existing, it is also true that that palpable male desire has its responsive counterpart in an equally palpable female fear. Similarly, if it is true that females are receptive "year-round," it is also true that that receptivity has its obverse in (the threat of) "year-round" penetration.

Taken simply as a violation of someone else's body, castration and rape are synonymous; they are each mutilating acts. Moreover, as something feared, they are synonymous; each gives rise to similar kinds of fundamental existential anxieties, that is, each is conceived as something more than an act committed against a *part*. Where castration and rape are *not* synonymous is in the fact that the former is not an actually practiced act;[32] neither is it a worldwide social problem. Furthermore, fear of castration, unlike fear of rape, has no everyday effects on behavior (except perhaps in psychotic males). It is, in short, a fear that exists wholly in the head—an *imagined* mutilating act. Fear of castration can thus only approximate in the most tenuous sense to rape anxiety and to the actual trauma of rape. The actual trauma is so extreme as to destroy the implicit foundations of trust upon which a raped woman has heretofore lived her life. It is worthwhile examining this loss of trust because it will make clear the basically aggressive nature of rape.

The most trenchant theoretical understandings of how bodily violations of rape come to have the seismic force they do emerge from what is called "crisis theory."[33] This theory places central emphasis upon the concept of homeostatic equilibrium—that faculty humans normally have for meeting crisis situations and for bringing them to resolution. What the act of rape does is destroy the often tacit structure of beliefs that normally makes crisis resolution possible. Rape takes away basic assumptions—of safety, of the world as an orderly place, of oneself as a person of merit. In other words, what is described as a "loss of equilibrium"[34] is a loss so intense as to sweep away, in one fell swoop, deeply held beliefs about oneself, about others, and about the world. Self-consciousness, an awareness of oneself not as a subject but as an object in the world, is in turn intensified: one is not invulnerable;

on the contrary, one's life is perpetually on the line; one is in consequence from now on, on guard. Given the intensity of the trauma, and the fact that threats of death—and fear of imminent death—are common to rape, it is understandable why the symptoms of rape victims have been compared to those of combat veterans; and further, why precisely because it is a *human* assault rather than an act of God—a tornado, an earthquake—or an unexpected accident—a plane crash—that recovery is prolonged.[35] Like combat veterans, rape victims "are often edgy, jumpy over minor incidents, and show startle reactions."[36] Because humans are social animals, and because they live in societies in which certain laws and modes of conduct are ordinarily commonly recognized—tacitly or not—one human's unpredictable and brutal violation of another human is understandably a self- and world-shattering experience that takes a temporal toll beyond that of nonhuman-induced traumas. Precisely because it too is an act of wanton brutality of one human upon another, mugging generates symptoms similar to those of rape. One comparative study showed that males who had been mugged began having nightmares, just as rape victims do, that they too "avoided people, and had prolonged periods of sexual dysfunction," and this even though mugging is a less complicated crime, socially and psychologically. Moreover, mugging victims too "became aware of fears seemingly unrelated to their mugging—namely fear of heights or confining spaces or of riding as a passenger in a car."[37] Human-induced traumas clearly undercut the very fabric of one's life. Considering the severity of their impact, it is not surprising that a 1980 study of the American Psychiatric Association showed that human-induced, in contrast to nonhuman-induced, trauma, "appears to involve a much longer recovery process, spanning months or years."[38] With respect to rape victims specifically, the trauma of the experience in fact not infrequently produces chronic stress.[39]

The above discussion and facts are sufficient to indicate why rape calls into question basic 'I can's' that have structured one's life, and how in this sense rape results in an intensification of self-consciousness, a loss of spontaneity, and a feeling of oneself as at the mercy of powers greater than oneself, powers altogether unpredictable and gratuitous. With rape, life itself is called into question. It goes on, but in a radically different mode. In effect, no matter how intense a particular case of castration anxiety, it cannot compare to rape anxiety. It cannot come close to equalling the actual trauma of rape. The nihilating power of the act is dramatically epitomized in a passage by writer Deena Metzger, a woman who was raped at gun point. In "Shadows/SILENCE: A Love Letter in the Form of a Novel," Metzger describes "a woman both overpowered and divested of her power. Her last words are *'I cannot'*. She is emptied out" (italics added). The nihilating charge of the act is set in heightened relief by Metzger's own comment on her fictionalized but accurate rendition of the experience. She states, "I decided to make this public in order to break some of the silence and isolation which reinforces the personlessness of women. The private voice in the public sphere confirms our common experience through which we begin to assert ourselves. Unlike my character, now, *I can*"[40] (italics added).

B. *Evolutionary Considerations*

Sociobiologists, though they do not speak of archetypal behaviors, would take as axiomatic the notion that, *unlike rape,* castration is not an archetypal behavior. There are, in other words, no evolutionary precedents to castration as there are, *so they believe,* to rape.[41] Hence, although sex and aggression are connected in an evolutionary sense, that is, although rape is a behavior they claim to identify in some species of nonhuman animals and to explain on the basis of its adaptive value, the idea that castration (or castration anxiety) has evolutionary roots is absurd. The question of whether rape, unlike castration, has an evolutionary basis, however, actually requires extended examination, particularly in the context of its being viewed as paradigmatic of an evolutionary connection between human sex and human aggression.

In many nonhuman primate societies, socio-sexual and socio-aggressive meanings are manifest by the same archetypal behaviors. Indeed, as exemplified and discussed earlier, an economy of nature prevails in nonhuman primate societies. Castration is clearly nowhere represented among the archetypal behaviors. Thus, while castration may be a fear for some human males—or for all human males at some points in their lives—it is a fear that has no actual basis in evolutionary life.[42] There is in consequence no disagreement between the foregoing discussions of sex and aggression and the tenets of sociobiology with respect to this point: castration itself and castration anxiety are unrepresented in an evolutionary sense. It should be noted as well that, *vagina dentata* myths to the contrary, castration anxiety cannot be said to be a cultural phenomenon in any fundamental sense. While castration anxieties might be tied to feelings of hazarding oneself by the act of penetrating a hole, or while it might more explicitly be tied to fantasies of being eaten, and thus to the culture-spawned archetype of females as witches, as incarnations of evil, and so on, castration anxiety is not so pervasive a phenomenon in any culture that it deters males from sexual intercourse with females. In sum, male fear of being *actually* dismembered by a female directly, or by her "vagina dentata," is neither an omnipresent, life-saturating fear everywhere apparent in present-day American culture or in Western cultures generally, nor is it a fear of such proportions in cultures where vagina-dentata stories are part of the cultural lore that it restrains males from heterosexual intercourse. If it were, the culture in question would likely die out.

As suggested above, the claim that rape *does* have an evolutionary basis is highly controversial. While it might seem a short step to posit an evolutionary relationship between rape and sexuality on the basis of an economy of nature, rape being merely a further if extreme consolidation of everyday primate socio-sexual and socio-aggressive behaviors, the linkage is conceptually faulty. In particular, the fact that in two quite different situations—socio-sexual and socio-aggressive—the same behavior obtains is not grounds for conceiving rape to be a conjunction of sex and aggression, an act in which—to use a relevant example—the positional disadvantage of a female primate in typical dorsoventral mating simply merges with the negative spatial valence of presenting in socio-aggressive situations. Leaving aside for the moment the fact that, with a single exception, rape is infrequent if not nonexistent in nonhuman primate societies, and that in consequence, no sound analogical grounds exist for positing human rape to be just such a sexual-aggressive act, we must first call attention in a preliminary way to the erroneous sociobiological conflation of *sexual* and *reproductive* behaviors when it comes to understanding certain human acts.[43]

Far from being an extremely hardy form of *sexual* behavior, in the usual way in which *sexual* behavior is construed in evolutionary terms, i.e., as equivalent to, or a measure of, *reproductive* fitness, rape is a *counter-reproductive* strategy. It severely damages the psychological well-being of those human females who are so assaulted; raped human females most frequently find carrying a child "sired" under such conditions to be utterly repugnant; a raped female's genitals and breasts are commonly mutilated in the act; and so on. The enumeration of counter-reproductive realities could be extended. Claims that rape has "a foundation in comparative biology" notwithstanding,[44] what is called "rape" in avian and elephant seal behavior has in fact no fundamental relationship to the human artifact. For example, neither female ducks nor female elephant seals are forced to perform sodomy or fellatio by their male counterparts; neither "raped" female ducks nor "raped" female elephant seals feel the male's violence to be their fault, nor is their society such that others would judge them as "bringing it on themselves." Furthermore, what has been overlooked in the case of elephant seals is the fact that "prolonged resistance" of the female is part of the courting behavior of these animals.[45] Not only this, but unlike male elephant seals who "rape," male humans who rape do not engage in strenuous competition prior to raping females such that, like male elephant seals, they are part of the slim 4 percent who sire 85 percent of the next generation.[46] As to ducks, like all avians—and unlike all mammals—ducks do not procreate by internal fertilization; hence to speak of forced *copulation* is imprecise to begin with. There is no *sexual intercourse*. Furthermore, there is a basic contradiction in a theory which ties rape to polygynous societies but uses duck behavior to exemplify forced mating since the relevant species are monogamous.[47] Forced mating does occur in duck populations, and it can result in severe injury and even death for the female.[48] It can thus hardly be described as increasing the reproductive fitness of either the male or the female so involved. Just as forced copulation that maims and even kills is not a *sexual* act in human populations, so forced mating that maims and even kills is not an enhancement of *reproductive* fitness in duck populations. There is, in fact, a similarity between the human and waterfowl artifact: violence—violence that is gratuitous, not adaptive, with respect either to sexual or reproductive ends. Apart from this similarity, *the use of the term "rape" to define certain reproductive behaviors of elephant seals, insects, and a few species of avians is more than egregiously inappropriate; it is conceptually obliquitous.* The strong tie between sexual and aggressive behaviors in the animal kingdom, though it most definitely includes reproductive access by force in some species, does not include the violent, *counter-reproductive* human act of rape.

Many sociobiologists nevertheless persevere in the belief that rape *is* a reproductive strategy. For example, Martin Daly and Margo Wilson, who co-authored a highly popular textbook titled *Sex, Evolution, and Behavior,* cite the studies carried out by sociobiologist Randy Thornhill on captive scorpionflies in their discussion of rape. They state that the male scorpionfly has three reproductive options open to him: he can offer the female what is called "a nuptial gift" (a dead arthropod); he can offer her secretions from his salivary gland; or he can rape her, that is, "the male may simply try to knock a passing female out of the air, immobilize her with specialized claspers and clamps, and inseminate her forcefully."[49] As pointed out above, however, *rape* is not a *reproductive* strategy. There is thus confusion here between human behavioral artifact and nonhuman behavioral fact. There is even something more than confusion. Daly and Wilson go on to say that "Males prefer the arthropod ploy . . . [and that] rape seems to be a last resort."[50] They add that while it is obvious why a male would want to use an arthropod rather than debilitate himself by offering his own secretions, "it is less obvious" why the first two options are preferable to the third. They *conjecture* the reason the arthropod ploy is favored: it "appears," they write, that the "priority" is "female-enforced." Indeed, they write that *"Females flee from and resist males who attempt to use force, and they manage to escape about 85 percent of the time*[51] (italics added). Not only this, but they acknowledge that females have a "further line of defense" in that while insemination is successful in "'honest' resource-transfer copulations," *insemination is "somehow blocked as often as not if the female has been raped"*[52] (italics added).

As is evident, Daly and Wilson's conjecture first of all calls into question the very idea of female *reproductive* behavior: by fleeing and resisting, females are in fact acting in a *counter*-reproductive fashion. If they were genuinely interested in reproductive fitness, as Daly and Wilson assume them to be in their *nuturant* as opposed to *prodigal* adaptive role,[53] then they would hardly flee or resist. In broader terms, if rape were genuinely adaptive, then it would necessarily be something for which animals would be positively wired, i.e., females would not flee or resist, no matter what the degree of force used by males. Daly and Wilson actually explain female flight and resistance as adaptive on the grounds that a male who rapes has not courted properly by offering food and thus would not likely help with any offspring that result from the mating. A major problem with this ready line of adaptive reasoning is that it pays no attention to rape as an act of violence. The idea that a human female is fleeing or resisting because she has not been offered food and thinks that the male will similarly not provide for any resulting progeny overlooks the brutalizing trauma of the experience of rape and the overwhelming fear that goes with it. It overlooks the actual violent coercion, the physical assault. Indeed, an appeal to actual experience brings the point home immediately: If someone is beating you up, do you not try to flee? Is not "spontaneous self-preservation" the motivating factor for resisting or fleeing? Is not escape from pain the leading thought whether you are a female scorpionfly or a human female?

The above questions highlight Daly and Wilson's oversight and lead to a second, intimately related point regarding their conjecture as to why rape is not a priority on par with arthropods and glandular secretions. If insemination is "somehow blocked as often as not if the female has been raped," then rape cannot possibly be a bona fide reproductive strategy, that is, it cannot possibly be adaptive toward reproductive fitness. On the contrary, it would appear to be so distressing and unsettling and traumatic an event that while insertion takes place through sheer force—while proper reproductive contact is made—insemination does not result. The female body rejects the sperm. If the actual trauma of the situation *for the female* is in fact taken into empirical account, an explanation emerges as to why and how insemination is "somehow blocked." The trauma of the situation—the experienced pain, fear, dread, revulsion, and so on—is coincident with an extraordinarily agitated physiology. It is eminently reasonable to suppose that this extraordinarily agitated, *pain- and panic-stricken* physiology is a major cause of a female's *not* becoming duly inseminated. Normal physiological functionings are thrown off by violent assault to one's person; normal physiological functionings are thrown off when one is in pain and fears for one's life. From this *female* perspective, the sociobiological explanation of female resistant responses to rape is as far off the mark as the sociobiological explanation of rape as a *reproductive* behavior is to begin with. In sum, described within the context of male reproductive strategies and female cost-benefit assessments, corporeal matters of fact concerning rape recede into a Madison Avenue statistician's sunset.

The use of the term *rape* to describe certain nonhuman animal reproductive behaviors is controversial for numerous other reasons, not the least of which is that the behaviors are those of insects rather than mammals, not to say ducks rather than primates and invertebrates rather than vertebrates. It is of interest to note in this context that while insects reproduce by internal fertilization, avians do not. There is thus a sizable gap with respect to both analogies: on the one hand are creatures who, like humans, mate by internal fertilization, but who, quite unlike humans, are invertebrates; on the other hand are creatures who, quite unlike humans, mate by cloacal contact, but who, like humans, are vertebrates. Analogies are of course useful and provide insight, but only to the extent that there are sufficient *basic* similarities to begin with, and to begin with, such creatures as scorpionflies, for example, though mating by internal fertilization, belong to a radically different phylum altogether. This radical difference might be of minimal importance were these creatures social animals, but they are not; neither are they parents in the sense of assuming responsibility for ministering to and nurturing their young, nor, in fact, do their young, being self-sufficient at birth, require personalized care. Moreover, if "raped," the female, though internally assaulted, does not suffer in such complex traumatic ways that she is no longer able psychologically to live as before.

Indeed, a stronger case could be made for rape in ducks on the grounds that females are not infrequently maimed in forced matings. But even here, there is a fundamental contrast with the purported human equivalent. Maiming is here all on the outside; there is no penetration, no brutal violation of *insides,* insides that are not impersonal, a merely "functional" space. Insides that are violated are the locus of pleasure, gestation, fecundity. *Mugging* might for this reason be the appropriate human term to describe the gratuitous violence of male ducks.

When we consider our own primate biological order, with one exception we find evidence of rape only in our own human case. Nonhuman primates, whose behavioral repertoires include acts which serve both socio-sexual and socio-aggressive ends, do not conflate the two contexts of utterance. Of even greater significance, however, is the fact that a socio-sexual act is not appropriated by nonhuman primate males to serve aggressive, *violent* ends. Mounting is not taken as a possible occasion for violence; nonhuman male primates do not brutalize nonhuman primate females by or in mounting them. On the contrary, mounting serves either straightforward socio-sexual ends, *or it serves to defuse violence.* Rather than being a ready means of terrorizing females *or males,* it is a *deterrent* to physical assault or abuse. That mounting is just such a positive, socially stabilizing act perhaps explains why rape is not a behavior found among social primates such as chimpanzees and gorillas.

Observations of rape by male orangutans are best interpreted in the above context, that is, in the context of the species's comparatively *un*-developed sociality. Orangutans do not live in groups as their most closely related relatives (chimpanzees, gorillas, and humans) do; they are not social animals. Even when feeding close together, there is none of the social interchange common to other higher primates.[54] Summing up his study of orangutans in Indonesia, one researcher writes that "The solitary nature of these orangutans remains a striking exception to the social rule among the Anthropoidea."[55] Another prominent field researcher states more broadly that "Whenever orangutan populations have been sampled, the consistent features of their social organization have included a highly dispersed population and a relative absence of social interaction. This absence is not only striking when compared to the social behaviors exhibited by other pongids but also anomalous when viewed in the context of the range of behavior found among higher primates as a whole."[56] Moreover, not only is there a basic asociality but adult females and adult males rarely come together at all. Indeed, as two primatologists have noted, "[G]ranted that adult animals must sometimes come together for reproductive purposes, the orangutan seems to be as solitary as is possible."[57] While adult females are sometimes accompanied by one or two young, adult males are quite solitary animals.[58] To consider too is the fact that female and male eating habits differ, to such an extent that they would have difficulty adapting to one another's lifestyle were they to attempt social living, and this in major part because their eating patterns differ both temporally and spatially. A male feeds heavily in the morning and moves about very little; a female feeds uniformly throughout the day and moves through a broad range.[59] The difference in eating styles is in fact related to the species's extreme sexual dimorphism: a male's body weight requires him to move less, i.e., spend less energy traveling to different food sources. Finally, most encounters between males—whether subadult or adult—and females do not end in copulation. Observations have on the contrary documented contact-avoidance: "The standard adult female response to an encounter with a male was either to avoid or to ignore him. Males likewise frequently ignored females, although on other occasions they would approach and follow any female encountered."[60] It is of particular interest to note that orangutan females do not show any visible signs of estrus. Thus a female's fertile period is not signalled in a readily apparent way. This does not mean that she is "receptive year-round"—at least field researchers do not label her as such, but it does mean that, like a human female, she has no readily apparent signalling organ. Oddly enough, this point of comparison receives no attention in the context of discussions of "rape" among orangutans.

The above sketch of orangutans' solitary living style indicates that, from the point of view of *human* rape, rape by male orangutans is most properly interpreted as the result of a fundamental asociality, in particular, the result of living outside of an economy of nature or other socially informed system of behaviors which make communal living possible. Such a system or economy includes not only whole-body social acts on the order of presenting and mounting, but a variety of facial expressions and calls as well—in short, a developed and complex intercorporeal semantics structured in corporeal archetypes. It is socially telling that orangutans have a quite meagre repertoire of calls—they are in fact described as "silent animals"[61] —and their facial expressions are far fewer than those of chimpanzees, for example. The decisive explanatory significance of a fundamental asociality, specifically, a lack of those social archetypal patterns that typically undergird the behaviors of social animals, is underscored even further by taking into account the species's sexual dimorphism. Orangutan males on average weigh more than twice what orangutan females weigh—170 pounds as opposed to 81 pounds.[62] Theoretically, one might hypothesize that given the male's greater size, rape is simply the natural expression of greater power. But all species of female gorillas similarly weigh approximately half of what their male counterparts weigh, and rape has never been observed in gorillas, nor does their sexual behavior even suggest such an act possible.[63] Thus *size* is clearly not the decisive factor in the incitement to rape. Accordingly, although male orangutans are overwhelmingly larger than female orangutans, what might be identified as the primary morphological archetype of power cannot be cited as reason for a male orangutan's brutal behavior.

Now with respect to the notion of rape as a *reproductive* strategy, evidence from research on orangutan reproduction points compellingly in the same direction as evidence presented earlier on other species; that is, rape is a most *unsuccessful* reproductive strategy and cannot in fact be properly considered as such at all. It has been written, for example, that when females were unwilling to copulate and males forced copulations that "only one (or no) completed copulation took place per encounter, although the male may have initiated copulations several times or even followed the female for a number of hours."[64] There are many such testimonies to the lack of completed copulations. Interestingly enough, male violence is resisted not only by the female being assaulted but by her infant: "In four aggressive attempts at copulation, males struck and bit struggling females. Infants accompanying these females also attempted to fight off the males concerned." The same field researcher goes on to say that "It is doubtful whether intromission was achieved on these occasions, and once the male's pink penis could be clearly seen thrusting on the female's back."[65] Given such field reports, the question of rape being a bona fide *"reproductive* strategy" surfaces on two counts: completed copulation and actual pregnancy. Consider the following unequivocal statements:

> The more aggressive forms of copulation with uncooperative females, which I have referred to as rape, would seem to be futile reproductively. I saw most of the aggressive sexual behavior north of the Segama River, and, indeed, that population was characterized by a very low birth rate.[66]
> A number of such [subadult male/adult female] forcible copulations or attempted copulations were witnessed. . . . During a period of 5 months, three such encounters were observed. The female's infant participated in the struggle with the male. The struggle was very fierce . . . After the forcible copulation was consummated, the male did not travel with the female. . . . Significantly, subsequent observation disclosed that the female did not become pregnant.[67]
> A large adolescent female . . . was forced to mate on 2 different days within 1 week by a subadult male and a large not fully mature male. . . . Again, although she later became pregnant, it is clear from the timing of the birth that this set of forced copulations could not have been responsible.[68]
> On one occasion [a male] located [a female], followed her for almost an hour and then initiated copulation. She began to struggle and he finally desisted in his efforts. Although intromission did occur, ejaculation did not.[69]

Prior to considering the significance for humans of the above facts about rape by male orangutans, it should perhaps be emphasized that subadult males rather than mature males are more sexually active and are in greater contact with females. When copulation is attempted, it is commonly young males who rape females.[70] Young males are, however, in what one researcher has termed "an untenable position" with respect to reproduction since on the one hand, "receptive females . . . seem to prefer the large fully mature males as sexual partners," and since on the other hand, "extensive association with any adult female—receptive or not—is dangerous[:] The abrupt appearance of a mature male could mean disaster."[71] A further fact should also be considered prior to assessing the relevance of rape in orangutans to human rape. An orangutan male has a small penis. Coupled with the fact that arboreal copulation poses its own problems, that is, copulation takes place in suspensory postures, mating ideally requires female cooperation since intromission cannot otherwise be easily, not to say, successfully achieved. As one researcher observes, however, "Such cooperation is usual only within the context of an established consortship, in which a female is familiar with and shows little shyness toward her partner."[72] Where such cooperation is not forthcoming, male violence can erupt in full force due to the superior size and strength of the male and his consequent overwhelming power to strike, bite, grip, and otherwise brutalize the female.

To give primary weight to orangutans' lack of intercorporeal archetypal behaviors, behaviors present in other primate species, is to suggest an evolutionary similarity between human rape and orangutan rape. The similarity might be buttressed by comparing humans' not uncommon sex-specific cultural archetypes—male dominance/female docility—to orangutans' extreme sexual dimorphism: both can serve to encourage rather than deter rape. In effect, cultural practices and institutions no less than size can cast females in a position of inferiority and proportionately augment the possibility of rape. The example of rape in orangutans is from this perspective doubly instructive. At the same time that it calls attention to the need for what ethologist Konrad Lorenz designated "ritual behaviors"[73] that serve to defuse violence, it calls attention to the ways in which male possibilities of violence upon females may be intensified: cultural practices and institutions that foster, encourage, or allow rape—practices and institutions that are of course largely the craft and product of male ingenuity[74]—do not just fill the void created by the lack of archetypal socio-sexual/socio-aggressive behaviors; they make rape explainable on biologically given grounds. In other words, precisely the lack of biological restrictions in the form of archetypal socio-sexual/socio-aggressive behaviors makes it possible for cultures to consider rape a viable, biologically given option. Male dominance is in consequence conceived not as a quirk of culture; male dominance is *natural*. The lack of biological restrictions correlatively accentuates and reinforces the *natural* inferiority and docility of females. In brief, the biological void is filled to overflowing.

The above facts and comparisons notwithstanding, we can question whether the term "rape" is being appropriately applied to orangutan male behavior. As emphasized earlier, rape is not a *reproductive* strategy. It is a violation of females, not an act in the service of reproduction. It has nothing to do with insuring progeny; it has nothing to do with "passing on one's genes." To claim that it does by speaking of it as a human male reproductive ploy is to claim that rape is basically of positive value: being in the service of procreation, males who rape are just doing their job. But as ornithologist Cheryl Harding has observed, in such a claim, the adaptive value of rape and its motivation have been declared before they have each been demonstrated by actual data.[75] This methodological-conceptual error is well exemplified in the interpretation of male scorpionfly reproductive behavior: rape is assumed a reproductive option before its actual reproductive value or success has been assessed and shown. Indeed, its reproductive value is on the whole nowhere in evidence since such matings seldom result in insemination or progeny. Moreover a further central problem with the claim—though one not given concomitant central attention—concerns the fact that direct evidence of paternity is often lacking. In other words, in many instances, particularly those in which it is a matter of observing behavior in the wild and those concerning species in which individuals, because of their number and form, cannot be separately identified, observers can observe as many matings as they might want, they can gather all the data they might wish to gather on who copulates with whom, but they have no guarantee, and often no reliable idea at all, of which mating actually produces ensuing offspring, thus no idea of just *whose* male genes are being successfully passed on. Daly and Wilson make light of this problem: "The correct attribution of paternity remains something of a problem in field studies . . . so that we must usually rely on an indirect measure of fitness, namely mating success, which is presumably highly correlated with eventual reproductive output."[76] Presumptions are not of course the stuff of which viable scientific knowledge is made; counting "mating successes" gives no evidence whatsoever of actual genetic relationships. Even in studies of *captive* monkeys, the evidence that mating success is equivalent to siring is not conclusive.[77] Daly and Wilson suggest as much when they state that "The selective consequences of the reproductive competition among males will not be fully elucidated until determinations of paternity are carried out in natural troops [of primates]."[78]

Now if the data are not yet in, in either field or captive populations, then present-day sociobiological theory outstrips sociobiological evidence by a very wide margin, particularly when it comes to truths about *human* "troops." In fact, precisely in conjunction with human troops, the sociobiologist's primary focus on male reproductive fitness—siring as many children as possible—becomes quite curious. In American culture, for example, it is most common for unmarried males (and adulterous married ones) to run from any liaison which results in pregnancy. A genuine desire for children apart (and/or specifically, a desire that *this* sexual act result in a child), the last thing human males—or females—want from copulation is a child. The motivation for the act is elsewhere. While there is no statistical evidence available for this claim, foster homes, abortions, homes for unwed mothers, patrimony cases, legal statutes, and the like provide all the hard facts that are needed to support it. In this context, Daly and Wilson's gratuitous remark that "If a marriage contract provided a man with a magical guarantee of paternity, the world would be a more peaceable place!"[79] is particularly striking. It insidiously proposes that women's sexual dalliances—their wanton looseness—and women's deceitful ways are at the root of human male sexual promiscuity, the *raison d'être* of the "indiscriminate," "prodigal" male, as Daly and Wilson describe him.[80] Their exclamatory statement epitomizes how one can be carried away by theory rather than adhere to facts and to the honest truths those facts reveal. "Doubtfulness about paternity" masquerades as an explanation of male promiscuity when in actuality it attempts to provide an adaptive justification for it. In just the same way, "rape" in nonhuman animals masquerades as an explanation of human male violence upon females when in actuality it attempts to justify it on adaptational grounds.

Many scientists would be sympathetic to the above critical analysis because they strongly disagree with the idea that rape is an evolutionary phenomenon. Indeed, in partial response to the question of whether rape is an evolutionary reproductive strategy, one psychologist wrote that "Humans tend to excel in ugly acts, which sets us apart from the majority of the animal kingdom."[81] Psychologist Dolf Zillmann similarly observed after a thorough review of nonhuman behavioral data that, "Despite behaviors in other species that resemble rape, the achievement of sexual access by violent means, or by the threat of such means, is uniquely human."[82] In sum, when one considers the actual corporeal facts of rape and the actual corporeal facts of nonhuman animal reproductive behaviors, there is no "evolutionary legacy" for rape.[83] There is only a gratuitous assault on the body of one person by another person. The assault is by means of penetration—in all of its possible forms. Penetration in these instances is neither the result of invitation nor a prelude to intensifications in "reciprocal incarnation." It is the result of a cultural linkage between sex and aggression that is based not on an economy of nature but on the myth of the Phallus, a larger-than-life cultural exaggeration of penile display. It is in consequence an act of degradation on the one hand and of unconditioned power on the other. In this fact lies a fundamental meaning of the conception of females as being "in the form of a hole."

C. *Brain Structures, Neurohormonal Factors, and Cultural Studies*

There are other scientifically investigated links between sex and aggression that throw light on the male threat/female vulnerability equation. Much of the work showing a connection between sex and aggression to lie in brain structure came originally from neurophysiologist Paul Maclean's experimental brain stimulation studies of squirrel monkeys and his subsequently developed, well-known thesis of a triune brain, a brain that evolved progressively from a reptilian-type upper brain stem structure into a paleomammalian cortex and then into a human-type neo-cortex.[84] Maclean's specific thesis regarding sex and aggression is that the limbic area of the brain—that paleomammalian part of the brain whose convolutions form around the upper brain stem—is the site at which both sexual and aggressive behaviors originate. In particular, two landmark structures mediate sexual and aggressive behaviors, the septum and the amygdala respectively. These structures are not only close to one another, but the stimulation of the one spills over into the stimulation of the other. In the ensuing discussion of Maclean's work, and of brain ablation and stimulation experimental studies generally, I will draw on Zillmann's near exhaustive review of the literature in his book *Connections Between Sex and Aggression.*

Numerous experiments have been conducted subsequent to Maclean's original research. They have all centered in one way and another on his findings on squirrel monkeys, *male* squirrel monkeys. His brain stimulation studies showed that penile erection could be produced, and produced to varying degrees, according to the limbic area stimulated. Stimulation of the septum, for example, produced erection and led to neural discharges in the hippocampus, the discharges themselves tending to result not only in the most pronounced erections but in penile "throbbing."[85] Moreover penile erection was produced along with elicited fear and anger behaviors. As Zillmann points out after considering the evidence from major research studies, the conclusion drawn from experimental findings was that "sexual and aggressive stimulation were . . . in immediate proximity of one another."[86] Upon stimulation of regions in the hypothalamus, for example, "Subjects exhibited cackling, piercing vocalizations, retraction of one or both corners of the mouth, and perhaps most importantly displays of their fangs" along with penile erection.[87] What is of moment to consider briefly here (and in more detail presently) is the fact that this "orgastic appearance"[88] did not terminate in ejaculation. In fact, as Zillmann comments, *"ejaculation has never been observed under these circumstances"*[89] (italics added). Ejaculation occurred only when stimulation was at loci along the spinothalamic pathway. This distinction between loci is important since *erection is not exclusively a sexual manifestation in nonhuman primates;* it is equally a manifestation of sociality, excitement, and fear, for example, and it occurs in a variety of contexts, a fact we will presently consider at greater length. It suffices to emphasize here that what is termed *sexual* behavior again needs clarification, in this instance with respect specifically to penile erection.

Maclean's early work also showed a coincidence between oral behaviors and sexual ones. Maclean himself viewed these oral behaviors as aggressive-defensive ones, thus the coincidence between biting, for example, in socio-sexual interactions and in socio-aggressive interactions. Of perhaps greatest interest in this regard is the fact that various experiments in which the amygdala has been removed have resulted in changes in the dominance hierarchy of the groups of animals so surgically treated.[90] The overall result of the experimental studies was to demonstrate that, brain abnormalities aside, social circumstances and social dispositions are essentially determinative of aggressive behavior. For instance, the animal who was initially the alpha animal later became submissive after bilateral amygdalectomy; the animal who was initially the beta animal was, after bilateral amygdalectomy, equally submissive, though still aggressive to some degree over the lower-ranked animals; the animal who was initially the gamma animal was, after bilateral amygdalectomy, strongly aggressive over the lower-ranking animals, and both previous alpha and beta animals were submissive toward him. As for the lower-ranking animals themselves, however, although they were in a position to challenge the newly ranked gamma animal, they made no move to do so. The explanation offered for their non-aggressive behavior is quite plausible: they had no experience of social aggression and control over others. Unlike their confreres, *"they had not developed dispositions to dominate others."*[91] Thus, although an opportunity was present to take advantage of the new situation, the animals remained complacent vis-à-vis their new leader(s).

A similar study using brain stimulation procedures rather than brain surgery demonstrated the effects of social conditioning even more conclusively. Animals in this experiment were subject to various freedoms and restraints in conjunction with being either isolated or in group situations. The behavioral effects of brain stimulation proved contingent upon the physical and social situation, and also upon the social rank of the animal concerned. For example, animals restrained to a chair responded to the stimulation by staring; animals isolated in their colony cage but free to move about did in fact pace about and were restless; animals who were unrestrained and in the company of others were aggressive in varying ways depending upon their rank. Stimulation of alpha and beta animals produced aggressive behaviors; stimulation of low-ranking animals produced no aggression at all, only restlessness.[92]

Experiments with human subjects corroborate the considerable influence of social factors. For example, stimulation of the amygdala was found to result in aggression "only in patients with a history of violent behavior; it had no appreciable effect on nonviolent patients."[93] In this same experiment, it was found that in those patients who responded violently, it mattered little the particular site within the amygdala that was stimulated.

All together, the above results are cause for extended reflection and questioning. To begin with, are socially low-ranking creatures generally undisposed to act aggressively regardless of brain surgery or brain stimulation? In more general terms, is social grooming in the ways of a particular group (or culture) determinative of behavior in the sense that if one has never developed an aggressive behavioral repertoire in the course of growing up, one does not engage in aggressive behavior simply upon being given the opportunity to act aggressively? Clearly a positive response would seem to be of considerable moment in considering the culturally induced notion of females as inferior beings and its related cultural archetype, female docility. It would suggest why, either having—or "being given"—opportunities for aggression, human females tend to remain less combative and generally far less aggressive than human males in the sense of using physical violence on others and of manifesting criminal behavior generally.[94] It would, in effect, lend considerable weight to the thesis that males and females behave socially as they have been led to think over their formative years that they *should* behave and that *they have sanction* to behave.[95] Habituation to these patterns of social behavior lead to normative, culture-specific sex roles, roles that, through lack of other types of social experience, are perpetuated as social behavioral matrices.

At a deeper level of analysis, the positive response opens the way to a further seminal insight: sex-differentiated archetypes mediating social behavior are not *necessarily* sex-specific ontological archetypes. In other words, social behavioral archetypes do not necessarily coincide with, or constitute, indelible *sex-specific natures*. This is not simply because cultural practices are not straightaway biological mandates. It is first of all because behavior is not animate form; behavior *assumes* animate form, which is to say it assumes bodies and bodily experiences, both corporeal and intercorporeal. In this sense, it assumes all those spatio-kinetic relationships described earlier, relationships that are sex-neutral as well as sex-specific. Second, while certain behaviors and sex-specific natures might be said to be causally reducible or equivalent to one another in animals such as bees (worker and drone behaviors are sex-specific), they are not necessarily causally reducible or equivalent to one another in animals such as primates—precisely for a reason specified earlier, namely, the existence of sex-neutral behaviors, ones which, though sex-specific in socio-sexual interactions, are performed by both sexes in socio-aggressive interactions. Third, a distinction clearly must be made between social behavioral possibilities as *possibilities of animate form,* where evolutionary archetypal patterns emerge and where sex-specific ontological attributes, if they exist (for example, in the form of holes), would be rightly founded,[96] and social behavioral possibilities as *possible options* within the context of a motivated individual life. What the distinction strongly affirms is that socio-sexual and socio-aggressive behaviors are for humans a matter of choice, choice itself being essentially founded upon pan-cultural, pan-individual invariants in the form of "specific facts about the body" that translate into a specific repertoire of 'I can's.'

The experimental results raise further questions about aggression since if low-ranking creatures do not take active advantage of a situation in which they might gain power over others, but ones higher up the power ladder do, then human social behavioral archetypes are of considerable moment. In particular, male dominance/female docility coincides with the notion that human males in their formative years develop "high-ranking," aggressive behavioral dispositions and repertoires, dispositions and repertoires that become as fixed as those of "low-ranking females." Violence toward females would on this account be a matter of each male's individual history in the form of precepts of power deriving from his social attitudes and values, and ultimately, of course, from his being the animate form he is (the latter to be more fully spelled out in the next chapter). The idea of "fixity," however, is of particular importance. The perseveration of non-aggressive tendencies is one thing; the perseveration of aggressive tendencies is quite another. Certainly there is nothing socially wrong in a low-ranking creature's not taking socio-political advantage of others when it has the opportunity to do so. There is, however, something distinctly wrong socially when a high-ranking creature, simply on the basis of its group or cultural rank, does consistent violence to others. Violence is not equivalent to hostility—or to anger, wrath, or rage. Violence, like threat behavior, is an *act,* not a feeling. As such, it is *chosen* in a way that feelings are definitely *not* chosen. As will become apparent in the ensuing discussion of the neurophysiological links between sex and aggression, where acts of violence are not appreciated as such, the act of rape can easily become morally detached from the agent of the act. The agent can be absolved of responsibility in these instances because in a crucial sense, he himself is seen as having disappeared.

Specific attention should be called to brain research studies in which human rather than nonhuman brains were stimulated. Reports of septal stimulation raise especially pointed questions. Of sizable significance, for example, is the fact that such stimulation "has been observed to diminish violent behaviors."[97] Moreover one research report shows not only that pleasurable sensations are induced by septal stimulation but that penile tumescence is also.[98] In more general terms, septal stimulation elicited feelings of "well-being and euphoria."[99] Interestingly enough, the reports of such pleasurable, *non*sexual feelings have been criticized on grounds of being "vague"; moreover, the analysis has been criticized as tending to be "highly informal."[100] In light of such reservations, it seems reasonable to wonder whether where overall rather than discrete bodily feelings are expressed or reported, present-day science is at a loss, and this because such feelings as pleasure and euphoria have no measurable or specifically *organ*-anchored, visually verifiable behaviors attached to them. Obviously, if one uses penile tumescence as an indicator of pleasure, a measurable, visibly evident phenomenon is present and ready to be consulted. *The question, of course, is whether in using such an indicator, investigators and/or their respondents do not confuse feelings of well-being with sexual arousal.* (The question of whether such confusion might surface equally with a female "indicator" will presently be considered.) Clearly there is a distinct possibility of conflating the very things one wishes to clarify. Furthermore, to discount feelings of euphoria in connection with septal stimulation as vague but at the same time count penile tumescence in connection with septal stimulation as pleasurable is to skew experimental findings. "Pleasurable sensations," rather than being recognized as feelings of "well-being," are all too quickly and exclusively equated with penile tumescence, i.e., with sexual excitement. The fact is that in everyday life, *penile tumescence is not limited to sexual situations; it occurs in nonsexual situations, and in both human and nonhuman primate species.*

With respect to the functionings of the autonomic nervous system, and in contrast to the differentiated brain structures Maclean isolated, researchers in fact report little distinction between sexual and nonsexual neural excitation. For example, sexual and aggressive behaviors are reported as virtually undifferentiated in physiological terms: "hedonically positive and negative emotions have been shown to hold similar sympathetic charges."[101] Thus, as Zillmann notes, "Sympathetic excitation associated with reactions to conflict, provocation, and physical pain . . . is commensurate with sympathetic excitation fueling sexual desire and consummatory action."[102]

In consideration of this fact, and in consideration of a thorough review of the research on "cerebral representation," that is, on cortical localization of function, Zillmann goes on immediately to say that,

> In this connection, it is worthy of note that what might be considered the unique feature of sexual arousal (viz., vasocongestion in the access organs [penile tumescence][103] is by no means a specific response capable of unambiguously differentiating states of sexual and nonsexual excitedness. Research [brain stimulation studies and anthropological evidence] . . . leaves no doubt about the fact that in many mammals, including nonhuman primates, penile erection may accompany numerous aggressive and nonaggressive actions in nonsexual contexts. Uncovered men [cultures in which males are nude] similarly exhibit erection in a variety of emotional reactions devoid of sexual connotations. Erection has been observed during anxiety, surprise, appeasement, anger, and aggressive behavior as well as during greeting, joy, and elation.[104]

Now if "hedonically positive and negative emotions" carry a similar sympathetic charge, and if "[e]rection has been observed during anxiety, surprise, appeasement, anger, and aggressive behavior as well as during greeting, joy, and elation," then clearly males are epistemologically challenged. How do they know whether they are angry (or hostile, or feeling irritable) or sexually aroused? Moreover if *both* anger and sexual arousal are *pleasurable*—they are both characterized as *hedonic*—then the challenge is proportionately heightened and *the issue of a human male's capacity for discrimination* is all the more pressing to investigate. We will return to consider this vexing challenge and its implications further. For the moment, let us note that there is ample confirmation of the confluence of erection and *non*sexual feelings in the literature on nonhuman primate behavior in the wild. Indeed, Maclean's experimental findings on male squirrel monkeys and all subsequent related experimental research documents in the laboratory what is observed in the wild. I have elsewhere discussed this literature and quote from my summary findings:

> Primate penile erection functions in the same type of circumstances as bipedal posture/locomotion . . . and often accompanies the latter behaviors in chimpanzee societies. . . . Primatologist K. R. L. Hall and ethologist Wolfgang Wickler separately report on conspecific-oriented warning behavior by male savanna baboons, behavior in which the animal sits facing away from its own group and displays its genitals toward a neighboring group. Detlev Ploog and other ethologists give accounts of penile erection in the thigh-spreading greeting and dominance behavior of male squirrel monkeys, the behavior occurring in both agonistic and sexual encounters. Jane van Lawick-Goodall reports penile erection (chimpanzee) in bipedal swaggering

(a behavior occurring in both sexual and socio-aggressive contexts), in situations of heightened excitement over food, and in greeting situations. Penile erection is also reported in the infant and juvenile play of Japanese monkeys and in intertroop encounters of colobine monkeys. In short, . . . it is clear that as with bipedality, penile erection occurs in both sexual and nonsexual contexts.[105]

Precisely in view of the confluence of erection and *non*sexual feelings in primate life—what Zillmann more generally calls "The Excitation-Transfer Connection"—penile erection cannot be taken as an indicator of exclusively *sexual* inclinations. While male genital tumescence is commonly associated with sexual excitement and readiness, erection is in fact not limited to sexual occasions. What might in the two preceding sentences appear unnecessarily emphatic or repetitious is perhaps not. Indeed, the confluence of erection and nonsexual feelings in many primates has far deeper significance than a simple factual reporting of data might suggest. The same negative claim is made above via two different locutions—*erection* and *genital tumescence*—in order to call attention to a distinction that Zillmann seems to want to make but does not spell out concretely. He appears to want to distinguish linguistically between erection in sexual and nonsexual situations, ostensibly using "erection" as a generic term and "genital tumescence" to designate the sexual species of erection. Although he does not explicitly distinguish the two locutions, index entries on the subject are peculiarly divided into "erection" and "tumescence," no explanation being given in the text for the two separate entries. Close examination of the relevant passages suggests the above interpretation. In fact, only with the above interpretation of the indexical division between erection and genital tumescence does it become understandable how, given his emphasis on cerebral "spill-overs" and particularly on physiological fusion (indeed, "excitation-transfer" highlights the role of the sympathetic nervous system), Zillmann can claim that penile tumescence is a "specifically sexual [response]."[106]

While a linguistic distinction can certainly be made—one can arbitrarily decide to call the same thing by two different names according to the situation in which it is found[107]—can an *experiential* distinction be made? Can and does a male readily distinguish between erection in anger (more generally, in aggressive moods) and erection in sexual readiness? As the earlier quotation ("What might be considered the unique feature of sexual arousal is by no means a specific response . . . ") from Zillmann shows, there seems to be some doubt. This is not simply because sympathetic arousal is non-discrete with respect to situations of anger on the one hand and situations of sexual readiness on the other. The quotation reflects more than this acknowledged *neuronal* ambiguity. To begin with, it reflects an *organ*-ic ambiguity. Vasocongestion, Zillmann states, "is by no means a specific response capable of unambiguously differentiating states of sexual and nonsexual excitedness." In this passage, neuronal ambiguity passes over into organic ambiguity since what is vasocongested is the penis, which, as respondent, is judged incapable of discernment. Indeed, it is as if "the owner" of the response were not there. What Zillmann means to say, of course, is that physiologically there is no distinction between sexual and nonsexual excitement; in effect, he is saying that one cannot rely on the body for a clue as to whether what is felt is sexual or aggressive in nature. The pressing question is, *who can one rely on—the owner?* Do males—all males?—regularly and properly discern sexual from nonsexual excitement?[108]

In later passages, Zillmann confirms the likelihood that males (presumably some males regularly, perhaps all males on occasion) may fail to distinguish erection-in-anger and erection-in-sexual readiness. With respect to his review of research on rapists in particular, he writes that "Essentially . . . the findings show that in their penile reactions rapists fail to differentiate between sexual situations that entail threats of violence and violent actions and sexual situations devoid of coercion and aggressive behaviors."[109] Commenting on the fact that "a strong preference for excitement-laden sex is undeniable" in contemporary Western cultures as it is in "a large majority of cultures," he goes on to note that "the transfer connection between sex and aggression" suggests that aggression may fuel and/or revive the desired excitement.[110] In this context, the ambiguity of penile response is of monumentally critical significance: "If erection is not unique to sexual stimulation and also occurs during aggressive action, men should be prone to misconstruing their inclinations and 'detect' sexual impulses in purely aggressive actions."[111] The impact of this suggestion is immediately softened by Zillmann's follow-up comment that "This [misconstruing], for one thing, might explain the male tendency to express violent intentions with sexual verbiage." Notwithstanding the possibility that Zillmann's explanatory surmise is correct with respect to sexual verbiage, the more logical follow-up given the topic of concern—the conflation of sexual stimulation and aggressive *action*—would be that *"This [misconstruing], for one thing, might explain why some males rape."* There are several points to be made in conjunction with this suggestion.

To begin with, recognition is due Zillmann in suggesting the possibility of males' confusing sex and aggression. The suggestion is out in the open, as indeed it should be on the basis of the neurophysiological evidence, not to mention the act of rape. Not only this, but a quite interesting contrast is made between male and female vasocongestive responses. Zillmann points out that while nonspecificity of "vasocongestive genital response" might be similar in females, "feedback of this response is not obtrusive, as in the male, and misconceptions about sexual responsiveness are less likely as a result."[112] Thus while females might be found in a laboratory or other experimental situation to respond with clitoral excitation to sexual and aggressive situations alike, clitoral erection is not focally experienced, i.e., at the sensorial center of awareness of the situation. This means that females experience anger and sexual excitement or readiness in corporeally distinctive ways such that, unlike males, they are not "prone to misconstruing their inclinations and 'detect' sexual impulses in purely aggressive actions." Clearly, tumescence as a *felt* bodily presence is different for males and for females. Clearly, an erection as a *visible* bodily presence is also different for males and for females. Those sensory differences may in part explain why females are not similarly prone to confusing sex and aggression. They may also in part explain why females are not violent in the substantively physical ways that males tend to be. Sensory differentials result in differential senses of power—a claim justified in the first chapter and further validated on the basis of animate form.

Second, if those who rape are those prone to detecting aggressive impulses in sexual actions and sexual impulses in aggressive ones, then research that shows rapists to be intellectually wanting in terms of discernment, judgmental abilities, and the like, must be taken with substantial seriousness. Consider, for example, A. Nicolas Groth's findings from his original study:

> Rape is symptomatic of defects in human development. Although there is a wide variety of individual differences among men who rape, there are certain general characteristics that men who are prone to rape appear to have in common. Although his cognitive abilities appear intact, his actual behavior appears inconsistent with his rational functioning. Although intellectually competent, he tends to exhibit poor judgment, especially when he is emotionally aroused or under stress. He does not anticipate the consequences of his behavior. He acts without thinking. . . . He tends not to be introspective and exhibits little capacity for self-observation. Insight and self-awareness tend to be lacking. He is not very much in touch with his own . . . feelings . . . except for anger. . . . He tends to experience himself as being controlled by his feelings rather than being in control of them. He does not know how to identify his feelings. . . . He . . . seeks immediate need-gratification.[113]

Groth, who at the time of his research—the mid-to-late seventies—was Director of the Sex Offender Program for the State of Connecticut, leaves no doubt but that rapists are generally prone to poor judgment and tend *not* to be aware of bodily feelings: self-awareness and introspective capacities are deficient. It is not surprising then that, anger apart, a rapist is out of touch with the feelings of others as well. Groth, in fact, notes specifically that the rapist "tends to misread the feelings of others and to misinterpret their motives, largely because of his tendency to project his own characteristics onto others. He does not differentiate well among people or separate his own interests from theirs. . . . Lacking empathic skills, he has no point of reference except himself and, consequently, attributes his own attitudes and motives to others, with the result that he feels threatened and victimized by them."[114]

It is noteworthy that Groth's generalized description of a rapist accords psychologically with the picture of Symons's pornotopian male: projection figures prominently in the social relations of each. Indeed, from the psychological perspective of projection, pornotopian males and rapists are at opposite ends of the same spectrum, a spectrum from which one can view images of females across a correlative spectrum from the lustful, orgasmic to the threatening, victimizing. Given their psychological correspondence, it is not surprising that feminists have voiced concern about a relationship between pornography and actual violence against women. Male projection nullifies females as autonomous subjects in their own right. At either end of the spectrum, females are simply externalizations of male fancy, the result of male psychopathology and male conceits turned inside out.

It should also be pointed out that, in an attempt to gain a better picture of rapists, strong emphasis has recently been given to separating profiles of incarcerated rapists from studies of "self-identified nonincarcerated" rapists.[115] In other words, the picture of the incarcerated rapist that Groth and others have given has been viewed as a picture of a male who is outside the mainstream. This picture of the rapist as "an exception to the male rule" has in fact been roundly criticized by feminists who contend that males who rape are not peculiar, out-of-the-ordinary people. While psychological or psychoanalytic studies of rapists might suggest otherwise, rapists are not radically deranged individuals. Neither are they fundamentally evil as historically they have been portrayed. Lorenne Clark and Debra Lewis, in their book about rape in Canada—in particular in their chapter, "Rapists and Other Normal Men"—write that in psychiatric studies, rapists were "found to be relatively normal. . . . Some researchers found rapists to display 'inadequate personalities' or 'sociopathic tendencies', but none discovered that gross abnormality which had been anticipated."[116] Cross-cultural studies bear out this claim; rape-prone cultures are neither the

product of morally aberrant genes nor culture-wide psychiatric patholo-
gies.[117]

The characterization of rapists as normal has a deleterious cultural effect
precisely because it *normalizes* rape. Normalization trivializes rape: "Because
the rapist does not display extreme abnormality, he has not been judged a
danger to the community, and he has not, therefore, been subjected to the
kind of controls reserved for those who are seen in this way. 'Real rape' may
itself still be seen as a serious moral crime, but when the rapist exhibits
nothing more disturbing than hostility towards women, he is not regarded
as seriously ill, or subjected to the restriction of liberty which we reserve for
those who are so afflicted."[118] The normalization of rape requires specific
discussion in the context of Groth's findings, for in spite of those findings
deriving from incarcerated rapists only, they provide deep insight into a
relationship between a patriarchal society and the crime of rape. Groth's
general characterization of rapists has in other words broad implications.

If a rapist is a normal male human being, his normality coincides with
what is normal for his society. In present-day American society, news
reports, films, television, entertainment fare in general, popular reading
material, and so on, all purvey violence in one form or another. A social
setting thus exists in which individuals are surrounded by, and daily
consume, violence. Such a society produces individuals not substantively
different from the individuals Groth describes. These individuals also tend
"not to be introspective and [exhibit] little capacity for self-observation."
Not infrequently they are too caught up in the values, entertainments, and
other commodities the society offers to be self-reflective in such ways.
Moreover, these individuals similarly tend not to delay impulses but "[seek]
immediate need-gratification." These individuals too, being part of a
fast-living, time-is-money, competitive society, tend "to project [their] own
characteristics onto others, . . . [find] it hard to appreciate that other people
may have needs, feelings, attitudes, and values that are separate and distinct
from [their] own, . . . [and], lacking empathic skills, [have] no point of
reference except [themselves]." In short, in a society in which people tend
not to be in touch with feelings other than those of aggression and
competition, and in which they tend to perceive others merely as projections
of themselves, a rapist's actions are normal. Trivialization reflects the
sociological tenor of the times; the psychological deficiencies of the
incarcerated rapist are a microcosm of the psychological deficiencies of the
society at large. A society so used to violence that it does not consider rape a
crime, much less a serious crime, is normal. This is why a picture of a
promising student—a young man who is listed as a member of the Varsity

Club, Drama Club, Dean's List, and Student Council—can be aptly displayed in an advertisement about rape. The promising young man is also a rapist—a rapist "you would least suspect." The advertisement attempts to drive home the fact that, *its "normalcy" notwithstanding*, rape "is a criminal offense . . . It's a felony. Even for the biggest man on campus."[119]

It is apposite at this point to make reference to Zillmann's own research which focuses on both the physiological and cultural factors that constitute the link between sex and aggression. Research focusing on cultural rather than physiological factors led Zillmann to consider, among other things, how callous attitudes toward women are developed. A number of interesting facts resulted from the research which bear indirectly but pointedly upon male conflation of sex and aggression and the possibility of misconstruing the meaning of one's own bodily sensations, i.e., penile erection:

> "Men who had been massively exposed to erotica [became] highly callous toward women."

> "Massive exposure to erotic materials . . . apparently trivialized rape as a criminal offense."

> "Massively exposed subjects prescribed far less severe punishment for rape than did control subjects."

> "As a behavioral disposition, sexual callousness undoubtedly promotes the sexual harassment of women."

> "To the extent that massive exposure to pornography, as has been demonstrated, fosters lasting dispositions of sexual callousness, such exposure can be considered to promote hostile behavior in the long run."[120]

These findings, of interest in themselves, are set in heightened relief by the fact that "Characteristically, the female in pornography yearns for sexual and pseudosexual stimulation from any male in the vicinity, responds euphorically to any stimulation from anybody, and shows no loyalties."[121] *What is characteristic sounds tantalizingly familiar.* A moment's reflection, and the original readily surfaces. Females in pornography are modeled on the self-image of those males who create, dream of, support, and in general are lovers of pornography. Their characterization of females is a projection of themselves: they are ready and potent "year-round" and show no loyalties. Males thus project onto females the correlative cultural archetype of year-round erection: in pornography, it is females who are always ready and potent for all that "easy, anonymous, impersonal, unencumbered sex with an endless succession of lustful, [handsome], orgasmic [men]."[122]

Pornotopia is indeed, "and always has been," as Symons affirms, "a male fantasy realm."[123]

Anthropologist Peggy Reeves Sanday's comments likening certain cultural practices with pornography are of interest in this context. In a study of rape-free and rape-prone societies, she singles out the Mundurucu peoples of South America as paradigmatic of the latter. In this society, as in other tropical forest societies of South America and Highland New Guinea, "it is fairly frequent to find the threat of rape used to keep women from the men's houses or from viewing male sacred objects."[124] The Mundurucu sacred objects are trumpets which Sanday notes are "like reproductive cavities, the source of all fertility." She speaks of the "uterine-phallic nature of the trumpets," describing them as "long and hollow" and containing spirits. Moreover she points out that "the idea that animal and human fertility depends on their being properly fed, suggests that the trumpets symbolize the wellspring of life." The trumpets are in fact " 'fed' with meat from the hunt to please the ancestor spirits that dwell within the trumpets and to ensure the fertility of humans and animals." Sanday concludes that "the playing and feeding of the trumpets constitutes a fertility rite, the only ceremonial activity in this society."[125]

Sanday's analysis of male dominance through symbolic usurpation of female reproductive powers is compelling. By symbolically usurping what is female in their cultural practices and by threatening females with rape (or rapes since it is a question of gang rape) should they trespass on male territory or view the trumpets, Mundurucu males preserve and protect what Sanday terms "their culturally constructed maleness."[126] The analogy to pornography thus becomes evident. In this Amazonian society as in Western pornography, what is female is recreated according to a preferred—and enhanced—male image. The result for females in each case is impoverishment, a loss of both personal value and a sense of oneself as an autonomous subject. On the one hand, female 'I can's' are appropriated by males—"If men and women both reproduce, each in their own way, there is nothing of *importance* that a woman can do that a man cannot"[127]—on the other hand, a familiar equation in intensities is instituted: male threat/female vulnerability. Their self-made reproductive rites aside, it can of course be claimed that Mundurucu males are themselves only asking to be treated as autonomous subjects. This response, however, misses the point. One need only suppose that Mundurucu females had sacred instruments and sacred houses that they wanted to keep to themselves. With what bodily assault could they threaten males? Could they threaten "gang-castration"? Would they not first have to take possession of the society and remold its patriarchal foundations?

(Lest it be thought that Mundurucu practices are in any way unique, quite similar appropriations of female powers by males are evident in other societies, in Sambia society, for example, and in perhaps the most literal sense. Recall that in Sambia society semen is likened to mother's milk. Thus the homosexual practices first enforced upon boys when they are seven to ten years of age and continuing until they are initiated into marriage are a way of weaning them away from what is female. Female value is thereby culturally expunged, and male value inculcated.)

As the above discussions in this section show, evidence of a connection between sex and aggression—of an excitation-transfer—is persuasive. The evidence comes from a variety of sources—brain stimulation studies, neurophysiological studies, cultural studies. The question is, what does this transfer—or "excitation confusion"—mean in the concrete terms of everyday human power relations, specifically, rape? However persuasive feminist studies have been in their attempt to show that rape is a violent act and not a sexual act, rape continues to be explained as essentially a sexual behavior, "fraught" though it may be with violence. The stronghold for this latter explanation is sociobiological theory. But "brain theory" can also be appropriated as a stronghold, not in terms of the above discussed research on the limbic system, but in terms of hormonal influences on the brain. The distinction is important. Drawing solely on the limbic brain research discussed above, one could say that the contentiousness of the issue of whether rape is a criminal or sexual act is epitomized by limbic "spill-over": rape can as readily be explained as a sexual act that happens to "spill over" into aggression as an aggressive act that happens to "spill over" into sexual expression. When it comes to brain chemistry, however, the ambiguity is thought by some to disappear. The topic of rape thus warrants further examination, this time in the context of the chemical link between sex and aggression.

D.   A Question of Brain Chemistry

A recent exposition of rape as sexual act, as in fact "in our brains"—differentially, of course, with respect to sex—is given by sociologist Lee Ellis in his book *Theories of Rape*. Ellis's "synthetic theory" proposes a *forced copulation threshold,* the threshold marking that point along a probability axis where individuals would be likely "[to use] physical force in attempting to copulate."[128] Ellis's chart showing this axis is underwritten by a linear "minimal-to-maximal" factor specified as "Brain functioning patterns associated with a strong sex drive and a desire to possess and control multiple sex partners, and with insensitivity to the suffering of others."

According to his "synthesized theory," which he says draws on the advantages of feminist, social, and evolutionary theories of rape while avoiding their disadvantages, "the peaks (or means) of both the female and male distributions represent what has been optimally favored by natural selection in terms of each sex's sex drive, drive to possess and control sex partners, and reticular formation functioning." He goes on to say that "in other words, the average male has been naturally selected for having a stronger sex drive, orienting his drive to possess and control multiple sex partners, and having less sensitivity to the suffering of others than the average female."[129] He affirms that "rape is largely sexually motivated and that it has had, and continues to have, significant reproductive consequences."[130]

In spite of his protestations to the contrary, namely, that his synthesized theory does not support the notion of specific "rape genes,"[131] Ellis states that "according to the synthesized theory, the motivation for rape is, so to speak, 'hard-wired' into the brain by numerous genetic programs that, in most cases, have survived countless generations of natural selection."[132] Indeed, according to Ellis, the only learned, i.e., *cultural,* aspect of rape has to do with the particular tactics the rapist uses. These tactics have little to do with attitudes or imitation — for example, with exposure to pornography or with pervasive cultural attitudes toward women which are abasing. Rather, these tactics, Ellis says, have to do with 'hands-on' experience. In particular, once males have had copulatory experiences, those experiences serve largely to determine *"how much force the males use"* (italics added).[133] Only for those males who have minimal sexual experience and are "beyond the forced copulation threshold" are "linguistically mediated attitudes" or "rape (and other sexually violent) depictions" of consequence in motivating rape.[134]

On Ellis's sociological view, rape is essentially a brain event that has been tooled over millennia by natural selection. An understanding of rape thus becomes an understanding of brains. While genetic factors are indispensable, they are in the background: "The brain is recognized as the direct controller of all behavior (whether the behavior is largely learned or unlearned), and any effects that either genetics or the environment have upon behavior must be mediated through the brain." Ellis affirms that if this is so, then "the brains of rapists, at least during the time they are in the process of planning and committing their offenses, must be functioning in ways that are significantly different than (sic) the brains of other persons."[135]

Multiple facets of Ellis's theory warrant comment. One of these concerns both the specialization and the scope of available chemical findings about bodies. Chemical investigations are such that, on the one hand, bodily functions appear more specialized than they are; on the other hand, chemical investigations are so extensive that for almost any given item, one can find some positive findings. In actuality, laboratory procedures such as those required for investigation of hormonal action yield data consistently more simplified than their normal counterpart. Writing about hormonal specificity with respect to sexual postures and behaviors in rodents, for example, biologist H. Feder states that "A current trend of thought in psycho-endocrine research is that steroid-specific limited-capacity receptors mediate the effects of steroids on behaviour. Although a research strategy based on this idea will almost undoubtedly result in at least a partial affirmation of it, proponents should bear in mind that the idea has simplistic assumptions in it. . . . For example, the assumption is implicit that 'activation' of a behaviour is a unitary process. However, as we have seen, activation of a behaviour such as mounting might have quite different bases than activation of a behaviour such as lordosis."[136] Feder also emphasizes the role of *non*-hormonal factors, and *non*-specific hormonal factors along with specific hormonal ones. Biologist John Bancroft similarly emphasizes that simplistic explanations are naive. He states in fact that "there is no reason why we should expect to find simple relationships." He furthermore concludes that even with greater laboratory sophistication and valid, relevant measurements of behavior, future findings are not guaranteed: "Whether such progress will allow us to consider hormones as important determinants of sexual behaviour in the human remains to be seen."[137] In short, the case that Ellis makes for brain-mediated sexual behavior is seriously challenged on the grounds of myopic simplicity.

A further serious challenge concerns the conflation of correlation with cause. This conflation has in recent years been somewhat prominently criticized. Primatologist Linda Fedigan, for example, in discussing the possibility "that in general a high ranking male mates more often with estrous females than a low ranking male," rightly points out that even if a correlation were conclusively found, it would not mean "that a high ranking male has the right to any female as a *result* of his dominance or his aggressiveness."[138] The causative pattern, as she goes on to show, might be just the reverse. A similar charge is made by Ethel Tobach and Suzanne R. Sunday against sociobiologists Randy Thornhill and Nancy Thornhill on behalf of the latters' claim that forced copulation is the result of polygyny. The Thornhills argue that forced copulation, i.e., rape, is an adaptive evolutionary human behavior: certain human traits, aspects of development, and types of behavior correlate with "an evolutionary history of polygyny shown by other polygynous mammals." In addition to pointing out that the characterization of ancestral hominid societies as polygynous is purely speculative, Tobach and Sunday rightly point out that to arrive at a bona fide causal explanation, "the theory would need to show how the processes responsible for monogamy, polygyny, or polyandry are also responsible for, or related to forced copulation."[139] Zillmann makes a similar point in his discussion of studies of the relationship between hormones and criminal behavior. While there is a correlation between high testosterone level and prisoners who have committed violent crimes as opposed to prisoners whose crimes entailed no violence, the correlation "does not implicate a causal connection between testosterone level and the propensity for violent crime."[140] The meaning of differing testoterone levels was earlier challenged on similar grounds by Bancroft, who, in reviewing a study of male impotence and testosterone level, emphasized that "Once again it remains uncertain whether or not the differing testosterone levels [between impotents who maintained sexual desire and masturbated, and those who did not] were the *consequence* of differing amounts of sexual activity or interest."[141]

Now where brains are conceived as "planning and committing their offenses," it is but a short step to the notion of brains being responsible for rape. Indeed, one could say that "people don't rape, brains do." This kind of thinking is insidious precisely in giving to brains—*or not being able to give to brains*—what properly belongs to humans: responsibility for their actions. For example, the claim that males are genetically less sensitive than females to the pain of others is in actuality an escape clause. It allows males to behave violently toward females with an ethical carte blanche. Their violence is readily explained by their brain chemistry. Ellis writes that "In moving toward identifying the detailed nature of brain functioning patterns that would be conducive to rape, one may recall that it was documented [in chapter 5 of his book] that, in all species where forceful attempts to copulate have been found, virtually all of the attempts were made by males. This single fact is impossible to explain neurologically without assuming that something about the functioning of male brains across a wide spectrum of species must differ from the functioning of female brains. In other words, in many species, male brains must be more prone to opt for forceful copulatory tactics than female brains."[142]

What Ellis's brain theory of rape forgets is that the act of rape is a *bodily* act. What Ellis forgets is, in a word, animate form. The "single fact"—that virtually all "forceful attempts to copulate" are attempts by males—is directly and parsimoniously explicable on grounds of animate form. Only males have the power to threaten sexually, to intimidate sexually, to force sexually; only males have a "sexual access organ" (as it is called); only they have a penis. Accordingly, if that organ's behavior is so similar in situations of aggression and sex that a male cannot distinguish between them, then those human powers of reflection that are so commonly touted, those powers of discrimination that are in fact most often deemed the unique province of humans, those powers so frequently aligned with and enshrined selectively in human brains, have failed. Indeed, males who do not have the capacity to distinguish between aggressive and sexual feelings and in turn between a violent act and a sexual act can hardly be considered rational. Even their closest evolutionary relatives regularly distinguish between presenting and mounting in socio-sexual and socio-aggressive situations. From an evolutionary viewpoint, one might in fact wonder whether, if "man is a rational animal," it is in part because his rationality was initially honed to allow him to distinguish sexual feelings from aggressive ones and thus to offset his deficiency, namely, that economy of nature found in his closest evolutionary relatives.

Ellis's brain-chemistry explanation of rape actually forces us to the startling conclusion that human males are biologically incapable of distinguishing sexual feelings from aggressive feelings insofar as copulation *always* requires force. The only discriminatory power a male has (or needs?) is the power to distinguish in each case *how much force to use*—presumably on the grounds of efficiency. From the perspective of the finely tuned equation of male threat/female vulnerability, Ellis's theory leaves no doubt as to how brain chemistry and power coalesce. His picture of human males is definitive. Human males are archetypally aggressive animals driven by sex, and driven not only to perform sex at whatever levels of force it takes to carry out the performance, but to control and possess as many partners as possible. The picture is morally—or more exactly, *amorally*—definitive as well. To protect his brain-driven compulsion to appropriate and dominate as many females as possible, the archetypally aggressive human male is not simply thoroughly absolved cerebrally from heeding any impulsion toward moral restraint; *by nature,* he is impervious to the suffering he causes others. He is the pawn of an all-powerful dictatorial brain.

Females, by this scenario, can do naught but take their cerebral lumps. Alas, they simply do not have the *brains* to rape.

The foregoing critical review of scientific claims regarding the relationship between sex and aggression in general and the act of rape in particular has shown from various points of view how archetypal formulations of females as receptive "year-round" relates to male threat/female vulnerability and in turn to female fear of male assault. In the following chapter, we shall examine these matters from the viewpoint of a psychoanalytic ontology.

6

Corporeal Archetypes: Penetration and Being "in the Form of a Hole"

The amorous act is the castration of the man; but this is above all because sex is a hole.

Jean-Paul Sartre[1]

Throughout his life [Freud] never abandoned the view that the thing that terrified men so much (especially "homosexual" men) was the "castrated" state of the female genitalia. That assumption is particularly enraging for women who refuse any longer to be persuaded that the vagina is an organ of "lack," while the prick is one of "fullness." They are right of course. Indeed, men who have loved and been loved by women, should be similarly enraged. In an act of pleasurable lovemaking, how is the vagina inferior or castrated? Is it any less potent, say, than the phallus?

Klaus Theweleit[2]

I. The Patriarchal Sartre

As noted earlier, the characterization of females as being "in the form of a hole" is one Sartre gives in *Being and Nothingness*. In the penultimate sections of that book, Sartre begins describing that area where ontology ends and existential psychology begins. He has just finished his analysis of the human condition and is at the point of exploring the path to which it leads: human reality as desire, specifically, the desire to do, to have, and to be. He repeats his basic claim that the human condition is always a living relation *to something*. His characterization of females as being "in the form of a hole" occurs in this context. The characterization has been strongly criticized by feminists who have found it an affronting, ontologically reductive view of females. In particular, Margery Collins and Christine Pierce, in an early article titled, "Holes and Slime: Sexism in Sartre's Psychoanalysis," persuasively show how Sartre consistently aligns female nature with negative values of existence—with all that is obscene and menacing, for example, and all that is less than a "plenitude of being."[3] In a later essay, Pierce singly points out that, while Sartre has defined female being as "the obscenity of the hole," he has "not seen fit to write an essay on the obscenity of the dangling."[4] The criticism is clearly warranted; a myopic and severely biased male perspective subtends Sartre's characterization of females. But the characterization and its psychological context must be taken together. In this broader view, there is both a patriarchal and archetypal account of female (and male) sexuality to examine. The best way of capturing the patriarchal Sartre is by direct quotation:

> The obscenity of the feminine sex is that of everything which "gapes open." It is *an appeal to being* as all holes are. In herself woman appeals to a strange flesh which is to transform her into a fullness of being by penetration and dissolution. Conversely woman senses her condition as an appeal precisely because she is "in the form of a hole." This is the true origin of Adler's complex [the complex of *inferiority*]. Beyond any doubt her sex is a mouth and a voracious mouth which devours the penis—a fact which can easily lead to the idea of castration. The amorous act is the castration of the man; but this is above all because sex is a hole.[5]

Now there is no question but that in the "psychoanalysis of *things*"[6] that has preceded the above characterization, the feminine is linked with the In-Itself. In Sartre's ontology, the In-Itself is the antithesis of the For-Itself. The latter being is free; it has possibilities in the world. The former being simply is what it is: it has a certain fixed nature; it has no projects. Sartre has in these earlier passages used an extended analysis of *the slimy* to show how a certain character is implicit in beings that simply are what they are. A certain quality immediately distinguishes them as soft, dull, or slimy, for example. Insofar as Sartre both directly and by suggestion aligns what is slimy with what is female, and what is female thereby with what has a fixed character, it is of special interest that at one point in his elaborations he says that "the slimy is *docile*" (his italics),[7] and speaks of its docility in terms of a "leech-like" softness. Though the relationship is not explicitly spelled out in his ensuing description of the slimy as docile, Sartre unequivocally connects female bodies with docile bodies. Whereas when holding a solid object, he says, "I can let go when I please," the slimy in contrast compromises his freedom:

I open my hands, I want to let go of the slimy and it sticks to me, it draws me, it sucks at me. Its mode of being is neither the reassuring inertia of the solid nor a dynamism like that in water which is exhausted in fleeing from me. It is a soft, yielding action, a moist and feminine sucking, it lives obscurely under my fingers, and I sense it like a dizziness; it draws me to it as the bottom of a precipice might draw me. There is something like a tactile fascination in the slimy. I am no longer the master in *arresting* the process of appropriation. . . . In one sense it is like the supreme docility of the possessed, the fidelity of a dog who *gives himself* even when one does not want him any longer, and in another sense there is underneath this docility a surreptitious appropriation of the possessor by the possessed.[8]

We should also note that prior to this description of how the slimy is docile, Sartre has attempted to show how all desire is appropriative. "The being of human reality"—desire—is a lived relation, he says, and lived relations are appropriative acts, acts of a For-Itself, a being who has possibilities and is free to choose among them.[9] It is relevant in this context to spell out Sartre's description of *knowing* as a form of appropriation, for it sheds light on his implicit grasp of the sexual modes of threat and vulnerability in which males and females respectively commonly exist in Western societies (as in the majority of present-day cultures), and which we have analyzed earlier. Sartre's account of knowing focuses specifically on scientific knowing and is not of course pejorative—although there is a faint hint of criticism of the ways in which scientists conceive their work. On the whole the account shows how knowing *as a desire to appropriate* is a male act, and how male appropriation—or male dominance—can and does express itself sexually in essentially nonsexual endeavors. In fact insofar as Sartre explicitly calls attention to a similarity between scientific knowing and carnal knowing (see below), the suggestion is that, particular contexts of appropriation to the contrary, male desire is everywhere the same: it is *always* self-arrogating. Sartre speaks furthermore of knowledge being "at one and the same time *a penetration* and a *superficial* caress" of the epistemic object of desire, and of "an appropriative enjoyment" in the context of describing knowledge as scientific discovery.[10] Moreover he describes his own philosophical work as just such a penetration and caress.[11] This consistent, all-embracing nature of desire to appropriate, and the enjoyment of appropriation itself, exists because male dominance is "in the nature of things," or more properly, *in the nature of the For-Itself.*

While Sartre offers no justification for there being a "First Sex,"[12] it is clear that he sees the supremacy of males and male dominance not as culturally or individually idiosyncratic but as natural. Hence his non-pejorative, simply descriptive tone. The absence of a critical tone notwithstanding, his description coincides exactly, one might almost say preternaturally, with present-day feminist critiques of Western scientific research as an essentially male enterprise with respect both to the schism drawn between the knower and the known, and to the essentially assaultive, exploitative, and visual nature of the research and its parallels with the degradation of all that is female.[13] One could readily take Sartre's 1943 descriptive account of epistemic appropriation in scientific research for a feminist tract written some twenty and more years later:

> What is seen is possessed; to see is to *deflower*. If we examine the comparisons ordinarily used to express the relation between the knower and the known, we see that many of them are represented as being a kind of *violation by sight*. The unknown object is given as immaculate, as virgin, comparable to a *whiteness*. It has not yet "delivered up" its secrets; man has not yet "snatched" its secret away from it. All these images insist that the object is ignorant of the investigations and the instruments aimed at it; it is unconscious of being known; it goes about its business without noticing the glance which spies on it, like a woman whom a passerby catches unaware at her bath. Figures of speech . . . like that of the "unviolated depths" of nature suggest the idea of sexual intercourse more plainly. We speak of snatching away her veils from nature. . . . Every investigation implies the idea of a nudity which one brings out into the open by clearing away the obstacles which cover it, just as Actaeon clears away the branches so that he can have a better view of Diana at her bath. . . . The scientist is the hunter who surprises a white nudity and who violates by looking at it.[14]

Where this uniquely male freedom to appropriate the world—this "male mode of knowing"—collides with the world itself is in its encounter with things in what one might call their "psychological aspect," for in such an encounter, an appropriative relationship to things is disturbed by the fact that quality reveals itself. As is apparent from the earlier description of the slimy, the For-Itself is thereby challenged: "I am no longer the master in *arresting* the process of appropriation," Sartre writes; there is "a surreptitious appropriation of the possessor by the possessed." Appropriation here folds in on itself because while the character of a thing—its sliminess, sweetness, or softness—gives itself, it at the same time remains wholly on the side of the thing. Sartre at one point in fact speaks of this worldly state of affairs in terms of male and female sexual relations: "the lover's dream is to identify the beloved object with himself and still preserve for it its own individuality; let the Other become me without ceasing to be the Other."[15] The lover's appropriative desire fails here of course in that the female is never *his*. This failure is mirrored by the experience of the slimy, an experience in which mastery slips from the side of the possessor to that of the possessed. Indeed, the lover loses his control in the same way that the "master of appropriation" loses his control in contacting the slimy; both are had by their "beloved object." Though concerned with the fact that "desire destroys its object,"[16] that, in other words, whether a matter of scientific research, amorous pursuits, or eating ("if I eat it, I do not have it any more"),[17] the object of desire always escapes assimilation to the subject of desire (in the manner of a Lacanian lack: see chapter 9), Sartre's more basic concern in these descriptive passages is to show how qualities we find in the world reveal being. In effect, his concern is to demonstrate how things in the everyday human world have an intrinsic nature. This intrinsic nature is their ontological meaning, a meaning which, according to Sartre, hangs between the physical and the psychic: it can be attributed wholly neither to the thing itself nor to what we make of the thing; rather, it is peculiar at once to them and to us. Sartre's association of females with worldly qualities (i.e., with holes and with the slimy) occurs in just this context. Females are thus necessarily conceived as something at once *in-themselves* and at the same time something whose meaning is bestowed upon them from the outside. They are indeed docile bodies awaiting male appropriation and validation.

Sartre nowhere explicitly acknowledges the fact that he accords females a different humanness from males, that, in extreme terms, he puts being on the side of males and nothingness on the side of females. He does not seem to realize that such a division, particularly since nowhere substantiated, vitiates his ontology. It denies to females precisely the consciousness that he is at pains to show is a *human* consciousness, a consciousness that is not simply a container of experiences but a plenum of possibilities. Moreover, while he does not openly censure females in aligning them with holes and with the slimy, or openly degrade females in aligning them uncritically with the immaculate unknown object to be deflowered in the name of scientific knowledge or with the beloved object in the lover's dream, he very definitely compromises female existence by robbing it categorically of autonomy. It is indeed remarkable that Sartre does not realize how his characterizations and analogies, by jeopardizing female existence, jeopardize his ontology. The two are inextricably bound. One cannot forge an ontology of human existence exclusive of female existence. But neither can one forge an existential psychology of females inconsistent with one's ontology of human existence. Clearly, in the last sections of *Being and Nothingness,* it is the patriarchal Sartre who is speaking, a Sartre whose customary probity and introspective clarity have at this point deserted him. Being blind to his prejudices, the patriarchal Sartre stops short of a full and balanced psychoanalysis of *things,* in particular, things that he himself chooses to focus on by way of examples, things that are not just objects in the world but corporeal aspects of persons—of persons in general, but also quite pointedly of female persons in particular. In consequence, he stops short of the possibility of understanding corporeal archetypes, male as well as female. What is wanting on the one side is a self-elucidation that would illuminate the "strange flesh" that penetrates holes, and on the other side a full rendering of being "in the form of a hole." We will see precisely in an examination of the archetypal Sartre that it is these synecdochal corporeal archetypes that need elaboration and clarification.

II.   The Archetypal Sartre

Just as the patriarchal Sartre was best presented by direct quotation, so also is the archetypal Sartre. Immediately following the statement, "The amorous act is the castration of the man; but this is above all because sex is a hole," Sartre writes,

> We have to do here with a *pre-sexual* contribution which will become one of the components of sexuality as an empirical, complex, human attitude but which far from deriving its origin from the sexed being has nothing in common with basic sexuality. . . . [T]he experience with the hole, when the infant sees the reality, includes the ontological presentiment of sexual experience in general; it is with his flesh that the child stops up the hole and the hole, before all sexual specification, is an obscene expectation, an appeal to the flesh.[18]

Thus, in Sartre's psychoanalytic of things, just as the fundamental ontological meaning of the slimy is to be "swallowed up by the in-itself," so the fundamental ontological meaning of a hole is its "appeal to the flesh," an appeal, be it noted, which is not neutral or which does not run a gamut from positive to negative, but an appeal which is equivalent to "an obscene expectation." Sartre exemplifies this psychoanalysis of things further by briefly elaborating alimentary symbols of being—"existential significances of . . . foods"[19]—and suggests that our preferences in tastes—our preferences for oysters, pink cakes, chocolate, shrimp, and so on—have psychological import: "they all represent a certain appropriative choice of being."[20] He asserts that it is the task of existential psychology to compare and classify human tastes and inclinations.[21]

A hole, according to Sartre's existential psychology, is a manner of being in the world which is obscene, and which, as obscene, reflects in a quite specific way on those who relish holes as well as on those who abhor them.

Sartre's characterization of holes as obscene is puzzling and is perhaps best explained not only by his patriarchal sexuality—his purblind, one-sex view of the For-Itself—but by his Cartesian-leaning ontology—his less than complete rendering of the body. By questioning, clarifying, and elaborating his equation of "an appeal to the flesh" to "an obscene expectation," we will be led to a deeper and more just psychoanalysis of holes, one which is not only experientially sound but one which leads to an understanding of a hole as corporeal archetype. What we must ask first, and quite specifically, is why an appeal to the flesh is *obscene*. Why is it not, for example, fascinating, or wonderful? The fundamental ontological meaning of a hole for Leonardo da Vinci, for example, would likely have been precisely something wonderful as well as something fearsome. Of his encounter with "a great cavern," da Vinci wrote,

> Urged on by my eagerness to see the many varied and strange forms shaped by artful nature, I wandered for some time among the shady rocks and finally came to the entrance of a great cavern. At first I stood before it dumbfounded, knowing nothing of such a thing; then I bent over with my left hand braced against my knee and my right shading my squinting, deep-searching eyes; again and again, I bent over, peering here and there to discern something inside; but the all-embracing darkness revealed nothing.
>
> Standing there, I was suddenly struck by two things, fear and longing: fear of the dark, ominous cavern; longing to see if inside there was something wonderful.[22]

Clearly if we take da Vinci's account as a phenomenological base, the fundamental ontological meaning of a hole is something quite other than obscene. To find holes fundamentally *inviting* at all is to find them fundamentally positive, intriguing; they impel us to fill them or bid us enter in some bodily way. It is curious that Sartre himself does not recognize this fact particularly since he affirms that to plug up a hole is "one of the most fundamental tendencies of human reality—the tendency to fill." Indeed, he states that "A good part of our life is passed in plugging up holes, in filling empty places, in realizing and symbolically establishing a plenitude."[23] If this is so, then we humans, males and females alike, are wedded to a world of obscenities. Moreover given what Sartre recognizes as our tendency to fill holes, to characterize that which invites us as *obscene* is possible only to the extent that one simultaneously characterizes human nature itself as obscene. The issue of the obscene aside, insofar as the human tendency to fill necessarily conjoins with an appeal, then holes are basically an appeal to the flesh—*an invitation*—and the tendency to respond to their appeal is equally basic; perception and act are intrinsically related. As Sartre himself specifically suggests, and as experimental studies of infants and young children directly attest,[24] a child is fascinated by putting (or pouring) things into holes, including putting its thumb into the hole which is its own mouth. Living bodies are moreover full of holes not only into which things can be put, but in which things are found (e.g., saliva, mucous, wax), and out of which things pass (e.g., urine, feces, phlegm when one sneezes, and so on). Holes are or can be clearly fascinating. They may also be fearsome. As da Vinci's descriptive account shows, one is not sure of what one might find in them. But as da Vinci himself indicates, that is part of their allure. They *are* an appeal to the flesh. Finally, where it is specifically a question of filling a bodily hole by bodily entering into it in some way, then unless one preconceives bodily insides as impure, fetid, or disgusting, there is nothing fundamentally obscene about an appeal of flesh to flesh, whether a thumb in a mouth, a tongue in an ear, or a penis in a vagina.

On the above critical analysis, holes appear to be archetypally inviting forms, and to invite is not in and of itself obscene. On the contrary, in beckoning us to fill them, holes make a primordial appeal characterized by fascination, wonder, fear, and/or awe, precisely the felt dimensions that da Vinci describes and that infant and child behaviors validate. Were we to take Sartre's characterization of females seriously simply at the level of the form of their genitalia—their *internal* genitalia at that—then we would have to say first that holes as psycho-ontological forms have not the wholly negative character that Sartre ascribes to them. Second, we would have to add that their appeal to the flesh must be taken into account. If female genitalia make an appeal which is obscene, then the flesh that answers to their invitation must itself have a character that is obscene. In other words, the flesh that answers to the invitation of female holes is not obscene by *contaminating* itself with the obscene. It is itself obscene because anything that would *answer positively* to an obscenity can only be an obscenity itself. Thus not only are perception and act intrinsically related as described above, but appeal and response are *qualitatively* linked: object and subject are *together* part and parcel of the same psychoanalytic of things in the world. Sartre himself avers this very fact when, as noted earlier, he tells us that ontological meaning hangs between the physical and the psychic. In sum, a psychological ontology of human genitalia demands far more thorough and neutral treatment than what Sartre offers us.

In fact, while Sartre's ontological characterization of holes compels us to think critically about the ways in which genitalia might be tied to archetypal forms, his emphasis upon fundamental ontological meanings, that is, upon the intrinsic character of things such as their softness, their liquidity, and so on, compels us to consider the possibility that the fundamental quality of things is indeed something other than a merely subjective feature or a mere function of the thing's materiality, each ontological option being exclusive of the other. If we know things in the world as having a certain character which is theirs but which at the same time indicates our own project or being—our own peculiar feelings in face of the thing—then a psychoanalysis of quality as a fundamental feature of the world must necessarily reveal something which is neither wholly on the side of the subject nor wholly on the side of the object, something which, precisely as we have seen in terms of possible genitalic archetypes, is not simply transient and personal but, as Sartre insists, is part of a "universal symbolism."[25]

There are several significant points to be made here. First, when Sartre writes that "What comes back to us [following extended reflection upon our experiences of a certain quality in the world] as an objective quality is a new nature which is neither material (and physical) nor psychic, but which transcends the opposition of the psychic and the physical,"[26] his description verges on the aesthetic; it coincides with descriptive accounts of aesthetic experience, experiences in which we give ourselves to the object as the object gives itself to us. Glossing on this correspondence, and elaborating Sartre's own description, we might say that the everyday world in its psycho-ontological mode presents itself not as a parade of *things* but as a parade of *physiognomies*. In just this way a painting, a piece of music, or a dance presents itself not as a thing but as a dynamic configuration, a quintessential quality or play of qualities that have certain evocative powers; their very character makes a certain appeal to us. Such dynamic configurations or quintessential qualities are a revelation of being in the same psycho-ontological sense that holes and the slimy are a revelation of being. Thus, in order to understand the psychoanalytic nature of things in the world—holes, for example, and things that can penetrate holes—one must first of all appreciate the fact that what is there reveals itself qualitatively to us and in so doing wields a certain power over us, attracting or repelling us—or both. The physiognomic aspect of what is there is in other words ascendant over all other aspects. At the same time, in Sartre's words, "we appropriate" what is there in some way as something *for us*. From this perspective, a hole is clearly neither a mere material emptiness nor a purely psychic "appeal to fill." Let us elaborate this physiognomic character of holes further in the way of a Sartrean psychoanalytic of things.

Holes are among the first discoveries if not *the* first discovery one makes both *of* one's body and *in conjunction with* one's body. Holes are orifices that lead to *insides*. Whatever the adult meanings one comes to attach personally to holes, those meanings are tied to fundamental experiences of one's own bodily holes as just such orifices. In all of these experiences, the felt body is primary, that is, tactile-kinesthetic awarenesses and tactile-kinesthetic explorations of one's own body are foundational to a conception of one's body as "full of holes." When Sartre writes in criticism of the traditional psychoanalytical theory of anal sexuality that a child cannot apprehend a part of his own body, i.e., his anus, "as an objective structure of the universe,"[27] he mistakes the visual appearance of a hole with the distinctly tactile-kinesthetic-kinetic sensations of excretion that reveal to the child (and adult) the bodily hole that is his anus, or with the distinctly tactile-kinesthetic sensations prompted by finger explorations that reveal the same bodily hole. He mistakenly believes that "it is only through another— through the words which the mother uses to designate the child's body— that he learns that his anus is a hole."[28] Proctologists aside, this would virtually make mothers alone privy to the fact that a hole exists where it does. It would furthermore make *all* knowledge of our unseen body both Other- and language-dependent, as if our own tactility and kinesthesia did not exist. On the contrary, in such acts as defecating and sneezing, in all-over bodily self-touchings, in sensations of warmth and cold, in experiences as diverse as babbling and having a stomachache, a child's sense of bodily insides and outsides develops. Its ultimate *linguistic* notion of its anus as a hole is contingent on these experiences. Indeed, children would have no idea of what their mothers could be linguistically referring to as a hole (or by the word *anus*) unless they had had the tactile-kinesthetic-kinetic experiences they have had. In short, the concept of a hole is first and foremost a corporeal concept. As I have shown elsewhere, insides and outsides are fundamentally not linguistic categories but corporeal ones; they derive from first-hand bodily experiences.[29] There is thus every reason to believe that the fundamental ontological meaning of holes in the visually perceived world derives from our original corporeal concept of holes; it has its origin in first-hand bodily experiences—of defecating, urinating, coughing, sneezing, sucking, drinking, eating—even *closing the eyes*. Indeed, given these everyday corporeal matters of fact, matters which, it should be noted, are *pan-cultural,* one can readily understand Leonardo da Vinci's feelings in face of "a great cavern." One's own body is "a great cavern" which fills one both with fear and wonder, fear perhaps that its functionings, unseen to the eye, are defective (or will some day be defective); wonder perhaps at its incredible and complex productions. So it appears to the Wik-mungkan, for example, an aboriginal Australian people who are fascinated by the difference between urine and feces and by the way each is formed.[30] Where no Western science "explains" feces, the latter are readily taken as mysterious emanations from the body. Clearly, the experience of holes as entrances to mysterious, awesome, wondrous insides derives from our own bodily experiences.

Put in the above perspective, it is clear that Sartre's pyscho-ontology of meaning with respect to holes is both idiosyncratic and archetypally one-sided. His view of holes as obscene derives from an archetypal fear or aversion to corporeal holes; his purportedly fundamental psychoanalytic ontology neglects altogether the positive aspect of the archetype so readily apparent in da Vinci's description. Indeed, at the most fundamental level, as Sartre's own account affirms, holes are clearly inviting: they fascinate, they have a positive bodily attraction, an affirmatively as well as negatively compelling valence.

The above critique of Sartre's psychoanalytic of holes actually brings to the fore an interesting psychoanalytical question about Sartre himself. The question is implicit in his own explanation of how it is that people have the specific feelings they do about things in the world. "I can love slimy contacts, have a horror of holes, etc.[,]" Sartre says, and then proceeds to explain:

> That does not mean that for me the slimy, the greasy, a hole, etc. have lost their general ontological meaning, but on the contrary that *because* of this meaning, I determine myself in this or that manner in relation to them. If the slimy is indeed the symbol of a being in which the for-itself is swallowed up by the in-itself, what kind of a person am I if in encountering others, I love the slimy? To what fundamental project of myself am I referred if I want to explain this love of an ambiguous, sucking in-itself?[31]

By this very token, we can ask Sartre to ask himself, "What kind of a person am I if in encountering holes I find them *obscene?* what is there about me in my relations to the world that fundamentally explains my finding the inherent *appeal* of holes to be obscene?" Whatever his answer might be, it is likely rooted in the archetypal fearsomeness of holes (a fearsomeness that is perhaps for Sartre linked to their *abysmal* character insofar as that character ties in with their *nothingness*) to the exclusion of their archetypal fascination or wonder. A thorough and deeper psychoanalytic of Sartre himself vis-à-vis holes would likely show further how his determination, "sex is a hole that is obscene," is tied to a larger but unexpressed determination of the whole body—with all its holes and itself conceived as a hole—as obscene. Sartre's determination in other words derives axiologically from the way in which he experiences his own body. The deeper psychoanalytic questioning of Sartre himself would undoubtedly point in the direction of a psycho-ontology in which the obscenity of holes in general and the obscenity of the hole that is sex in particular are tied to a nausea of the physical.[32]

Beyond asking for a self-questioning and for the deeper psychoanalytic that self-questioning suggests, we can furthermore ask Sartre to explain how "a *pre-sexual* contribution" can possibly result in *"obscene* expectations." When Sartre states that "the child stops up the hole and the hole, before all sexual specification, is an obscene expectation,"[33] he errs in attributing to the child what cannot possibly be the child's: *obscene* expectations cannot in any reasonable way be attributed to a two-year-old much less to an infant. Obscenity is an *adult* assessment. Obscene expectations are in fact not a *pre*-sexual contribution at all. They are not only decidedly adult, but in the context of Sartre's descriptions, they are also decidedly male.

Put in the broader perspective of his work as a whole, Sartre's psychoanalytic of holes is consistent with his general account of the human condition as *angst;* there is no joy, only anxiety. As is apparent, however, his account is richly heuristic. It has opened the way toward an analysis of the hole as corporeal archetype. Not only corporeal acts and intercorporeal positionings and behaviors but also aspects of the body—bodily parts—may figure as archetypal forms.[34] His psychoanalytic moreover strongly supports the claim that corporeal archetypes function in an essential way in our conceptions of things; archetypal meanings enter into our *personal* tastes and inclinations.[35] If this is true, then both individually and culturally spawned archetypes are derivative forms; that is, not only cultural tastes and inclinations are variations on fundamental corporeal themes, but individual preferences and dispositions are as well. In turn, this means that archetypal aspects of the body are as susceptible to personal exaggerations and reductions, distortions and elaborations as to cultural ones. Just such a derivative relationship was intimated earlier in noting invariant human bodily acts such as defecating and sneezing. These and other everyday corporeal matters of fact are the foundation of particular personal as well as cultural values. To understand male threat/female vulnerability in the context of a psychoanalytic of things would thus require understandings of how corporeal archetypes subtending the equation in intensities—be they positionings, behaviors, or bodily parts—are semantically intensified and elaborated not only by cultural practice but by personal psychoanalytic ontologies.

By providing insight into the hole as corporeal archetype, the above critical analysis of Sartre's characterization of females as being "in the form of a hole" has provided insight into one side of the equation in intensities. Docile female bodies, at once cultural products and products of a particular personal psychoanalytic of things in the world, are rooted in an archetypal aspect of female sexuality, but one so stripped of its full living values— negative as well as positive—as to silence its full complement of fundamental meanings. Its appeal is reduced to its use; it exists to be appropriated. The archetypal hole is in short degraded to a body part that "gaps open" and is there not simply to be filled, but that, in gaping open, is receptive "year-round" and incites male desire simply by existing. At this level, of course, where there is neither fascination nor fear, incitement may easily be to violence. "The castration of men" of which Sartre speaks is an inverted projection of this incitement to violence.

III. Penetration

We turn now to the other side of the psycho-ontological equation and ask what corresponds to being in the form of a hole; what is the nature of the "strange flesh" that is in the form that fills holes? Necessarily the answer is "a fleshly form having the power to penetrate," which in turn means a form having *instrumental* power. Now what corporeal aspect more aptly fits this criterion—at least potentially—than something elongate and relatively inflexible like an erect penis? Indeed, what aspect of human bodies lends itself more to a conception of instrumental power than something which of itself not only changes in size, shape, but texture, and which furthermore dynamically changes position, something which, in brief, starts out inauspiciously being soft and flaccid—squeezable but not solidly grippable—and ends by becoming hard and erect? Unlike fingers which can do other labors, the penis "self-designs" itself to do a special work: the work of filling a certain hole, not only with itself but with its emissions. Instrumental power is concretely potentiated in just such flesh. The instrumental power of an erect penis is in no way merely figurative.

What we must distinguish in a general way about this strange flesh are actually two powers: on the one hand the power of a bodily member to make itself into an instrument—by changing size, shape, texture, and position—and on the other, the power of the realized instrument—the bodily member *as changed*—to penetrate holes. While the latter power is dependent upon the former, it is independent of it in the Sartrean sense of a project. Penetration is in other words the act of a living subject; morphological change in the form of penile tumescence is not. There is in consequence a difference between the two powers in terms of autonomy. We will return to this fact at length in chapter 10. The point of moment here is not autonomy per se but rather that, however miraculous its instrumental potential, a flaccid and soft penis is impotent when it comes to filling holes. In order to transform a female "into a fullness of being by penetration and dissolution," a male himself must first be transformed. Sartre's characterization of a penis as *strange* flesh is precisely suggestive of its unique self-transforming power—from mere swag of flesh to instrument.

In addition to distinguishing two powers, we must also offer a clarification. Because it figures in a critical way, it is important to set forth clearly at the start what Sartre means by dissolution when he speaks of the strange flesh that transforms females. Taken out of context, the above-quoted phrase might suggest that in the sexual act, a female is dissolved at the same time that she is penetrated, i.e., brought to a fullness of being by having her hole plugged up. Indeed, at first glance, the statement might appear paradoxical. What dissolves in the sexual act, however, is not the female but the male. Sartre has in an earlier paragraph described the thumb-sucking of an infant as an act by which the infant *dissolves his thumb,* transforms it into "a sticky paste" which will seal the hole of his mouth.[36] What penetrates and what dissolves are thus conceived by Sartre to be one and the same thing. In the sexual act, an erect penis dissolves into "a sticky paste" that seals off the hole that is the female. It is easy to see how through this psychoanalytic account of the nature of sexual body parts as *things* in the sexual act, Sartre arrives if not at the full-blown notion of the obscenity of holes, at least at a robust repulsion of insides. The everyday act of eating might have been exemplary in this respect. In our culture, at least, people are repulsed to see dissolutions in the form of partially masticated food; they are repulsed to see others spit food out of their mouths. They are even repulsed by their own expelled dissolutions. Infants who squeeze food out of their mouths are of course exempt from this repulsion as are the adults who feed them and who in fact reinsert the squeezed-out food. No doubt infants are conceived to be so new, their insides so pristine, there is no possibility of finding their expelled dissolutions repulsive. In Western cultures (at least), bodily insides corrupt only with age.

With the above clarification, we can immediately see a connection between Sartre's idea of dissolution and his characterization of the amorous act as one of castration. Insofar as it is a question not only of penetration but *dissolution,* a male hazards himself. He himself is subject to transformation within the female. If his identification of himself is with what is upright and hard, this flaccid, moist swag of flesh—this sticky paste—is his undoing. Indeed, it is the erect penis, not the flabby, slack swag, that archetypally in a psychoanalytic of things—and in a synecdochal way—typically defines what is male. In just this sense, a male who conceives his penis to be equivalent to himself loses himself within the hole. By inserting his penis inside a female, he allows himself to be negatively transformed into a being that is no longer a fullness of being, a being who can no longer fill holes, a diminished human specimen. In effect, by penetrating, a male dissolves himself not just sexually but, in a synecdochal sense, ontologically. Given this descriptive rendition of strange flesh, it is readily apparent why some males come to conceive of females as having power. In the amorous act, they perceive themselves to be dispossessed of their power by females. It is just this transposition of power that Sartre implicitly pinpoints when he writes of the slimy that there is "a surreptitious appropriation of the possessor by the possessed."

What semiotician Daniel Rancour-Laferrière terms "the personification of the penis" can be understood in just this context. In his discussion of the synecdochal role of the penis for males—of its standing for their whole body, indeed for their very person—he first of all presents evidence supporting his claim that personification of the penis is a cross-cultural phenomenon. We should note that to say that personification is cross-cultural is not to say that it is *pan-cultural*—a cultural universal—and further, that what the identification and analysis of corporeal archetypes allow over cross-cultural comparative studies, such as those of Rancour-Laferrière or Donald Symons, is precisely the discovery of a common denominator—an invariant form—together with its cultural variations, i.e., its elaborations, exaggerations, reductions, and even suppressions. The superiority of an archetypal orientation aside, Rancour-Laferrière interprets the cultural evidence of penis-personification in sociobiological terms: what is personified, he says, is "likely to be involved in the maximization of inclusive fitness. . . . To personify an entity, i.e., to treat it as a person, is to attempt to parlay that entity into interactions which are good for genes."[37] He proceeds to make an evolutionary case for castration anxiety on this basis, showing how castration anxiety has been "selected for."[38] Males are rightfully fearful—psychologically on guard—about the well-being of their "person." Their person is their passport to the future, their assurance of genetic representation in successive generations. The connection between what Sartre terms "dissolution" and "castration" is thus seen by Rancour-Laferrière to lie in a male's fear that, by engaging in intercourse, he puts not just his life but his genes on the line. Castration means not being able to pass them on, and thus to damage irreparably his inclusive fitness. This sociobiological interpretation is peculiarly reminiscent of the plight of females as described in chapter 3: females are at risk of being regarded loose if they spread their legs; they are at risk of being raped if they do not. Males, on the above view, are similarly "damned if they do and damned if they don't": if they do engage in intercourse, they hazard themselves (castration); if they do not engage in intercourse, they hazard themselves (no progeny). The difference, of course, is that for males, there is nothing to fear but fear itself. Indeed, how could anything which simply "gapes open," to use Sartre's phrase, inflict damage? What gapes open is absolutely empty, a nothingness, until filled. Thus while vulnerability is ostensibly present on both sides, the penetrator's vulnerability is empirically groundless. Once again it is easy to see why female fear and male fear, as described earlier, are categorically different fears. They are categorically different not simply because of a *sex* difference—one is properly male and one is properly female—but because of the nature of the fear in each case. Castration as a possibility within the amorous act is all in the head, one might even say an *irrational* head insofar as it is a thoroughly fantasized liability. Moreover, castration is not and has never been an actual act of females committed upon the bodies of males. The evidence Rancour-Laferrière has gathered shows that where castration has actually been practiced, it has been either self-imposed, or it has been imposed *by males upon other males.* In sum, castration fear is an existential anxiety that depends upon a male's reducing himself to his penis in the same way he reduces females to being "in the form of a hole." An interesting psychological study might indeed be made to determine the degree to which the latter determination is dependent upon the former. It would be surprising if empirically gathered evidence did not support male psycho-ontological logic. One's psychological conception of one's own body is the semantic source of one's psychological conception of the body of others. Freud's renditions of female sexuality offer incontrovertible support to this claim. The renditions are in fact a form of projection, of which more will be said in later chapters.

Analysis and acknowledgment of corporeal facts demand recognition of a second point. If dissolution is part of the act of intercourse, then the subject himself who engages in the act, if he is not guilty of bad faith, must assume full responsibility for engaging in it. In other words, dissolution is not the work of a female; it is the result of a penis that is spent. Rather than admitting to this fact, an anxious, psychologically distressed and/or myopic male blames the female. Thus, the myth of *vagina dentata.* Thus also, perhaps, a male's fear that his penis will involute, dissolving itself into his abdomen and disappearing. In short, to the extent that a male personifies his penis, he is at ontological risk. Freud himself, though hedging in the extreme, says as much when he writes that "The effect which coitus has of discharging tensions and causing flaccidity may be the prototype of what the man fears."[39] While it might be said that the funneling of all psychoanalysis into infantile sexuality has deflected attention from a thoroughly adult corporeal matter of fact, namely, the vicissitudes of the instrumental organ that is the penis, the funneling of all sexual responsibility onto the female deflects attention from the same quite adult corporeal matter of fact. In the same paragraph of the same essay, an essay in which he is concerned with "primitive" sexual behavior, but behavior in which "there is nothing obsolete, nothing which is not still alive among ourselves," Freud conjectures that because "woman is different from man, for ever incomprehensible and mysterious, strange and therefore apparently hostile . . . the man is afraid of being weakened by the woman, infected with her femininity and of then showing himself incapable." He then conjectures further that the "hypothesized" prototypical fear of flaccidity, *being the woman's doing,* may be justifiably augmented on just that account: "realization of the influence which the woman gains over him through sexual intercourse, *the considerations she thereby forces from him,* may justify the extension of this [prototypical] fear" (italics added).[40] The fear of which Freud speaks initially is a male's fear of being *incapable,* a fear that *it*—*he*—will not rise to the occasion. This fear is augmented into a condemnation of females on the grounds that females, being "strange and therefore apparently hostile," take forceful advantage of their sexual "power" over males. Ultimately, however, if a male is responsible, he must face the fact that penetration and dissolution go hand in hand, that with the act of intercourse there is a *double* transformation to be lived through: transformation into an instrumental power is at the price of a later transformation into a mere swag of flesh. However much the mere swag is anathema to a male's conception of himself, a physical contradiction of his manhood, it is a corporeal matter of fact. From a Sartrean view, for a male to blame a female for his dissolution is to compromise responsibility for his "project," to be guilty of "bad faith."

In addition to personification and moral responsibility, corporeal matters of fact warrant the singling out of an erroneous aspect of Sartre's notion of dissolution. The notion that a filler of holes—strange flesh—is turned into sticky paste by contact with *the slimy,* that is, with female insides, runs counter to experience. Dissolution here is actually an epiphenomenal effect rather than a property peculiar to a bodily hole. As briefly noted above, the dissolution of strange flesh is not the work of females but the result of a spent penis. Dissolution is in fact an *aftermath* of the filling. In the filling itself, that is, in the penetrative act, there is no sticky paste to be found, certainly not in the substantive sense of *dissolution.* It is only in *withdrawing* a penis, an *emptied* penis, from a vagina that one meets with sticky paste. While one might say that in the same way, it is only in withdrawing thumb from mouth (or perhaps foot from shoe on a hot day) that one meets with sticky paste, one must also acknowledge the fact that thumbs (and feet) neither ejaculate nor are they substantively transformed by their own powers.

A central psychoanalytic aspect of strange flesh is readily apparent precisely in the context of self-transformative powers. In its initial transformation, strange flesh is erect and hard, and what is erect and hard is or can be in itself threatening. Clubs, steles, shafts, lances—all attest to this fact. Archetypally, they are forms that can inflict harm, that can violate others. With respect specifically to holes, an instrument that is hard and erect can intrude forcefully, lacerate, wound, brutalize. This foreboding psychoanalytic aspect of what is erect and hard has significant affinities with two corporeal archetypes discussed earlier: penile display and making a spectacle of oneself. Penile display, not as sexual invitation but as threat to other males, is a practice among some nonhuman primates. Squirrel monkeys in particular are described as abducting the thighs, i.e., spreading their legs apart, to display an erect penis close into the face of another male in an attempt to dominate him. As one researcher has described it, "If the recipient does not remain quiet and submissive during the display, it may be viciously assaulted."[41] In a variety of human cultures, past and present, penile display has figured and still figures similarly as threat. I have elsewhere given multiple examples of these displays as part of an account of how male sexual signalling behavior figured in a central though by no means exclusive way in human evolution.[42] The examples are sufficiently diverse and historically broad enough to lend weight to the thesis that socio-aggressive penile display is an evolutionary, culturally elaborated phenomenon. Suffice to note here the example of the god Priapus (from early Greek and Roman times) whose penile erection is legendary and whose statue was used in gardens to ward off possible thieves.[43] (It is parenthetically of interest to note that priapism is an actual affliction which is extremely painful.[44] While not wholly equivalent to "year-round" erection, priapism suggests that year-round erection would in fact be an excruciating misfortune, something *actual* to be anxious about in comparison with castration.)

By making a spectacle of oneself—by *brandishing* something, for instance—one can similarly intimidate or subdue others. As discussed in chapter 2, male chimpanzees and other male primates pick up branches or other readily grasped natural objects to call attention to themselves and/or to threaten. They wave these objects about in situations where they want to get the upper hand in some way. Human penile display as exhibitionism is a form of brandishing in just this sense. A male exhibits his penis to females not simply to make them take notice, but to distress them, to hold sway over them, as it were. Psychiatrist Robert J. Stoller writes that "[the exhibitionist's] purpose in displaying his genitals is not to seduce a woman into making love with him but rather to shock her. If she is upset—is embarrassed, becomes angry, runs away—and especially if she calls the police, he has, he feels, absolute proof that his genitals are important. . . . For him, this sexual act serves as a kind of rape—a forced intrusion (at least, that is how he fantasizes it) into the woman's sensibilities and delicacy. If he cannot believe that he has harmed her, the act has failed for him. . . . Therefore, we find that the exhibitionist displays himself to strange women, not to his wife, who could hardly feel assaulted by a view so ordinary."[45] In another descriptive context, Stoller writes more pointedly that "The uninformed person may think the man who exhibits his genitals to women does so as a form of invitation to sexual intercourse. That is not so. . . . [The exhibition] does not achieve its object unless [the exhibitionist] believes he has caused a furor. It is as if, in exhibiting his penis, he must upset the observing woman and later the civil authorities; only then has he forced society to attend to the fact that he unquestionably is male, is masculine, and has a penis. Then he can be certain, temporarily, about his status as a man."[46] Synecdochal personification of the penis clearly grounds such behavior.

Psychologists Ivor Jones and Dorothy Frei, in their medical study of exhibitionism, present evidence supportive of Stoller's findings that exhibitionism is most usually an expression of hostility rather than an attempt to initiate sexual or social contact. They furthermore find grounds for believing exhibitionism to be an innate male disposition.[47] The evidence they cite in support of what they call a "biological hypothesis" comes from a variety of both anthropological and biological sources, i.e., from both human and nonhuman behavioral studies. In light of the evidence, they suggest that "a precipitating event"[48] triggers the expression of the disposition, the mechanism being similar to imprinting in birds. They note too, however, that any simple aetiological explanation must be discounted since one must also explain the continuance of the behavior. Whatever might be the psychological factor(s) motivating exhibitionism, the biological hypothesis that exhibitionism is an innate disposition of males is consistent with a psychoanalytic ontology of male genitalia: males who procreate by internal fertilization are beings "in the form of a powerful instrument," an instrument that, in its display, can be used to intimidate as well as to invite.

It is also interesting in this context to compare Stoller's psychoanalytic account of an exhibitionist's personification of his penis with Freud's statement that "maleness combines [the factors of] subject, activity [,] and possession of the penis; femaleness takes over [those of] object and passivity."[49] For the exhibitionist as for Freud, a penis—a possession, something a man *has*—is quintessential to being a man. Moreover Sartre, in spite of disavowing Freudian psychology with its central focus on the unconscious, would agree at least in part with Freud's tripartite characterization: to be male is to be a subject rather than an object, to have possibilities in the world rather than to be passive, and to have a penis that can transform things in the world (i.e., females) into a fullness of being. It is furthermore of interest to note Freud's use of the definite rather than indefinite article with reference to a penis in his definition of maleness. Males do not possess *a* penis, as they possess *an* arm or *an* ear; they possess *the* penis. What can this mean but that a penis is not up for grabs, so to speak. It is *definitively* spoken of, one could even say definitively spoken of as a privilege or prize that males alone have; only they "possess" it. It is notable, even odd, that femaleness carries with it no comparable possession. In his characterization of femaleness, Freud accords females not even a vagina, much less *the* vagina.[50] Perhaps it is because a hole is an emptiness not worth mentioning—except, that is, as a refuge, a phallic sanctuary, at which point it gains *definition:* "The vagina is . . . valued as a place of shelter for the penis; it enters into the heritage of the womb."[51] Freud and Sartre obviously have much in common here again. Females, as Freud would claim along with Sartre, are merely in the form of a hole. Moreover, Freud initially elevates female genitalia to a place of psychoanalytic significance only in speaking of that critical developmental psychological point at puberty where, finally, maleness *and* femaleness exist. Until that point of development, Freud affirms, only maleness exists because until that point, there is only the antithesis between "having *a male genital* and being *castrated.*"[52] Clearly, in such an account, the potential instrumental power of an erect penis is unquestionable, as is its potential as an instrument of dominance and possible aggression.

IV.   A Deepened Psycho-Ontology of Male Sexuality

As a way of deepening the above psycho-ontological elucidation of male sexuality, it will be helpful first to focus on the original side of penile power, the side from which "strange flesh" derives its very status as an instrument, namely, its power to be transformed. Indeed, as noted earlier, a mere swag of flesh cannot transform anything without first being transformed itself into a fullness of being. Its first office is thus to make itself into an instrument by filling itself up. In this sense, an emptiness must exist; a hole must be present and in turn made to disappear. This descriptive account is far from fantastic—either physiologically *or* historically. For centuries, a penis was considered to be a vagina turned inside out. The notion originated with Galen (second century A.D.), who stated that "All the parts . . . that men have, women have too, the difference between them lying in only one thing . . . namely, that in women the parts are within [the body], whereas in men they are outside, in the region called the perineum." Galen went on to suggest that the reader "[c]onsider first whichever ones you please, turn outward the woman's, turn inward, so to speak, and fold double the man's, and you will find them the same in both in every respect."[53] Galen's "directions" can only mean that the penis itself was considered a hole. The descriptive account is furthermore far from fantastic psychologically. As we have seen, synecdochally construed, it is precisely and only by virtue of a transformation into a *fullness* of being—having an erect penis—that a man considers himself a man.

The way in which a penis is transformed into an instrument of power by filling itself up is similar to the way in which a creature, by inflating a body part or the body as a whole, makes itself into an instrument of power. By increasing its size, and at the same time solidifying itself, the creature makes itself or a part of itself hard and erect. This transformation, in turn, is not dissimilar from the transformation effected by a creature who, in appropriating and brandishing branches or sticks, creates itself as a threat. The actual transforming project, however, in a strictly Sartrean sense of the term, is not the same in each instance. Unlike picking up a tree branch and brandishing it, and unlike the inflation of laryngeal pouches (male orangutans), whole-body inflations (common European toad), or the assumption of a bipedal stance (male chimpanzees), for example, penile transformation is not actually a project at all since erection is not a male's choice. It is not something that is *his* possibility, an upsurge of *his* freedom. In exactly this sense, the power of a swag of flesh to be transformed is awesome, fascinating, extraordinary. It is describable in exactly the same terms as "the great cavern": it is something fearsome, and it is also "something wonderful." This double aspect adumbrates a range of possible meanings, meanings that are at the same time signs of a range of behavioral possibilities: a transformed swag of flesh has the power to invite as well as to threaten, to give pleasure as well as to inflict pain. Since the felt character of the transformation is the psychological key to its specific employment, the critical significance of a discerning intelligence—first described in chapter 5—is obvious. The capacity to distinguish between feelings of agonistic belligerence from feelings of sexual arousal or more generally, *socio-aggressive bodily feelings and dispositions from socio-sexual ones,* is paramount. The task and necessity of perceiving this distinction is in fact critical to a full understanding of how penile erection can come to cultural eminence as a Phallus and thus become an object of worship, a symbol of power, an organ of dissemination, or "a floating signifier." The latter designation comes from French psychoanalyst Jacques Lacan, with whom we shall be concerned in the following chapters. What is of moment here is how that designation is archetypally instructive. In particular, two quite other significations attach to "a floating signifier" than those specifically given it by Lacan. On the one hand, an erect penis is a floating signifier in the sense of having dual archetypal meanings and of necessitating a discerning intelligence as described above. On the other hand, in Lacan's strictly phallic sense, "a floating signifier" is indeed equivalent to "le signifiant sans pair."[54] Lacking any corporeal moorings, a floating phallus lacks the capacity to join—with anything. It in fact lacks the capacity to invite in the first place. What invites has the power to attract, to fascinate, to give pleasure to touch. A floating signifier, unlike an erect penis, can hardly appeal in this physiognomic way; what is devoid of flesh has no sensuous character. Moreover what invites in this physiognomic way has the power to connect rather than intrude, to bring together rather than objectify, to unite rather than merely penetrate. Penile display as *invitation* clearly engenders this archetypal possibility. A floating signifier, in contrast, is not only devoid of flesh and complete unto itself; as *"le signifiant sans pair,"* it rules magisterially *über alles.*

In its invitational as opposed to threatening mode, and in its consequent power to bring together, an erect penis is something of a non-pareil in the natural world; it has no ready worldly counterparts as does a hole. This is because the power to bring together is in this instance not simply the power to fill a hole, a question of spatial logistics, but of both an initial self-transformation and a subsequent conjoined sensuality. In this respect, its closest worldly counterparts are in fact other male creatures who also procreate by internal fertilization and for whom the pleasure of connecting flesh to flesh is similarly a potential instrumental power of an erect penis. When sociobiologist David Barash describes animals' finding "sweet" the activities in which they engage, he necessarily includes the pleasure of mating activities.[55] More broadly, when biologist William Eberhard describes penes as "internal courtship devices," he is recognizing penes as instruments that give pleasure.[56] Instruments that merely fill holes are, by comparison, simply plugs of one sort or another. They have nothing particularly inviting about them. On the contrary, they are not uncommonly purely functional things like sink stoppers. Such fillers are ordinarily human-made rather than natural objects.

Where natural *in*animate objects in the world might be tenuously likened to fillers, and where they make an appeal to the flesh, their appeal is quasi-aesthetic. A sensuous aspect is present and it is this very sensuous aspect that is the source of the appeal. In a psychoanalytic of things, it is what turns the object into "something wonderful." Water that rushes into an inlet, or that cascades into an expanse below, or that simply fills the crater that is a lake, can have just such a sensuous character. Rushing or cascading waters resound with a kinetic dynamics we can feel in our own bodies; the placidity of a crater's water invites our touch, its smoothness resonates within us. Because it is a sensuous event, because it does not function as a plug, water in these instances constitutes an appeal of flesh to flesh. The sensuous experience recalls Sartre's description of "the flesh of objects": "My shirt rubs against my skin, and I feel it. What is ordinarily for me an object most remote becomes the immediately sensible; the warmth of air, the breath of the wind, the rays of sunshine, etc.; all are present to me in a certain way, as posited upon me without distance and *revealing my flesh by means of their flesh*" (italics added).[57] When we are open to such invitations in the natural world, we find ourselves connected with a sensuous presence, we are "at one with nature."

While in their similar sensuous appeal the relationship between inanimate and animate fillers of holes might be considered strong, the analogy between them is obviously weak. In none of the above instances, for example, is water perceived as akin to an instrument of any kind. In fact, although the water in each instance has power, it is not something one touches or can touch, handles or can handle, as one can touch and/or handle a tree branch, spear, or penis. It is not so much a *thing* as an event, and this because of its kinetic character. Whatever the disanalogies, however, the similarity in terms of appeal remains. This is because the sensuous character predominates. In other words, *in terms of a psychoanalytic ontology,* the miraculous power of worldly things to make an appeal of flesh to flesh and to unite flesh with flesh has far less in common with the literal power of a thing to fill or to penetrate another thing than it does with both the qualitative power of the thing to invite our attention and the power of animate life itself to apprehend the invitation—to apprehend the world as flesh, as physiognomic presence.

The quasi-aesthetic connection focuses our attention on the potentially strong invitational aspect of the strange flesh that is a filler of holes. The strange flesh that makes an appeal of flesh to flesh reinforces Eberhard's description of a penis as "an internal courtship device." An erect penis can penetrate another body in ways that bring mutual pleasure. There is indeed great truth in Sartre's claim that we are "on another plane of being" in flesh to flesh encounters. Though the actual sensuous domains go unacknowledged in Sartre's account, it is clear that in such moments, the body gives itself over to more than a passing tactility, kinetics, and kinesthesia. Sartre intimates as much both when he says that in making himself flesh, "I discover something like a *flesh* of objects," and when he comments on the extraordinary ways in which things in the natural world can affect him, "revealing my flesh by means of their flesh."[58] Sensuous experiences such as these belie the generality of his later insistence that doing, having, and knowing are all forms of desire which is through and through appropriative. The appeal of flesh to flesh is not necessarily appropriative. When "I discover something like a *flesh* of objects," I discover a form of sensuality. The sensual is a dimension of desire. When the body is "lived as flesh," the world it encounters is indeed correlatively lived as flesh: "The world . . . come[s] into being for the For-itself in a new way."[59] In effect, a proper psychoanalytic ontology would recognize what Sartre himself affirms. The outstanding character of that which penetrates holes lies not in appropriation but in both the unique sensuality of its flesh and its potential to bring pleasure by joining flesh to flesh.

Sartre's psychoanalytic ontology of sex—"sex is a hole"—actually contradicts his earlier phenomenological ontology of sex—sex as desire—on this very point. His earlier ontology, discussed in chapter 1, shows how, through desire, we realize ourselves as sexed beings. Desire is expressed by the caress; its aim is reciprocal incarnation. When Sartre in spite of this experientially supported claim goes on to characterize desire as through and through appropriative, he falsifies the complementary, even interchangeable, nature of the intent of desire "expressed by the caress." Such desire clearly has nothing to do with holes—nor with the ensnarement of another's freedom. What is *reciprocally* incarnated cannot be appropriative in character. On the contrary, in a reciprocal interchange, something is shared, not seized, commandeered, or otherwise self-arrogated. Reciprocal incarnation in fact exemplifies in an attenuated way the double powers indicated above: the qualitative power of things in the world to invite our attention and the power of animate life itself to apprehend the world as flesh. It similarly exemplifies, again in an attenuated way, Sartre's characterization of quality as "a new nature which is neither material (and physical) nor psychic": the Other is neither a mere body nor merely a personal projection. While reciprocal incarnation, an intensification of being in the flesh, is not a matter of material objects in the world, thus not directly part of Sartre's psychoanalytic ontology of things, it nevertheless evokes a psychoanalytic ontology. We do—male and female alike—enter a different world, a world in which utility gives way to sensuous presence, and sensuous presence to a heightened revelation of flesh by flesh. Sartre is right, of course. Desire is not the effect of which an erection is the cause; neither is it a beginning of which the act of ejaculation is the end.[60] But these negatives do not nullify the possibility of any connection whatsoever between caress and sexual act. If desire is not a desire of "doing," as Sartre claims, then perhaps the sexual act itself is not a "doing"—anymore than it is a "having." Certainly it is not *necessarily* or even preferentially a doing. In turn the sexual act is not necessarily the appropriating act Sartre conceives it and describes it to be.

There is in fact an ontological problem in Sartre's sexology bordering on a Dr. Jekyll and Mr. Hyde transformation. A male who makes himself flesh in desire has somehow to unmake himself in order to appropriate; he must de-flesh himself in order to penetrate and, in penetrating, possess. Now the reason Sartre gives for separating desire and sexual act is that one must learn amorous techniques; he states that there is a whole apprenticeship to the sexual act "that is added on to the desire from outside."[61] But clearly, a lack of continuity attends this ontological scenario. The making and unmaking of flesh is inconsistent with the flesh-to-flesh encounter of the caress *and* of the sexual act. It is consistent only insofar as penetration is appropriation, only insofar as the being that is in the form of filling holes is a being who, at a certain moment, simply penetrates them. In its duly appropriative mode of being, this being is indeed one incapable of conceiving of an instrumental power to join. It is solely in search of holes to penetrate, which in terms of positive and negative archetypal valencies means that its instrumental powers are not experienced as miraculous but as an impulsion to possess and to dominate, or more radically, to coerce and to harm. Perhaps its possibility of finding a continuity between a caress which is the expression of desire and an act of penetration which is an act of joining threatens its sense of autonomy. Perhaps its possibility of bringing together is felt as a compromise of its freedom. This is in fact what Sartre implicitly suggests when he writes that "desire *compromises* me . . . one is invaded by facticity."[62] Desire clogs one, paralyzes one; one is "swallowed up in the body."[63] To experience oneself as flesh is to be *vulnerable*.

It is little wonder, then, that the (male) "For-itself" wishes not to prolong the agony of being a body. Sartre's ontological inconsistency and his ambiguity as well—his clear rendition of the caress as the expression of desire and his troubled consciousness in face of that flesh to flesh encounter—are testimony to a being who is uncomfortable in the "different" world of flesh, and in consequence testimony to a being who has difficulty realizing positive meanings of his instrumental power.

Two final aspects remain to be noted in this deepened psychoanalytic of male sexuality. First, it is not surprising that, like a vagina, an erect penis has a double aspect, that is, negative and positive poles of meaning, and that the poles of meanings are archetypally similar with respect to both female and male genitalia. Viewed as qualitative forms within a psychoanalytic ontology, penis and vagina are both intimidating and wonderful, in the same way that things in the natural world, viewed as qualitative forms rather than as objects having a certain utility or everyday triviality, can be overwhelming, even fearsome, and at the same time inviting, fascinating, exciting. Indeed, things in the natural world can appear inexplicable as if housing an impenetrable mystery. Thus, at the same instant they magnetize us by their splendor, they can cause us to stop and draw back. It is only to be expected that male and female genitalia, taken as things in the world, partake of these same double archetypal powers.

Second, it is not either surprising that in any culture given over to violence in its entertainment and politics, as in its daily news, that negative valences predominate, and that the connective significance of "that which penetrates holes" is suppressed. We saw in the last chapter how exposure to pornography, for example, increases the tolerance for violence. It is thus once again not surprising that the wonderful instrumental power of an erect penis to unite flesh with flesh, to create pleasure by bringing together is, in our present-day culture, hardly apparent much less even conceived. The same is true of any culture—past as well as present—in which penetrators are hierarchically valued above the holes they penetrate. Indeed, statements valorizing "possession of *the* penis" never hint at miraculous 'I can's'. They proclaim only the 'I can's' of an oppressing patriarchy. For a male in such a society, the felt character of an erect, hard penis has all too typically a myopically one-dimensional signification, perhaps more aptly, a myopically non-dimensional signification since it is nothing more than a matter of filling holes. There is a deep irony of course in the fact that in these very cultures females are typically conceived as "the *relational* beings," beings who relate closely to others and to nature, beings whose concerns are not distant but empathic. While the source of their relational proclivities can be tied psychologically to child-bearing and child-caring with all of the deep bodily involvements—whole-body involvements—those activities engender, there remains nevertheless a profound irony in the fact that maleness concretely embodies a miraculous power to couple, to *pair,* and that this positive side of male instrumental power remains unacknowledged and unelucidated. It is in fact suppressed by a cultural *eidos* wedded not only to violence and devastation but to a view of males as insatiable sex machines who want nothing but an ongoing bevy of holes to fill. The instrumental power to bring flesh together with flesh by a closing of the distance between two bodies can hardly be expected to surface in such circumstances. It is completely eclipsed by a device, an implement that penetrates, a filler of holes.

V. A Critical Summation of Sartre's Psychoanalytic and a Substantiation of the Claim of a Universal Symbolism of Genitalia

Clearly, to adhere to Sartre's psychoanalytic ontology, one would have to say that coequal with the conception that sex is a hole is the conception that sex is a filler of holes. In consequence, one would also have to say that in Sartre's psychoanalytic, strange flesh is as shorted of its full possibilities as its female counterpart. Precisely for this reason, the idea that strange flesh transforms a female into a fullness of being is a conceit. Moreover if "the amorous act is the castration of the man . . . because sex is a hole," then the amorous act is no less a violation of the woman insofar as sex is nothing more than a filler of holes. In a strictly Sartrean psychoanalytic, filler as well as hole are charged with negativities. *Both* are wholly intimidating, offensive, or nihilating. While it is true that, taken simply as things in the world, a hole and that which fills a hole have each a certain human nature that is invariable—vagina and erect penis are stable archetypal forms in terms of past and present if not future human evolutionary history—that nature in each case is not a simple quality, much less simply a negative quality. It is a nature that in fact requires full qualitative elucidation, a setting forth of positive as well as negative meanings. If a just psychoanalytic ontology is to be formulated of male and female genitalia as things in the world, then first of all *each* of those things must be given its due, and second, each must be given its *full* due.

A second critical aspect of Sartre's psychoanalytic of genitalia follows from the first. It concerns a further dimension of the relationship between sex and aggression insofar as it calls attention, and forcefully so, to the act of reducing persons to one of their bodily parts. As preceding analyses and discussions have shown, the implications of that act are formidable. The synecdochal move to symbolize or to personify by way of genitalia is de-humanizing, whether a matter of personifying others or oneself. In either case, the individual so personified is silenced as if anaesthetized and at the same time distilled into a figment of himself or herself. Such anaesthetized, partial bodies are in no way equivalent to full-bodied animate subjects. They are without genuine autonomy. A male who personifies himself in the form of his penis accords himself an autonomy that he himself is actually without, moreover an autonomy for which he assumes no responsibility.[64] His penis's will is simply his will; he does what it bids him do: seek a hole. Short of peeing, seeking a hole is the synecdochal penis's only aim because filling holes is the sum total of its 'I can's'. Indeed, it has only the power of a *thing*. It lacks the power of reciprocal incarnation, the power to connect bodies, *because it itself is not a body*. It is simply "filler." Because it lacks the power to instantiate a flesh to flesh sensuality, touch is unfulfilled; female sexuality is unfulfilled; and male sexuality is equally unfulfilled. A mere penetrator of holes is an emaciated figment of a full-bodied sexuality. In fact, given this emaciated figment, it is not difficult at all to understand how a personified piece of flesh becomes infatuated with its own being to the point of being neurotic, insisting on calling attention to itself, fixating on filling holes at all costs, raving on about the possibility of castration—in short, of intimidating, of exploiting, of nihilating if necessary. It is all it can do. The swag of flesh has transformed itself materially into an instrument, but stopped short of transforming itself into an extraordinary intercorporeal power. Just as it has no sense of its miraculous powers, so it has no sense of a discerning social intelligence, and no sense of moral agency. On the contrary, it is a travesty of responsible autonomy. The hole(s) it fills are as detached a body part as it itself. A synecdochal swag is in this sense a caricature of male sexuality no less than the synecdochal hole it seeks to fill is a caricature of female sexuality. The two caricatures are intimately related. Fillers in a strictly synecdochal sense call forth holes to invade, to violate, to vanquish. Both caricatures are a gross distortion of being, but the caricature of male sexuality is of far greater consequence precisely because of its *instrumental* possibilities. A synecdochal swag is always at the brink of throttling what it penetrates. At this point, of course, it is no longer a question of male sexuality at all but of male violence. In such instances, what is performed is a socio-aggressive act. The swag of flesh has transformed itself into an instrument of brute power, the term *brute* referring not to nonhuman animals but to unthinking, insensate human ones whose only connections are made for ravaging.

    In tracing out the roots of power, how much weight are we to give to a psychoanalytic of genitalia as archetypal forms in the world? Because human genitalia resemble certain things in the world and because the things they resemble are in all cultures similar if not the same; and conversely, because certain things in the world resemble human genitalia and because human genitalia are in all cultures similar if not the same, the psychoanalytic is pan-culturally invariant. Sartre is right in thinking of a universal symbolism. Holes and shafts, like caverns and sticks, are archetypal forms that have fundamental psychoanalytic meanings. These meanings may be culturally elaborated or suppressed, emphasized or neglected. They may, with respect to specific male/female power relations within any particular culture, be set at odds or harmonized. Whatever the cultural variation on the archetypal themes, the meanings take their place in that wider context of fundamental meanings that emanates from animate form and in which animate form with varying degrees of emphasis figures as a semantic template.[65]

To substantiate further the claim of a universal symbolism of male and female genitalia, cultural variations on archetypal themes will be exemplified in three ways. In particular, we will show first how archetypal meanings can be neglected; second, how they can be emphasized; and third, how, within a psychoanalytic of male/female power relations, they can be distorted. It has already been shown, of course, in the above analyses and discussions how they can be suppressed.

The attire of Grand Valley Dani males of New Guinea was initially of considerable psychoanalytic interest to Western anthropologists until it was discovered that the attire carried none of the expected, typical Western meanings. Dani males wear a penis gourd, which does not simply cover the penis but extends up to the navel and sometimes even up to the chin of the wearer. The *holim,* as it is called, is a hollowed out gourd—essentially of a tubular form—"of varying lengths and sizes [and] fit[s] over the penis . . . [Holims] are anchored at the base by a string around the scrotum, and are held upright by another string under the arms."[66] Holims may be straight or curved; they may be curled at the end; they may be "festooned with furry marsupial tails sticking out of the tip."[67] A male begins wearing a holim at the age of four or five. He wears it "at all times except when urinating or having sexual intercourse."[68] Karl Heider, the anthropologist who first studied the Dani, comments that the gourd "conceals (but does not protect) the penis, and people are never hindered by this apparently awkward thing. Boys soon learn to sleep in them, run in them, work in them, and even wage war in them." He adds that "Through 1970 at least, few Dani had any interest in exchanging them for cotton trousers."[69]

The above description, even without accompanying photographs, would lead a person steeped in present-day Western culture to believe that by their attire, Dani males are intensifying the fact that they have a penis. The gourd is, after all, a facsimile of perpetual erection. By Western standards, the male is presenting himself in the form of an ever-ready erect, hard penis, indeed, an every-ready erect, hard penis enlarged to formidable degrees. He is thus announcing not merely his masculinity but his *phallic* masculinity. Moreover, by wearing one gourd one day and another on the next, he might be announcing certain day-to-day variations in his feelings about phallic masculinity. In other words, and as Heider-as-Western-observer acknowledges, one might expect the penis gourds "to be a Freudian library of projective information giving immediate insight into basic personalities, rather like wearing one's Rorschachs on one's sleeves."[70] After approximately four and a half years of intensive study spread over a period of approximately ten years, however, Heider was forced to conclude, "Alas, as data they were next to useless. People would have whole wardrobes of penis gourds of different lengths and shapes, and I could find no correlation between the gourd of the day and either long-term and [sic] short-term personalities." In brief, the idea that Dani males are concerned with "phallic masculinity" is mistaken. The gourd is not a focal point of masculinity at all. Males simply wear them. They are a "neutral covering."[71]

It is clear from Heider's descriptions and discussions that the penis is not either being shrouded in mystery by a holim. It is simply tucked away. Since it is tucked away, a Dani male does not and cannot play with his penis; he does not and cannot compare its size with the size of his friends' penises; he does not and cannot make it the focus of his life; he is by and large out of touch with it. Being out of touch with it, it is neglected. It is in fact neglected to an extent most Westerners would consider unbelievable and even impossible. The Dani space childbirths and observe a prolonged period of postpartum abstinence. Children are born four to six years apart and to judge from a table showing the number of children per woman in a particular Dani neighborhood, most women have only one or two children. This means of course that males remove their gourd for sexual purposes with an infrequence unheard of in Western society. Notwithstanding the fact that some of the men have more than one wife, the approximate five-year period of postpartum abstinence is consistently observed. Heider writes that "in those few instances which I could follow closely, the husband spent his time with the wife who had just had a baby, and ignored his other wives."[72] The relatively small number of Dani children corroborate his observations.

In contrast to the neglect of archetypal meanings of a penis by Dani males is the unmistakable archetypal emphasis given the penis by several different southern New Guinea peoples in their respective "penile display" dances. Anthropologist Carleton Gajdusek writes, for example, that "when frightened, excited, elated, or surprised, groups of Asmat men and boys spontaneously meet the precipitating event by a penile display dance, which involves much the same sequence as the presentation display of the squirrel monkey." He goes on to say that "This behavior is performed on the arrival of strange visitors, on the departure of strangers who have been received with friendship, or in response to excitement or anxiety-producing events, such as the burning of a house, victory in fighting, a severe thunderstorm, completion of a communal effort involving exertion. In more formalized ritual form, the vocalization, thigh spreading, genital grasping and rubbing, erection and pelvic-thrusting behavior pattern has been introduced into the traditional night dance of the Asmat and Auyu peoples, which at times may become even more overtly erotic and copulatory."[73]

Gajdusek gives in particular a brief description of the ritual penile display dance of the Waina-Suwanda, a people living north of the central ranges of New Guinea. He writes that the dance "leaves no ambiguity, explicitly emphasizing the display nature of the performance. . . . The usual glans penis-covering gourd is replaced by a much larger and longer gourd, which throughout the dance is flipped from between the legs up against the abdominal wall by undulating movements of the thighs and pelvis."[74] He also gives a brief description of a New Guinea people whose ritual penile display dance is more subtle and complex. "The Anga," he writes, "who modestly hide their genitalia under two or three dozen grass sporans, wear so many sporans that they produce a phallic-like anterior projection from the body wall of the piled sporans themselves. In their dance they jump about, with their feet slightly spread, so as to flip their voluminous skirts up and down. In its most exuberant climax they do this in such a way that the heavy pile of sporans flaps against their abdominal wall and exposes their genitals."[75]

Clearly, a penile display dance in virtue of its movement—both body movement and clothing movement—is an intensification of penile erection. Clearly too, where the dance is performed on varied occasions, the performances validate not simply the fact, *but an experiential awareness of the fact,* that penile erection occurs in a variety of social contexts, nonsexual as well as sexual. The dance is in this added sense an intensification of penile erection. It calls attention to the fact that penile tumescence is not limited to sexual occasions but is a transient feature of everyday male life. Thus, it is not surprising that the dance is performed as "an expression of aggression, . . . greeting, appeasement, or rejoicing"[76] as much as an expression of sexuality.

In all that has been said of holes, it was suggested only in passing that such an account is of human female sexuality is far from adequate. One might indeed say that the account is precisely a distortion of female sexuality insofar as it misconstrues a vagina as "the female genital organ." While the well-known French feminist Luce Irigaray has written at length of this misconstrual,[77] psychologist Harriet E. Lerner's earlier article on the mislabeling of female genitals is the more cogent example here for it succinctly points out the cultural aspects and consequences of the mislabeling and demonstrates as well how consistent misconceptions of female genitalia have grave psychological effects. "If one interviews parents," Lerner writes, "or reads literature on sex education it is evident that the girl child is told that she has a vagina and nothing else. Even literature written for an adolescent population typically communicates this same undifferentiated picture of female genitals."[78] Though Lerner does not use such language, her criticism is clear: construed simply as a vagina, female sexuality *is* simply a hole, something to be appropriated, a handy place to put a penis. Lerner points out that "Such an incomplete, poorly differentiated and anatomically incorrect picture of female anatomy may have its most critical effect during the preoedipal and early oedipal phases of development when the girl discovers her clitoris as the prime source of sexual stimulation and gratification." In Lerner's view, "it is of serious psychological consequence to the child that she discovers an organ of pleasure that frequently is not acknowledged, labeled, or validated for her by the parents [or by the culture in which she lives], and which is thus inevitably experienced as 'unfeminine' (only boys have something on the 'outside')."[79] Lerner's clinical evidence strongly supports her claim that penis envy is less a wish for "what boys have" than a wish "to validate and have 'permission' for female sexual organs, including the sensitive external genitals."[80] A culture that does not recognize female external genitalia—or which in fact *excises* or *ablates* them—distorts female sexuality. In the one case, distortion is effected through conceptual manipulation, in the other by actual physical manipulation. The result in either case is to nullify a full-bodied female sexuality; a hole is not the whole story. It should be briefly noted in this context that there is in fact a further aspect of genitalia that must figure in a psychoanalytic of power relations. Insides and outsides, alluded to only in passing above and in the foregoing analyses, are of central significance both to an understanding of the binary oppositions that commonly structure human societies, in particular the opposition of male and female, and to the asymmetrical axiology in which those oppositions are commonly structured.[81] In this respect, there is quite obviously more to be said of holes and penetrators of holes, and of cultural distortions as well. The inquiry that opened on behalf of archetypal understandings of sex and aggression, however, closes with this critical analysis and elaboration of Sartre's provocative psychoanalytic of things in the world, which is best summed up by emphasizing once more that that which penetrates holes must itself first be transformed and that so transformed, it has instrumental power: to realize itself either as an appeal to flesh, an appeal in the service of a consummate sensuality of flesh on flesh, or as a weapon, an aggressive instrument born in hostility and bent on violence.

7

Corporeal Archetypes and the Psychoanalytic View: Beginning Perspectives on the Body in Lacan's Psychoanalytic

The whole history of ideas should be reviewed in the light of the power of social structures to generate symbols of their own. These symbols deceivingly commend themselves as spiritual truths unconnected with fleshly processes of conception, thus obeying the purity rule.

Mary Douglas[1]

. . . [T]o understand and to combat woman's oppression it is no longer sufficient to demand woman's political and economic emancipation alone; it is also necessary to question those psychosexual relations . . . through which gender identity is reproduced. To explicate woman's oppression it is necessary to uncover the power of those symbols, myths and fantasies that entrap both sexes in the unquestioned world of gender roles.

Seyla Benhabib[2]

I. Introduction

Multiple and complex issues surrounding the notion of corporeal archetypes of power must be considered within the framework of Western psychoanalytic views of human nature. These issues are best introduced and examined by placing Jacques Lacan's radical poststructural/postmodern psychoanalytic theory at the center, and this because, however much Lacan views his work simply as an ideological elaboration of Freudian theory,[3] it has important other theoretical and philosophical links which tie it to prevalent twentieth-century Western socio-cultural understandings of the body. Lacan's psychoanalytic is in fact a microcosm of twentieth-century practices and tendencies that preclude an appreciation of animate form and tactile-kinesthetic life, and in turn, preclude concerns about, and understandings of, the roots of power. These practices and tendencies are reducible to three major dispositions: to scientize life, to pedestal human language, and to silence the living body. The latter disposition, as will be evident, is fundamental to the first two. An extended critical examination of Lacanian psychoanalysis will not only allow us basic vantage points upon the roots of power within twentieth-century Western society. It will furthermore exemplify how culturally elaborated corporeal archetypes deriving from our evolutionary history may be individually elaborated, elaborated in subtle and intricate ways that entrench even further archetypal forms of power already subtending major cultural dispositions, in effect solidifying even more strongly the regnant conception of power as control. Introspection, infant-child developmental stages, the unconscious, and the phallus will be particularly thematic in the present chapter. Persistent inquiry into their status within Lacan's psychoanalytic will provide the foundation necessary to the more detailed critical and constructive inquiries undertaken in chapters 8, 9, and 10. It should be noted that the purpose in all of what follows is not to skewer Lacan—as if one could skewer so fleet and erudite a figure—but to bring to light the complex conceptual infrastructure of a psychoanalytic that denies the living body and thereby blinds us to the roots of power.

To begin with, Lacanian psychoanalytic theory is ontologically linked to Sartre's existential psychology. Lacan's work in fact appears deficient in acknowledging its Sartrean debt for it is an adroit and intricate psychoanalytic makeover of Sartre's existential ontology.[4] Lacan's work is equally coincident with the structuralist theory of Claude Lévi-Strauss. Lacan commentators John Muller and William Richardson summarize the coincidence in their reader's guide to Lacan's *Écrits,* specifically in the context of discussing two of Lévi-Strauss's essays, "The Effectiveness of Symbols" and "Language and the Analysis of Social Laws," essays in which Lévi-Strauss presents analyses supporting his claim that the unconscious operates according to structural laws that are linguistic in nature and basically invariant across cultures.[5] Other commentators as well have demonstrated how Lacan's psychoanalytic claims on behalf of the unconscious are inspired by and conform with Lévi-Strauss's anthropological claims about the unconscious.[6] Finally, and as unlikely as it might at first appear, Lacan's work is theoretically coincident with that of Carl Jung. His claim that the unconscious is ordered in invariant structures aligns his psychoanalytic with Jung's analytic psychology as much as with Lévi-Strauss's structural anthropology. Notwithstanding the fact that the "collectivity" of the unconscious is for Jung a built-in of the unconscious itself, i.e., a matter of diverse archetypal psychic forms that are part of our human psychic heritage, and that the "collective" unconscious for Lacan is a built-in of language, specifically, of the inherent Otherness of language, or generally, of the inherent sociality of human life,[7] the idea that the unconscious is ordered according to built-in invariants and that these invariants are the focus of psychoanalytic work strongly links the theoretical psychoanalytic of the two men. Most significantly, and not surprisingly in view of the theoretical linkage, Lacan's psychoanalytic is actually archetypally based, as we will see in detail in chapters 8, 9, and 10. It in fact rests ultimately upon a single, overarching *corporeal* archetype. To lead up to an initial unveiling of this archetype and the power structure it both generates and engenders is a major aim of this chapter.

The above-noted linkages show that in addition to thinking along Freudian lines, Lacan's thinking also runs along the same lines as other intellectuals of his day. Those lines are, in particular, the lines of desire, of language, and of invariant structures or forms of the unconscious. They are apparent, respectively, in Lacan's ontological dovetailings with Sartre's existential psychology, in his linguistic dovetailings with Lévi-Strauss's structural anthropology, and in his psycho-theoretic dovetailings with Jung's analytic psychology. The lines not only converge in Lacan's work; they provide it its theoretical-philosophical underpinnings. What is remarkable is how separately—and how together in an all the more emphatic manner—the lines create a disjunctive space in Lacan's psychoanalytic. That is, both singly and together they create a space in which the living body is marginalized, even pointedly unrelated to the psychoanalytic. In the schema drawn by the lines of desire, of language, and of invariant unconscious structures, the living body is something of a surd. Moreover, for all its material presence as a flesh and bone reality, it is regarded an obstacle to clarity. Accordingly, the living body rarely makes an appearance in Lacan's psychoanalytic. It is precipitated out of his initial and focal conceptual equations: it is precipitated out of desire conceived fundamentally as an emanation of language, precipitated out of language conceived fundamentally as an emanation of the unconscious, and precipitated out of the unconscious conceived fundamentally as a realm of signifiers.[8] In effect, a closed circle is drawn. The space initially marked off by and for desire, language, and the unconscious is a virtually bodiless space.

To consider Lacan's work as derivative of the work of Sartre and Lévi-Strauss in particular is to see in each case—and in the original as in the Lacanian makeover—that what occludes an appreciation of animate form is the inordinate conceptual and theoretical preeminence given consciousness (including the unconscious-made-conscious through psychoanalysis) and language, respectively. In each instance, the living body is undervalued or devalued and in consequence either left behind or unattended. With Lacan it is indeed all the more emphatically left behind, for in his psychoanalytic the wakes of consciousness and of language are together exponentially cumulative. The result is that power—power Lacan accords to desire and to language, and to the unconscious insofar as desire is born with the birth of language and language speaks the subject—is in essence not only power altogether separate from a living body; it is power that overwhelms the living body. It consistently reduces the latter to fractions of itself. It is hardly surprising then that the living body is emphatically outside the field of discourse; it is consistently *hors de la question*. Neither interrogated nor introspected, it is precipitated out of whatever solution constitutes the matter under investigation.

Now feminist texts that try to set the Lacanian record straight— especially the texts of French feminists attempting to establish *un écriture féminine*—may be seen precisely as attempts to reclaim the living body. That they fail to reclaim it in any generalizable way and thus mobilize feminist socio-political power is in good measure the result of their clinging to Lacan's (and more generally, postmodernists') basic tenet: that as subjects we are structured in and by language. So long as the body is conceived as a by-product of language, so long as it is viewed in essence as a linguistic concept—hence as a socio-political construct—it will continue to appear simply as a function of language, regardless of whether, as construct, it is open to socio-historical conditioning and hence to revision. It will in consequence generate discussion, research, and writing *only* at the level of language. It will not engage us at the level of concretely realized corporeal expressions of power and concretely realized corporeal manifestations of power relations. An excellent and telling example of this linguistic reductionism is the ongoing and seemingly incessant question of whether French feminist Luce Irigaray, in speaking of "two lips," is speaking of "two lips" metaphorically or anatomically, and whether, collaterally, to speak of "two lips" is not to essentialize woman.[9] The twin questions have occupied many a feminist and they continue to do so. Discussions, however, have produced not only no consensus, but have generated no new and further insights, either into the nature of what it is to be female—or male—or into the roots of power. The lack of progress strongly suggests that at a mytho-poetic level at least the linguistic pools into which followers of Lacan have been looking in the hope of gaining knowledge of themselves, or even finding themselves to begin with, are neither basically metaphoric nor essentially biological constructions but linguistic mirages. Animate form can hardly be expected to surface on such surfaces; neither can the roots of power. Ironically, and as the evidence gathered in this book shows, an elucidation of corporeal archetypes and, correlatively, of the roots of power, is precisely what is needed to gain fundamental insights into the oppression of females and in turn insights into viable modes of lifting it. If the original watchwords of the women's movement still ring true, then the personal must come to the fore, clothed in language, yes, but not so as to *forge* the reality of the personal; rather, to illuminate and communicate it.[10] In short, elucidating the living body necessitates detaching oneself theoretically from linguistic-centered theories. It means attending to the body itself, directly and patiently, which is why a consideration of Lacan's work is ultimately essential. It will show us in multiple ways, albeit by counterexample, what must be considered, both evidentially and methodologically, if we are to understand how we twentieth-century Westerners come to have the concept of power we do, and how we have come in turn to order our behaviors within certain concepts of power relations.

II. Introspection and Lacanian Psychoanalysis

In order to approach the complexities of the relationship between corporeal archetypes and power in the context of Lacanian psychoanalytic theory, it is relevant first to consider the central role of introspection in the formulation and development of the field of psychoanalysis. Psychoanalysis as initially formulated by Freud was in fact described as "a new species of introspection."[11] What Jung later termed his own "voluntary confrontation with the unconscious"[12] was similarly "a new species of introspection." Just what is this "new species" and what is its importance?

Psychoanalysis entered into a many-stranded science of psychology which, by the early 1900s, aimed more and more toward emulating the experimental laboratory of the physicist or chemist, and which correlatively, in America at least, moved further and further away from introspective techniques. The new psychoanalysis was not only interpersonal as opposed to impersonal and not only clinical as opposed to "laboratorial"; it was a psychology that was through and through methodologically tethered to introspection. Indeed, as indicated above, it was a psychology whose founder and whose most esteemed and illustrious disciple devoted years of their lives to self-analysis. *Self-analysis* was in fact the methodological key to *psycho-analysis*. Experimentation played no part in this new psychoanalytic work except in the sense that certain techniques were tried out—notably, free association by Freud, "active imagination" (deliberate fantasizing) by Jung—in the course of developing methods of psychoanalytic treatment. It is important to emphasize that the techniques were tried out by Freud and Jung themselves in the course of their own extended self-examinations and that they in turn tried out the techniques on their psychologically distressed patients in order to determine their efficacy.[13] In this kind of experimentation, as in their self-analysis generally, introspection was not only critical; it was the methodological anchor point of their experimental studies. Only through introspection could basic psychoanalytic techniques be developed and, in turn, could developing theories of therapy be validated. Thus, introspection was at once the *modus operandi* of psychoanalytic treatment itself and the testing ground of the theoretical structure of psychoanalysis. At one stroke, one authenticated on the basis of introspective evidence—on the basis of one's own experience—the efficacy of a technique and the validity of tenets within psychoanalytic theory. Let us examine the fate of this central methodological peg in Lacanian psychoanalysis.

To introspect is to look within: on the one hand in the common sense of turning attention to the inner terrain of thoughts and feelings, sensations and inclinations, actions and motivations that constitute everyday present experience, to the "stream of consciousness," as William James called it;[14] and on the other hand in the psychoanalytic sense of turning attention to gaps in the stream—to what Freud and those following him would see as upsurges of the unconscious in free associations, in dreams, in slips of the tongue, and so on—and being led therefrom to reflect upon a maze of heretofore repressed pains, conflicts, frustrations, desires, guilts, regrets, and hostilities, thus ultimately led to introspecting past experiences in the form of images, fleeting reminiscences, and vivid memories. To the new psychoanalytic breed of psychologists—those individuals who would conceive the psyche altogether differently from physiological, gestalt, or behaviorist psychologists—we humans are not just the stuff of our waking life but the stuff of which our dreams are made. Accordingly, the world is not conclusively captured by an altogether luminous human psyche in its immediate perceptions. Quite the contrary. In classical Freudian terms, at one extreme is a psychic time lag from which we continually suffer as from a dream that continually holds us in its grip. Our past clings to us in myriad ways, dragging us back to infantile levels of feeling and comprehension. At the other extreme is a psyche that does not know itself. Our unconscious has access to our consciousness in moments we do not control, such as when we make a slip of the tongue or when we "inadvertently" forget something. These moments aside, our consciousness is ordinarily dumb to our unconscious; we have no easy or immediate access to it, and without special effort, we remain oblivious of it. In classical Freudian terms, the special effort we must make is precisely an introspective one, for our unconscious, being out of our control, can never itself be the subject of our thoughts or reflections. It can only be caught in retrospect, that is, in its products, hence in dreams, moments of "inadvertent" forgetfulness, and the like. Moreover, these products of our unconscious are not themselves immediately meaningful. To yield their meanings, they must be subject to the technique of free association. Furthermore, free association is itself a stepping stone. It serves as a methodological anchor point for discovering repressed experiences—of fear, guilt, lust, and so on. Within the web of associations, the patient must search for that experience or complex of experiences that occurred in the past and that holds the key to the meaning of the associations. In particular, the patient must *look within* to find the experience or set of experiences that makes sense of them. Though not described or spoken of in such terms, that semantic match might well be described as a bodily moment of felt resonance; an experience in the past bears the cumulative felt weight of the associations. Alternatively, the patient might validate an experience that the psychoanalyst proposes as matching the associations. Regardless of whether the experience is remembered by the patient or suggested by the psychoanalyst, the critical factor in this matching is introspection. It is through introspection that the patient *verifies* meaning. If an offered interpretation does not feel right, it is jettisoned. A bodily felt sense of rightness obtains or does not obtain. In sum, and in classical Freudian terms, helping the patient to become conscious of her/his repressions requires precisely going back to the past, a past that is called forth through free association and that, through introspection, is found to match—or not to match—the evoked associative material.

Now in Lacanian psychoanalysis, one does not have to go back to past experience. One does not have to introspect at all. Lacan in fact scoffs at introspection; one does not need this "mirage of narcissism," as he calls it.[15] How does Lacan categorically distance himself from introspection? Not by discussing its liabilities and proposing an alternative, but, as we will progressively see in detail, by denigrating it and banishing its empowering source—the living body. In theory as in practice, Lacan does away with introspection on the grounds that one can *induce* the desired validating psychological experience in the here and now of the analytic treatment itself. In the analytic situation, there is only speech, the speech that takes place between analyst and analysand. Thus the experience that validates Lacanian psychoanalysis, in theory as in practice, is created in and through language, a particular kind of language, or more precisely a speech that is not just any speech. By inducing paranoia in the analysand—"a controlled paranoia" (*"une paranoïa dirigée"*)[16]—the analyst brings the analysand to paranoic speech. This special kind of speech validates at one stroke both the technique of delusional free association as *the* way of liberating the unconscious—*the* method by which the unconscious will manifest itself—and Lacanian psychoanalytic theory. The latter holds that the unconscious is manifest in psychosis and that the psychotic is "spoken" by the unconscious, i.e., by language.[17] Accordingly, the unconscious of the subject becomes the subject of the unconscious in the course of Lacan's developing psychoanalytic.[18]

In both theory and practice, Lacan's psychoanalytic thus short-circuits an introspective technique by making the past present. There is, in other words and to begin with, nothing to introspect, for there is nothing to remember, and there is nothing to remember because the key lies directly in the here-and-now relationship of analyst and analysand, i.e., in the transference.[19] The patient's past is not then something of which the details must be sought, for it exists in a veritable here-and-now present. By inducing paranoic speech—what at an early point in his career Lacan termed the *"method of symbolic reduction"*[20]—Lacan not only makes the past present; he reverses temporal order in a further sense. Experience follows theory: that is, by inducing paranoic speech on behalf of his patient, he induces experience on behalf of his theory. Thus at the same time that the usual temporal focus of psychoanalysis is reversed, the usual order of scientific theorizing is reversed. To realize this is to understand why, in his writings describing the stages we pass through in our infancy and childhood—from our initial stage of a *corps morcelé* to the mirror stage in particular—Lacan's verbal tenses appear mixed up and, indeed, self-contradictory. Postmodern theoretician Jane Gallop has puzzled over and complained about Lacan's future anterior (or future perfect) tense,[21] but that tense, as is evident, makes sense in the light of what his psychoanalytic treatment involves. His practice, in brief, must be seen in the light of his theory. When we gain insight into his psychoanalytic methodology—his actual method of treatment—we realize that by creating a present experience that will validate his theory, Lacan has no need of stabilizing or solidifying his theory in an actual past—a past validated by introspection in all its experienced (or fantasized) affective intensity. On the contrary, according to Lacan, all past experience, ever since the subject's introduction to language—to the Symbolic order, as Lacan terms it—is subsumed in the present paranoia-induced analytic experience(s). The invariant infant-child stages Lacan specifies in his psychoanalytic are in effect theoretically subsumed in what may aptly be termed ontological epiphanies, epiphanies experienced by the analysand in his/her "controlled paranoia." What needs validating, of course, in this psychoanalytic scheme is the linkage between "controlled paranoia" and the infant-child stages Lacan claims we have all experienced. Lacan supplies this linkage in his own idiosyncratic and unequivocal way,[22] as we will note in progressively greater detail in the course of examining his psychoanalytic and considering the archetypal infant-child stages through which he says we all pass.

What is of particular moment to point out here is that we are *pre-determined* by those stages through which we are all said to pass. According to Lacanian psychoanalytic theory (and practice), we are all of us temporally skewed and basically psychotic. For example, we have all initially passed through a mirror stage wherein we originally apprehended an "Ideal-I."[23] In the wake of this experience of ourselves in a mirror, we realize ourselves as at bottom irremediably split subjects: an Other is at the heart of our being. But we repress this knowledge. We delude ourselves, believing instead that we are thoroughly autonomous egos. No matter how quintessentially whole we appear in the glassy mirror of our presumed essence, however, paranoic speech will lead us to see ourselves in our true light. Through this speech, we will say with Lacan himself that "what is realized in my history is not the past definite of what was, since it is no more, or even the present perfect of what has been in what I am, but the future anterior of what I will have been for what I am in the process of becoming."[24] The future perfect is, as Gallop points out, both anticipatory and retroactive.[25] It converges on a present that is never there, that evades pinpointing in much the same way that, in the chain of signifiers that Lacan says is language, there is "an incessant sliding of the signified under the signifier."[26]

Now the manner in which Lacan describes the mirror stage—"[it] will be the source [rootstock] of secondary identifications"[27] insofar as it is the antecedent to what will be and the successor of what will have been—mirrors the ontological epiphany in which the patient, through his or her paranoic speech, both reawakens all that he or she has been and all that he or she will have been. The mirror stage that has been originally lived in the past is the stage as it will have been lived repeatedly in the future. Hence, we can conclude that the patient will be able to say and will have said:

> I am everything I have been and everything I will have been at this moment of desire within the psychoanalytic transference. This delusional moment links me to that mirror stage and to the earlier stage that preceded it, the stage in which my body was a *corps morcelé*; and it links me equally to that later stage which was my introduction to language, to the Symbolic order. This delusional moment is a replay of my original ontological epiphany which I have heretofore repressed, an epiphany which I cannot recover by any act, lateral or otherwise, of introspection. It is an undecipherable

moment of my history in that, while I have lived through it, I cannot remember it. Since it preceded my introduction into the Symbolic, I am powerless to dredge it up even indirectly. All of my speech—all of my free associations and interchanges with my analyst—are linked not to it but to the endless chain of signifiers which I am in the process of becoming.

We are all, in effect, ripe for Lacanian psychoanalysis. That, in fact, is precisely Lacan's message.

The above initial explications of Lacanian psychoanalysis show that an act of speech is not a point of departure for an act of thought: free associations are not a point of departure for an introspective glance at, or ruminations upon, the past, whether a past given in memory or in a dream. Any act of speech is simply a point of departure for further acts of speech—whether further associations, or further dialogues between analyst and analysand. Introspection, in Lacan's own words, is not only a *mirage éculé*—a worn-out psychological sham—but a *mirage* that is a *monologue*, a monologue that never achieves and can never aspire to *discourse,* in particular, to the discourse Lacan characterizes as the *travail forcé* of free association.[28] In short, introspection is a ruse; it is not the stuff of either psychoanalytic treatment or psychoanalytic theory.

Lacan's resolute distancing of himself from the mirage of introspection is linked to a fundamental motif of his work, perhaps, as has been suggested by a French psychoanalyst (see below), to *the* fundamental motif, namely, the project of making psychoanalysis a science. The resolute distancing is, in other words, not a trivial, idiosyncratic rejection of introspection. Neither is the project of making psychoanalysis a science a trivial, idiosyncratic scheme. For Lacan, to succeed in making psychoanalysis into a science would be to realize not only his, but Freud's—dare we use the term— desire. This desire is consistently evident in both men's writings and it is consistently recognized by commentators of both. There is a considerable thickening of (the) desire, however, in Lacan's work. There is at the same time a progressively more transparent desire to the degree one takes the history of psychology into account. That is, Lacan recognizes that to be a science in the mid-twentieth-century sense of that term, there are certain requirements. Psychoanalysis must, like experimental psychology, do away with "inner consciousness." In particular, it must eschew introspection. The task, obviously, is far more difficult for psychoanalysis than it is—or was—for experimental psychology. After all, in psychoanalysis, one is dealing with people who are emotionally disturbed. How is it possible to treat them if feelings, wishes, beliefs, and so on are not at the center of attention, if, indeed, such aspects of experience are inexorably pushed aside and ignored? It is not even as if such "inner" concerns were *secondary* to language in Lacanian analysis; on the contrary, they are nowhere to be found among the algorithms that define Lacanian psychoanalysis. Francois Roustang, a former Lacanian analyst who underwent training with Lacan himself and who later turned apostate, offers particularly lucid insights into Lacan's attempt to scientize psychoanalysis—whatever the psychoanalytic costs of the project. Roustang affirms that many points of view are possible on Lacan, but that one viewpoint can be privileged over all others in that it "allow[s] us to furrow a way through the entire *oeuvre.*"[29] This privileged viewpoint coincides with "the project Lacan pursued relentlessly: To turn psychoanalysis into a science."[30] An initial consideration of Roustang's key observations and conclusions will put the issue of introspection in the larger context it deserves and at the same time will point the way toward an understanding of how introspection is but one aspect of what must ultimately be put out of play in making a science of psychoanalysis: the living body. In the Lacanian psychoanalytic, that body is first shredded as it were (the initial stage of *le corps morcelé*), then idealized (the mirror stage), then finally riddled with holes (catapulted into the Symbolic order). It is thereby divested of its 'I can's'; it is rendered powerless.

III. Scientization and the Living Body

A true twentieth-century Western science is a Galilean science, one that is through and through susceptible to mathematization. Were psychoanalysis to fulfill its promising Lacanian press and become a science, then not only must introspection be censured and banished; everything that does not lend itself to mathematical formalization, or at least to the aura of such formalization, must be effaced. Roustang points out in commenting on Lacan's grandiloquent assertion that Freudian psychology is akin to Einsteinian physics[31] that "mathematical physics will in fact never cease to be the model *par excellence* for Lacan."[32] Following discussion of diverse instances of Lacan's attempt to position psychoanalysis as a science, Roustang notes that after aligning the psychoanalytic "real" with science,[33] "Lacan goes on to deduce that both the object and method of psychology are not subjective but relativist (a further allusion to Einstein) because they are founded on interhuman relations." Combined with his "energetic concept of libido," such "mathematical" reasoning provides Lacan with a basic psychoanalytic that "has rid itself of any dependence on the qualitative."[34] With these stage-setting moves in place, Lacan's subsequent procedure, according to Roustang, is to divest psychology—thus psychoanalysis

—of its subject, that is, of "what is most irrepressibly subjective." It must do this because, "unlike art, science precisely *must* efface such moments . . . in order to subject itself to the strictest rigor and rationality, stripped not only of all emotion, *but of any trace that might refer it to a specific person*" (second italics added).[35]

Clearly, Lacan's effacement of introspection is part and parcel of a much larger project. The latter swallows up not only introspection but the entire subjectivity that is its source: "what is most irrepressibly subjective" is an individual life. What is at stake is thus life itself. A brief look at Roustang's account of Lévi-Strauss's influence on Lacan will open a beginning vista on the specific cost of Lacan's attempted scientization of psychoanalysis.

It was noted at the beginning of this chapter that the unconscious is theoretically patterned alike for Lacan and Lévi-Strauss: the unconscious obeys structural laws in the form of invariant linguistic rules. Roustang's discussion of Lévi-Strauss's *Introduction à l'oeuvre de Marcel Mauss* and of its influence on Lacan illuminates in exacting ways the multiple dimensions of subjectivity that Lacan must efface—and in each case attempts to efface—in order to uncover and articulate those invariant linguistic rules for psychoanalysis. In a general sense, of course, Lacan's program effaces those dimensions by reducing them to the symbolic and ultimately, to "the autonomy of the signifier with respect to the signified."[36] Utilizing Lévi-Strauss's program of mathematizing the unconscious through language, Lacan, like Lévi-Strauss, more exactingly excludes all that pertains to "feelings, wishes and beliefs."[37] He will thus also "no longer bother with whatever has to do with the affections, the imagination, lived experience, the unspeakable, and the unfathomable."[38] Three years after Lévi-Strauss's publication of his *Introduction à l'oeuvre de Marcel Mauss* in 1950, Lacan himself comes to assert in his famous "Discourse of Rome," "Symbols . . . envelop the life of man in a network so total . . . that they bring to his birth . . . the design of his destiny."[39] In short, Lacan's proclamation is that language is not only everything we are; we are *pre-determined* by language. As indicated above, the future perfect is the grammatical key to our pre-determination: we are linguistically locked in to our fate; we simply fail to realize it. Most significantly, being merely spokespersons of language, we can hardly lay claim to "'force' or 'power'."[40] Such notions have no place in Lacan's psychoanalytic. Ignoring force and power, however, as Roustang points out, "means ignoring everything that pertains to feelings, wishes, and beliefs—not to mention anxiety and the imagination."[41] Lacan commentator David Macey calls attention to this same fact when he writes that language has been hoisted to such heights by Lacan that "he has frequently been criticized for privileging it to the almost total exclusion of affectivity, preverbal states and nonverbal communication."[42]

What is the effect of this linguistic totalization? In the course of discussing Lacan's Lévi-Straussian programmatic in which the Symbolic — language — reigns alone and supreme, Roustang states that "If the Symbolic is cut off from its social foundations, if it is forced to become autonomous in order to meet the needs of the scientific cause, no society at all could possibly be salvaged from it."[43] We need only paraphrase the remark to get a beginning grasp of the psychoanalytic distillation of the subject resulting from symbolic totalization. We need only substitute *bodily* concerns for social ones to realize that if the body is cut off from the Symbolic in order to meet the needs of science, no living creature could possibly be salvaged from it: to eliminate all that is irrepressibly subjective is to eliminate the living body. What Lacan puts in the place of a living subject are fabled stages on the way to the Symbolic, itself the *sine qua non* stage in the fable Lacan writes. With linguistic totalization, there is a divestiture of power in the living subject and, in effect, a divesture of 'I can's', of a felt life, of the immediacy of experience of the living subject. With this divestiture, of course, there is nothing left to introspect. No wonder then that Lacan speaks of introspection as a mirage. With the change of registers from living bodies to pure speech, living bodies themselves become a mirage. They are in fact "linguistified" at the same time that language is "substantified."[44] Lacanian psychoanalysis is indeed not a process but a *processing,* a reassemblage of life into language. When Roustang shows how feelings, wishes, and beliefs are excluded, when he shows how nonverbal communication is ignored, when he shows how experience is put outside the signifying field, when he shows how descriptive accounts are denigrated and banished, when he shows how force and power are reserved to language alone, he admirably captures the formidable impact of Lacanian psycho-processing. What he does not in turn observe — perhaps because his focus is on scientization, perhaps because of his closeness to the material, perhaps because he does not take a backward conceptual view over "The Science of the Real"[45] — is that these denials, each of which he so deftly identifies in the course of tracking Lacan's central project, converge in a denial of the living body. There is, in brief, a common denominator to all that Lacan is writing off and out of the psychoanalytic picture: how can there possibly be feelings, wishes, and beliefs without a body that winces, that gazes wistfully into space, and that solidifies itself in its convictions? how can there possibly be nonverbal communication without a body that gestures, orients itself, and moves in multiple and flexible ways? how can there possibly be experience without a body that lives it? how can there possibly be an account of life apart from descriptive renderings of animate form? how can there possibly be *power* short of dynamically moving creatures who are its source? In sum, how could there possibly be a living body if "[What] we are trying for [is] an algebra which would answer . . . to what . . . the kind of logic called symbolic accomplishes when it delimits the laws of mathematical practice"?[46]

A living body nevertheless haunts and continues to haunt the Lacanian psychoanalytic scene. When Lacan writes, for example, that "in analysis, there is a whole section of the real in our subjects which eludes us" . . . or of "[the] element of immediacy, of weighing someone up, . . . of deciding whether or not the subject has some substance, . . . ," he intimates by the very words he uses that it is a *bodily* presence that "eludes us."[47] When he furthermore continues by saying that "in order to pose the question of what is at stake in analysis, [one can ask] is it this real relationship with the subject which needs to be recognized?" it is clear that this *real* relationship is at the very least a bodily one, that is, that an intercorporeal dimension is part of the analytic situation, indeed, that in terms of corporeal archetypes, there is an intercorporeal positioning that is a decisive part of the immediacy of which Lacan writes. This intercorporeal immediacy is evident in an affective sense. For example, the classical analytic situation is not infrequently an initially intimidating one for the analysand insofar as he/she is supine while the analyst sits. The analysand is not only lower than the analyst but is in a vulnerable position with respect to the analyst. Moreover the analysand cannot see the analyst since, classically, he/she is lying on a couch *behind which* the analyst sits. There are, in fact, certain resemblances between the intercorporeal positionings in classical psychoanalysis and socio-aggressive presenting. The analysand, like the presenting animal, is in a vulnerable position; the analyst, like the presented-to animal, is in a position of power. Now Lacan is not only clearly interested in the question of the analyst's power. He sees the analyst's power as absolutely pivotal to the analytic situation, all the way from the analyst's being "the person supposed-to-know" to being the person who determines just how long each analytic session should be—whether three seconds or two hours.[48] He does not, of course, take into account intercorporeal presence. Yet to overlook the intercorporeal aspect of the analytic situation is to overlook the corporeal facts of the situation and the range of possible felt senses the situation evokes for analyst and analysand alike. Predictably enough, Lacan's response to his own question of whether the *real* relationship between analysand and analyst needs recognition is resolutely negative. The real relationship is not "what we're dealing with in analysis. . . . Definitely not. Incontestably, it is something else. Indeed, there we have the question that we and all those who try to theorize the analytic experience ceaselessly ask ourselves . . . the question of the irrational character of analysis."[49] It is startling to find this ceaselessly asked question immediately pushed aside in deference to what Lacan designates "the primordial question"; namely, "What is this experience of speech?" It is *this* question that poses "the question of the analytic experience, the question of the essence of speech, and of its exchange."[50]

"The question of the *irrational* character of analysis" is in actuality the question of an unexplored and unarticulated intercorporeal semantics. That Lacan, wittingly or not, characterizes *the real* in bodily terms—in terms of "immediacy," of "weighing," of "substance"—in the same context that he speaks of "the irrational character of analysis" points in the very direction of the archetypal intercorporeal semantics suggested above. What is irrational in a mathematical sense is what is not expressible in terms of a fraction; or in more precise spatial terms, not expressible in terms of *one integer over another*. But the classical analytic situation, reduced to mathematics, is indeed a fraction. It is a relationship between two bodies—one over the other. What is inexpressible—what cannot in principle be put into Lacanian psychoanalytic speech, and what has not yet been put into the speech of any classical psychoanalytic—is the relationship between these two bodies and the power relations they instantiate. Lacanian psychoanalytic speech has no words for this intercorporeal relationship because it is outside discourse. In mathematical terms, there is no formula by which it could be made to fit into what Lacan would designate the rigorously ordered algebraic system of psychoanalysis. The above quotation says as much. Whenever Lacanian psychoanalysis is threatened by *the real*, i.e., by animate form—living bodies—there is a change of "register." We should point out that Lacan uses the word "register" frequently to designate the categories of his psychoanalytic—the real (or Real), the Imaginary, the Symbolic—and to suggest, perhaps even denote, official records of documented facts. Lacan's psychoanalytic categories are in effect invested with authoritative status; they are each a compendium of what Lacan would have us believe are scientifically validated principles. Thus "the question of the irrational character of analysis" can be and is trivialized by "the primordial question." The trivialization, however, is in actuality no more than a change in register, from living body to analytic speech. The problem of course is that the register of the living body is nowhere properly filled in. Roustang's analysis supports this claim. He points out that in contrast with the Imaginary and the Symbolic, the real (and/or Real) is nowhere adequately spelled out in Lacan's *Écrits* or elsewhere. He emphasizes this fundamental lapse in a note in which he cites Maurice Dayan's book on Lacan, *Inconscient et réalité*.

Commenting on the book he states, "Lacan's notion of the Real has discouraged even the most patient and lucid minds"; as Maurice Dayan observes, "You never know what can happen in a Lacanian discourse on the Real and reality."[51] Presumably that is what makes the real (and/or Real) irrational in an epistemological sense.

Roustang documents his charge that Lacan's register of the real is improperly kept in his analysis of three kinds of Real in the Lacanian corpus, each kind being foundational in some respect to the real. There is first the Real of the Psychotic with respect to what we take to be normal everyday reality; there is next the Real of the Other with respect to what we take to be the everyday world of others perceptual reality; and there is finally the Real of articulated language with respect to the Imaginary and the Symbolic. But Roustang's charge of an improperly kept register can be documented even further, namely, in Lacan's affirmation that "Every analytic phenomenon, every phenomenon participating in the analytic field, in the analytic discovery, in what we deal with in symptoms and neurosis, is structured like a language."[52] The affirmation aptly encapsulates a denial of the corporeal and intercorporeal real. The denial, however, is vulnerable; it is open to question, as Lacan himself implicitly indicates in his acknowledgment of "the irrational character of analysis." Living bodies clearly haunt the psychoanalytic scene. Moreover, the mathematical edifice that Lacan attempts to construct in the name of psychoanalysis consistently crumbles now at this point, now at that point, from bodily infestations. In other words, it is not simply that living bodies haunt the psychoanalytic scene as Lacan implies when he discretely acknowledges the body and the problem it presents as an irrational element within the psychoanalytic. It is that living bodies haunt his theory-making. For example, in recalling Bertha Pappenheim's (Anna O.'s) "nervous pregnancy" in his chapter on sexuality in *The Four Fundamental Concepts of Psychoanalysis,* Lacan asks "What did she show by this?" He answers, "One may speculate, but one must refrain from resorting too precipitously to the language of the body. Let us say simply that the domain of sexuality shows a natural functioning of signs. At this level, they are not signifiers, for the nervous pregnancy is a symptom, and, according to the definition of the sign, something intended for someone. The signifier, being something quite different, represents a subject for another signifier."[53]

The first question to be asked is, "What *is* the language of the body?" What precisely does Lacan mean by "the language of the body?" If there is such a language, and if it enters into, or *could* enter into, the psychoanalytic situation, should it not be considered within psychoanalytic theory and be precisely and rigorously differentiated from a language of signifiers? Osten-

III. Scientization and the Living Body

sibly Lacan might be said to have done this. From what he
language of the body is to be regarded as no more than a syn
language. But again, from what he writes, a symptom belong
"natural functioning of signs," which is to say a symptom is "inte
someone." Symptoms are thus intentional, according to Lacan.[5]
Pappenheim's "nervous pregnancy," in effect, shows that a symptom
intentional even at the level of repressed desires, i.e., at the level
unconscious. No one, after all, can directly, through an act of con
willing, somehow *make* a bodily symptom appear. If this is so, then
language of the body," if there is such a thing, would seem to require ca
attention and examination in the context of psychoanalytic theory. On
one hand, if the body speaks in an intentional way, and if it can speak in
intentional way even at the level of the unconscious, then surely boo
"symptomatology" warrants a properly kept register; indeed, the register
properly a part of psychoanalytic theory. Perhaps even *a universal symbolism*
that is, archetypal symptoms, would emerge from a properly kept register
On the other hand, if a corporeal language exists, then such a corporeal
language has enormous interpersonal significance, precisely as suggested
above by intercorporeal archetypes and by an as yet unexplored and
unarticulated psychoanalytic intercorporeal semantics.

We might note that the popularity of "body language" some years ago
attested to the possibility of a properly kept register as it attested to the
significance of the body in everyday interpersonal communication. But this
body language quickly became a pop-culture attraction, and the language
itself was simplified, trivialized, and taken to be a thoroughly cultural
phenomenon. No concerted attempt was made to anchor this body
language within either an evolutionary framework or a psychoanalytic
perspective. Lacan's ill-kept register of the symptomatic together with his
denial of the living body might thus be seen simply as a reaction to the
pop-culture status of body language. But certainly both register and denial
may also be seen as a way of relegating all talk of the body—including
bodily symptomatology—to a place outside the science of psychoanalysis
where all signification is reserved to verbal language.

The problem with Lacan's dismissal and with his ill-kept register is that if
"the domain of sexuality shows *a natural functioning of signs*" (italics
added), then surely psychoanalysis, which is without doubt keyed to the
domain of sexuality, must take the "natural functioning of signs" into
account. A *natural* functioning of signs within the domain of sexuality
clearly suggests an *evolutionary* perspective. It furthermore suggests that that
evolutionary perspective on sexuality is essential. In other words, a *natural*
functioning of signs is a functioning that is in some unexplained but tacitly

fulfilled way part of our heritage. But here again, Lacan banishes what might count as a *natural* functioning of signs to a place outside the psychoanalytic. He changes registers, from the improperly kept one to the properly kept one. The "nervous pregnancy" is displaced from its possible interpretation within "a natural functioning of signs" to significatory status: in other words, it is a full-fledged signifier. Bertha's "nervous pregnancy," Lacan tells us, is really her psychiatrist Josef Breuer's desire for a child. Lacan would thus have us believe in linguistic action at a distance, namely, the power of desire to leap across bodies, causing a nervous pregnancy in another. The words he puts in Freud's mouth, though ostensibly such as to reassure Breuer that it is *Bertha's* and not *his*— Breuer's—desire for a child (for Breuer is presented as terrified of the transference that has taken place), are actually ambiguous to Lacan's point. According to Lacan, Freud says to Breuer, *"What! The transference is the spontaneity of the said Bertha's unconscious. It's not yours, not your desire, it's the desire of the Other."*[55] When Lacan puts the words "it's the desire of the Other" in Freud's mouth, he is actually speaking in two tongues, Freud's and his own. The Lacanian Other being Breuer's—not Bertha's—unconscious, Lacan can interpret the nervous pregnancy as Breuer's unconscious desire to have a child by Bertha. In this way, Lacan's Freudian words become a testimonial to Lacanian theory: "the symptom is resolved entirely in an analysis of language."[56] As Lacan himself goes on to say of "Freud's" words, "I think Freud treats Breuer as a hysteric here, since he says to him: *"Your desire is the desire of the Other."*[57] The testimonial aside, and the interpretive imputation to Freud aside also, the idea that the nervous pregnancy is actually Breuer's desire to have a child turns a natural functioning of signs into a signifying act, but one that is highly peculiar in that it involves a kind of psychokinetic leap. No matter that Lacan is insisting that the desire of the analyst cannot be discounted in the analytic situation, thus that Breuer's desires cannot be discounted. There is clearly a corporeal gap in Lacanian psychoanalytic theory, a gap in which the "natural functioning of signs" is all too swiftly and inopportunely entombed.

The pinnacle example of bodily infestations in Lacan's psychoanalytic is actually found in conjunction with corporeal gaps—interstitial places where the unconscious rises up and "expresses itself"—and this because the psychological gap in the Lacanian subject is topologically one with the corporeal gaps that mark human corporeal being. The living body that infests Lacanian psychoanalytic theory and haunts the actual Lacanian psychoanalytic scene is in other terms a function of the fact that the *lack* that is introduced into being by language, i.e., by the unconscious, is topologically one with *"le sujet troué,"* the subject riddled with holes.[58] Lacan speaks of drives (which he carefully distinguishes from instincts) being "expressed in the anatomical mark of a margin or border—lips, 'the enclosure of the teeth', the rim of the anus, the tip of the penis, the vagina, the slit formed by the eyelids, even the horn-shaped aperture of the ear."[59] (There are even more slits, Lacan assures us parenthetically: he could include "embryological details.")[60] The body thus *does* figure in Lacan's psychoanalytic, but it figures as a purely *visual* entity and as a purely *material* specimen. These margins and borders, as Lacan speaks of them, are not experienced in any bodily felt sense. There is no tactile-kinesthetic life to these surface openings. In effect, the subject is not a living body at all but simply a series of anatomical holes. Lacan says as much when he writes that the named anatomical openings are "no less obviously present in the object described by analytic theory: the mamilla, faeces, the phallus (imaginary object), the urinary flow," and when he goes on to ask rhetorically whether it is not obvious that partial features—anatomical gaps—are "rightly emphasized in objects [mamilla, faeces, and so on] . . . not because these objects are part of a total object, the body, but because they represent only partially the function that produces them."[61] By such claims, Lacan affirms that bodily slits are equivalent to desire that is always a lack—or in other terms, that analytic objects are simply the " 'stuff' "[62] of the drive that is desire. Corporeal gaps are hence relevant to Lacanian psychoanalytic theory not because analytic objects—feces, urine, and so on—that call attention to them are part of the body but because, like the analytic objects that emanate from them, the gaps represent the consistently negative character of desire, i.e., lack. Just as the objects have no "alterity," so the slits are the site of "nothing."[63] They *are* Other in this sense, Lacan tells us, in the same way that the specular image one sees of oneself in the mirror is Other; and, as he also tells us with respect to a lack of alterity and to sites of "nothing," "there is no Other of the Other."[64] In brief, partial objects align themselves with lack, with desire manqué, with the sliding of the signified under the chain of signifiers, with the Other that is always speaking us because we are *sujets troués,* subjects lacking integrity, subjects who are not really there at all.

The decimation of the subject in Lacan's attempted scientization of psychoanalysis is furthermore evident in his concept of a drive. His statement, "It is difficult to designate [the subject of the unconscious] anywhere as subject of a statement, and therefore as the articulator, when he does not even know that he is speaking," makes clear why Lacan calls upon "the concept of drive" in the first place. According to this concept, "[the subject of the unconscious] is designated by an organic, oral, anal, etc., mapping that satisfies the requirement of being all the farther away from speaking the more he speaks."[65] Drives are impersonal in the sense that they emanate from a "collective" unconscious. They are the expression of a body that is full of holes, i.e., full of desires. They are not *felt* realities but *linguistic* ones. Whatever their particular bodily site at any particular time, they are not on the register of the real; they are on the register of the Symbolic; they are signifiers (of desire). Thus it is that the body is reduced along with the subject him/herself to a signifying function, to various slit, rim, and border mappings which Lacan graphs in progressive fashion as mathematical formulae complete with retrograde vectors, algorithms, and so on.[66] Any vestige of a felt life within this mathematical formalization is nowhere to be found.

In sum, the body is incorporated into Lacanian psychoanalytic theory as a series of holes. One can hardly introspect something that is not there. Lacan's difficulty in pinning down the real (or Real) to anything remotely resembling a living body is symptomatic of a theory that is equally difficult to pin down: evidential support for it is lacking. The theory is, by any normal psychoanalytic standards, unverifiable. There is no way to "cash in"—as Husserl would say[67]—the mathematical formulae Lacan offers. Lacan would likely answer that this is because we are dealing with the unconscious and, accordingly, our procedure must be different. The unconscious has virtually nothing to do with the real (or Real). It is, after all, structured like a language, a language, Lacan would say, that we have to learn as apprentices in order to master. The power of Speech, he would insist, is properly reduced "to the form of an algebraic sign," just as things in the world are properly reduced to mathematical formulae, and the subject him/herself properly reduced to a signifying function. Given these formalized reductions, one could hardly expect the real to be anything but corporeal gaps. Indeed, Lacan actually states that if the situation were otherwise, if people were to insist on pursuing an experiential approach to psychoanalysis, "I dare say the last word in the transference reaction will be a reciprocal sniffing."[68]

Lacan's refusal to acknowledge corporeal life and his denigration of intercorporeality so fragment and mock the living body that it is no longer even alive: it is simply a registry of symptoms that plague language. "Speech," Lacan avers, "is in fact a gift of language, and language is not immaterial. It is a subtle body, but body it is. Words are trapped in all the corporeal images that captivate the subject; they may make the hysteric 'pregnant', be identified with the object of *penis-neid* [penis-envy], represent the flood of urine of urethal ambition, or the retained faeces of avaricious *jouissance*."[69] He avers further in a thoroughly sardonic appreciation of "two-body psychology" that analysis "[is] becoming the relation of two bodies between which is established a phantasmic communication in which the analyst teaches the subject to apprehend himself as an object; subjectivity is admitted into it only within the parentheses of the illusion, and speech is placed on the index of a search for the lived experience that becomes its supreme aim, but the dialectically necessary result appears in the fact that, since the subjectivity of the analyst is free of all restraint, his subjectivity leaves the subject at the mercy of every summons of his speech."[70] Lacan proceeds then to inveigh against the psychoanalyst who "interprets the symbol and, lo and behold, the symptom, which inscribes the symbol in letters of suffering in the subject's flesh, disappears. This unseemly thaumaturgy is unbecoming to us," he says, "for after all we are scientists, and magic is not a practice we can defend."[71]

Clearly the body that is the subject's flesh is not the subtle body that is language. The body of language is indeed so subtle that it can hardly be said to be a body at all, or if a body, is so consumptively torpid as to be moribund. When Lacan calls into question "the singular existence of the subject" and asks, "Why is he there? Where does he come from? What is he doing there? Why is he destined to disappear?" he answers that "The signifier is unable to give him the answer, for the very good reason that it situates him, exactly, beyond death. The signifier considers him already dead, and immortalizes him in his essence."[72] What more incisive statement could be had attesting to the absolute dismissal of the living body, to "the impossibility of the real," as Roustang terms it, or, to the fact that "signification is never a matter of reference to the real," as Macey describes it?[73] At the same time, what more conclusive proof could be had that it is fundamentally the living body that must be denied in order that, as with Lévi-Strauss, all power can be accorded to language?

The problem is that while power is put in the hands of language, language has no hands. This problem is not metaphorical. It is not even metaphysical. It is the simple problem of according power to something incapable of wielding power. The signifier that is powerless to explain the singular existence of the subject is powerless to explain animate form. It is thus powerless *from the beginning*. As Lacan himself states: "nothing in the Symbolic explains creation."[74]

However conclusive the above judgment against Lacan in his dismissal of the living body, it is vital to consider seriously the hypothetical claim attributed to him above, *viz.,* that because it is a question of the unconscious, it is precisely *not* a question of the body. In turn, if we are to verify his psychoanalytic theory, we must be open to verifying it *in its own terms*. The issue turns precisely on what those terms call upon us to do. Because documentation from self-analysis or case histories which would offer introspective corroboration is nowhere to be found, Lacanian theory is nowhere supported by the facts of analysis, the classical analysis that is linked to Freud and that Lacan consistently invokes. There are furthermore no experimental studies which we could carefully evaluate, studies which, through their quantifications—for example, of the comparative successes and failures of Lacanian psychoanalysis—would attest to the validity of Lacanian psychoanalytic theory. There are only Lacan's mathematical formulae that attest to the validity of Lacanian psychoanalytic theory, which is to say there are only Lacan's words. Roustang's cumulative evidence from Lacan's own writings, gathered to demonstrate Lacan's aim at making psychoanalysis into a science, eventuates at one point in an illumination of the startling—one might truly say *unconventional*—way in which Lacan himself seeks to validate his theory. The illumination is significant because, with the jettisoning of introspection, psychoanalytic validation is left hanging. The way in which Lacan deals with the formidable problem of validation is by invoking psychoanalysis itself—that is, *Lacanian* psychoanalysis itself. The experience of (Lacanian) psychoanalysis, he says, can be "universally verified" by everyone going through (Lacanian) psychoanalysis. Roustang tersely comments on this mode of validation, saying "Argumentation like this leaves you truly dumbfounded."[75] Perhaps because his central focus is not on scientific validation per se, Roustang stops short of noting that in fact there is a parallel, even an obvious if somewhat bizarre parallel, between standard scientific practice and Lacanian scientific practice. Lacan is merely emulating normal scientific procedure: what science recognizes as truth is what is verified and verifiable by the work of others. Just so with Lacan's psychoanalytic theory: it is validated by Lacanian psychoanalysis. The same validating procedures thus obtain in Lacanian psychoanalysis as in the natural sciences: others carry out the same procedures and in so doing validate—or invalidate—the original findings.

We will presently consider the double bind inherent in Lacan's program of validation; the question of universal verification will in other words surface again. For now, suffice to emphasize that in his attempt to scientize psychoanalysis by patterning it on the model of mathematical physics, Lacan can claim that his procedure of "transmission through recurrence"[76]—the analysand becomes analyst for another analysand, and that analysand becomes analyst for another analysand, and so on—does not in principle deviate at all from normal scientific validation procedures. It deviates only in the fact that in such a procedure Lacanian psychoanalytic theory becomes dogma. What each analysand learns in the process of psychoanalysis is not simply to leave the living body behind and unattended, as Freud did. It is to close the door on the living body as tightly as possible so that only "the subtle body of language" remains. Lacanian psychoanalysis is thus a kind of brainwashing in which animate form and the realities of corporeal life are reduced to nothing. The reduction leaves a permanent gap in the Lacanian record. There is, as indicated above, no point in looking within: there is nothing there. The extent of effacement of the living body in Lacan's larger project is, in a word, devastating.

IV. The Lacanian Unconscious

Insofar as psychoanalysis deals with the psyche, and preeminently with the underpart of the psyche at that, why would the living body enter into its analysis? What reason is there for not affirming that because it is a matter of the unconscious, the body, and most especially, the *living* body, is rightfully disregarded. Even given the fact that Freud's psychoanalytic studies began with hysterics who corporeally symptomized their anxieties, why would psychoanalysis focus on the living body in any central way?

Precisely because Lacan's psychoanalytic so decisively closes the door on the living body, it offers a paradigm example of the living body's focal significance to psychoanalysis. It shows that the living body cannot be successfully capped; a lid cannot be put over it such that, in effect, animate form is made to disappear. Animate form haunts Lacan's psychoanalytic not only in interstitial ways—in the slits and gaps of human anatomy to which Lacan draws attention—it haunts it in the form of an overarching and indomitable presence, *viz.*, an archetypal body part that Lacan raises to the power of privileged signifier. This linguistified body part is the phallus. If it is asked why a *corporeal* archetype is chosen for this privileged place in the Lacanian psychoanalytic, there is no reasonable and compelling answer forthcoming.[77] Indeed, if it is simply a matter of designating a special signifier, we can ask why a phallus and not a pickle?

This is a bona fide question we will progressively answer in the sections and chapters that follow, and in terms that both call into question the meager support Lacan offers as to why he chooses the phallus and show how his choice necessitates a denial of the living body.[78] We will orient ourselves in the direction of this extended answer first by sketching out more closely the Freudian unconscious and its method of access; second, by examining how Lacan travels from the Freudian unconscious to an unconscious structured in and by language and made accessible by paranoic speech; and finally, by exploring the differences and commonalities between this latter unconscious and Kant's "productive imagination." In this way we will hope to demonstrate how the hypothetical Lacanian claim—that because it is a question of the unconscious, it is precisely *not* a question of the body—is not only untenable, but how Lacan's psychoanalytic is tethered to the very concept of power that both explicitly and implicitly structures twentieth-century Western society. The concept of power as control can be culturally and evolutionarily traced out in an unconscious that "speaks the subject."

A.   First Approximations to the Lacanian Unconscious

It is important to emphasize from the start that we cannot directly introspect the unconscious, but can only introspect—or "retrospect"[79]—what we are actually aware of or have been aware of in experience. In consequence, to begin fathoming our unconscious, we need to pay attention to what are classically taken to be its spontaneous productions—to what are to us, in our conscious mode, inadvertent or involuntary acts and events. We thus need to pay attention to what slips into our experience *unbidden,* to what we do *without conscious consent or direction*—at least if we proceed as Freud first directed us. When we introspect *these* experiences, we begin to grasp—with or without psychoanalytic help—a domain of ourselves that is not completely lucid and sensical. As described earlier, these classic introspective investigations lead ultimately to revelations of desires and feelings harbored in our unconscious since infancy; they lead to the traumas and conflicts of infantile sexuality. In particular, all roads of the unconscious lead back to the Oedipal complex and castration anxiety, or to castration anxiety and the Electra complex, depending upon whether one is male or female. In either case, introspection leads to an illumination of the unconscious as a cauldron of repressed desires, a cauldron whose painful fires may be extinguished only through "the talking cure." In broad and summary terms, what makes people psychically sick, i.e., neurotic, is repression, and what turns psychically sick people into psychically normal, mature adults is both their capacity to put into words everything they remember and associate with their childhood, and their capacity to make sense of these remembrances and associations in terms of infantile sexuality. Language, in effect, is a means to an end, but introspection is the key to verifying "the talking cure." Insofar as introspective findings secure what is said, the anchoring act is not a speech act but a reflective act—a looking within. The turmoil of feelings besieging a neurotic, a turmoil plainly evident in Freud's case studies, makes clear why this act is necessarily central to psychoanalytic theory and to the psychoanalytic process, and why language is at bottom the means rather than the key to classical psychoanalysis. A sense of the body as locus of feelings, wishes, and so on—in general terms, as the generative source of desire—comes not from language but from concrete bodily experience. However much the living body is left behind in Freudian psychoanalytic theory and practice, what is articulated (even in a Lacanian sense: see below) subsequent to free association is what is or has been experienced.

If we look at Lacanian psychoanalysis and theory in terms of classical psychoanalytic methodology, we find ourselves without any such experiential anchorage. This methodological deficiency, as pointed out earlier, contrasts strikingly with the methodology of both Freud and Jung, who not only validated their own introspective findings by adducing corroborative introspective evidence from their patients, but who also insisted that psychoanalytic theory must be just so verifiable and verified. Bice Benvenuto and Roger Kennedy in their joint book on Lacan point out that Lacan's dissertation thesis contained the only detailed case history he ever presented, and that even here, one has neither "a record of psychoanalytic treatment," nor an "account of individual sessions." They point out in addition that "there is little interpretation of detailed clinical material, such as one can readily see in Freud's case studies."[80] They suggest quite clearly that in his development of psychoanalytic practice, as in his development of psychoanalytic theory, Lacan proceeded in a scientifically idiosyncratic manner. As previous sections of this chapter show, his idiosyncratic approach is fundamentally tied to his denial of the living body.

While it was noted that in analytic practice Lacan's method is to induce paranoia, there is another way of understanding his methodology: it can be theoretically as well as practically examined. On the practical side, his method was identified and shown *in principle* to claim support by appeal to clinical evidence in the form of "paranoiac alienation"[81] and paranoic speech. On the theoretical side, however, it is not supported by any evidence. This is because in essence his theory is buttressed solely by a backward projection of paranoic speech onto infantile "mental" development. "What I have called paranoic knowledge," Lacan writes, "is shown . . . to correspond in its more or less archaic forms to certain critical moments that mark the history of man's mental genesis, each representing a stage in objectifying identification."[82] The passive voice should serve as a sufficient warning about the actual lack of empirical evidence. Indeed, modest reflection on the theoretical side of Lacan's writings readily reveals a psychoanalytic done with word-play and mirrors. There is no actual pudding by which to prove the theoretics of Lacan's psychoanalytic, for there is no way of actually verifying the described infant-to-childhood stages that he describes. For example, the "within"—the *Innenwelt* which Lacan specifies as half of the basis of the mirror stage[83]—is part of the Imaginary. It purportedly grounds the Imaginary in the Real, but it is evidentially nowhere to be found. The mirror stage, the register of the Imaginary, is thus literally a work of the imagination. Lacanian theory, in turn, offers an optico-linguistic theoretics that has no foundation in fact, but that is affirmed "to correspond" to delusional paranoia.

We might in consequence ask whether a sound theoretical foundation might not be found in Lacan's writings themselves since, as his commentators have remarked, Lacan chose a personal style of writing that shows "many similarities in its form to that of psychotic writing." These commentators point out that Lacan's style "obeys the laws of the unconscious as they were formalized by Freud," i.e., it is replete with puns, contradictions, metaphors, ironies, and so on.[84] Psychoanalytically speaking, psychotic writings and certain manifestations of the unconscious are closely related, the distance from the normal to the psychotic being classically regarded a measure of the distance of the subject from his/her unconscious. Jung, for example, writes that "The definiteness and directedness of the conscious mind . . . are often impaired in the neurotic, who differs from the normal person in that his threshold of consciousness gets shifted more easily; in other words, the partition between conscious and unconscious is much more permeable. The psychotic, on the other hand, is under the direct influence of the unconscious."[85] Put in this perspective, Lacan's "psychotic" writings might appear to be immediate gleanings from his unconscious, that is, they suggest that Lacan has *communed* with his unconscious, and communed not mediately through its products—through dreams, slips of the tongue, and so on as with Freud, or through fantasies or artistic renderings of the "active imagination" as with Jung—but communed directly by speaking its tongue. They suggest further that he proceeded to theorize its tongue by decoding its *langue*.

The idea that Lacan communed directly with his unconscious by speaking its tongue and that he theorized its tongue by decoding its *langue* is supported by both sympathetic and critical commentators, who consistently point out the density and opacity of Lacan's writings. Commenting upon Lacan's use of the terms Symbolic, Imaginary, and Real, one reader notes, for example, that "Lacan's use of these terms is highly idiosyncratic, and at times one feels that they can be used to explain anything."[86] A similar charge is made against Lacan's style by an advocate of his psychoanalytic. In her attempt to explain "the preposterous difficulty of Lacan's style," Juliet Mitchell states that "the difficulty of Lacan's style could be said to mirror his theory."[87] Difficulty aside, the statement strongly suggests that Lacan's style *is* his theory, or in other words, that Lacan is writing out his unconscious. Another advocate, Catherine Clément, affirms that "Lacan's work is not a work of invention but of transmission."[88] Clément thus affirms that in formulating psychoanalysis, Lacan acted as a medium for the unconscious. He is indeed pictured as an extraordinary spirituo-linguistic agent, Clément likening him to "the mystical pelican that tears open its own liver and offers itself to its offspring."[89] Finally, that Lacan's writings "obey the laws of the unconscious as they were formalized by Freud" suggests not only a certain style but a certain content, which is borne out in Lacan's own view of his psychoanalytic as a strict return to Freud.[90] Indeed, the focus of his central aim—"[to] restore psychoanalysis to life"[91]—is the original Freudian unconscious in its pristine sexual splendor: "The reality of the unconscious," Lacan declares unequivocally, "is sexual."[92] In sum, first approximations to the Lacanian unconscious show it to be a veritable hot-bed of eccentric signifying powers which, while possibly restoring psychoanalysis to life, breed lasting mental derangement and an obsession with sex.

Now whether Lacan is really in communion with his unconscious, whether Lacan's unconscious is *the* unconscious, or whether his unconscious is a fabrication—in effect, a product of his imagination—are questions that underscore the central methodological issue. In negative terms, they are related to the repeatedly emphasized point that Lacan has not put together, neither in practice nor in theory, a psychoanalytic on the basis of self-analysis and of corroborative introspective evidence from others. But the questions also bring to the fore the question of the relationship between the unconscious and the imagination. The latter question can be most cogently addressed by putting Lacan's psychoanalytic in Kantian perspective. It is of moment to ask in preface to that placement, what is at stake? Why the concern with the question of whether, or to what degree, Lacan's psychoanalytic rests on his communion with his unconscious? What is at stake is precisely the credibility of Lacan's stages of infantile development, which Lacan links to delusional paranoia. The question provides a beginning context for evaluating that credibility. If Lacan's writings are the result of communion with his unconscious, i.e., with delusional paranoia, then the stages of infantile development that he specifies are similarly the result of communion with his unconscious. If Lacanian theory is valid, then it follows that were *we* to commune with *our* unconscious, we would find the very same infant-child developmental stages at the core of our own lives. It follows also, then, that "looking within" is unnecessary: we simply have to let the subject speak, and in speaking show us the invariant algorithms of our irremediably split human selves that originate in our irremediably troubled infancy. *What is at stake is thus clearly a matter of power.* If the roots of power lie in language, and if it is not we who speak but an invariant human unconscious that speaks us, then we are pawns of forces that are not simply outside our control, but that shape us, relentlessly and inexorably, from beginning to end. Only by acceding to Lacanina psychoanalysis, thus ultimately teaching the truth of these forces do we presumably come to power.

B.   Lacanian-Kantian Commonalities and Differences

What Lacan does with mirrors and words in his theoretical and practical restoration of psychoanalysis is to refract the original Freudian unconscious into what can, in a Kantian sense, be called "optico-linguistic schemata"; that is, Lacan structures the unconscious in a manner strikingly similar to the way in which philosopher Immanuel Kant structures the productive imagination: in principles or rules. The productive imagination, according to Kant, is a synthesizing faculty that orders experience and thus grounds our understanding of the world. It orders experience by producing rules— schemata—that are tied on the one hand to the sensible and on the other hand to the purely conceptual. Already there is a hint of an affinity. The Lacanian Imaginary is on the one hand linked to the Lacanian Real (however faultily in terms of sensible evidence) and on the other to the Lacanian Symbolic. While the Lacan-Kant comparison might appear far-fetched—particularly in view of Kant's judgment (and its sizable irony in the present context) that psychology will never separate itself from philosophy and be a separate science because it can be neither mathematized nor subject to experimentation[93]—a certain theoretical relationship obtains. Because it does, it is instructive to examine Lacan's psychoanalytic in terms of schemata and in the context of Kant's original formulations concerning *schematization*.[94]

To begin with, Lacan's optico-linguistic schemata are literally Kantian in the sense that they are taken by Lacan to be the forms or rules that anchor (psychological) experience and knowledge. More than this, they are the forms or rules that specify the Lacanian subject him/herself, the subject who is irreparably split, who is psychologically bifurcated, who is irremediably an ontologically damaged piece of goods. In the briefest of terms, Lacan's schemata all specify *lack*, lack that has its anticipatory and retroactive origin in the specular image (the "mirror stage" of infant development) and ultimately in the signifying nature of language. At the same time that Lacan's schemata conform in their ordering function to Kantian schemata, however, they are also purely Lacanian creations insofar as in their mature form, i.e., as schemata of language, they derive not from a Kantian 'productive imagination', but quite specifically from the Lacanian unconscious; that is, psychoanalytically speaking, they derive not from an *imaginative* faculty but from a *symbol-making* faculty, the unconscious, and, by extension, from *language*. In fact, as Lacan reminds us many times over, the subject him/herself is parasitic on language. Most importantly, in contrast with the Kantian power of the imagination, which is to put things together—to connect on behalf of human understanding—the Lacanian power of the unconscious is utterly disconnective, precisely as schemata specifying lack would suggest. The unconscious is the source of an ontological rupture at the core of being. Thus Lacan's optico-linguistic schemata are psychoanalytic formulae that specify and mark out the lack within every desire, the schism within our every relation with Others, the gap that is the perpetual condition of our being. These psychoanalytic formulae are laid down in and by our unconscious. They are brought to evidence only in the words uttered in the psychoanalytic session, that is, when, in our symbolically reduced, i.e., paranoic, state, we speak out "the gaps" of our being to the analyst-Other.

In light of the above comparison, one might say that what the Kantian productive imagination joins together in the service of human understanding, the Lacanian Symbolic rends asunder. The statement at first glance seems to accord in a remarkable way with the infant-to-child stages Lacan describes as charting our unconscious. But the Lacanian Imaginary stage marks the point at which the impoverished *corps morcelé*—the fragmented body that defines the *real* stage of the infant, a stage Lacan describes in more dynamic terms as one of "uncoordinated turbulence"[95]—is superseded by an intact image. With that intact image, the infant totalizes him/herself; he/she sees him/herself in the mirror and perceives a whole. The Imaginary thus *joins together;* it is the work of a 'productive imagination'. The schemata for this stage are rooted in what Lacan terms "the *imago*"—a relation between the organism (*Innenwelt*) and reality (*Umwelt*).[96] At this stage, Lacan tells us, the infant assumes his/her specular image, that is, his/her "reality." Whatever he may mean by saying the infant *assumes* its image, Lacan affirms that a transformation "takes place in the subject when he assumes an image."[97] More specifically, he states that "This jubilant assumption of his specular image by the child at the *infans* stage, still sunk in his motor incapacity and nursling dependence, would seem to exhibit in an exemplary situation the symbolic matrix in which the *I* is precipitated in a primordial form, before it is objectified in the dialectic of identification with the other, and before language restores to it, in the universal, its function as subject."[98]

Clearly the Imaginary is midway between an impoverished bodily life and the life of a language-bearing *I*. Its seemingly positive aspect—the jubilance which marks its introduction—is, however, short-lived. The Imaginary form in the mirror, Lacan declares, "would have to be called the Ideal-I, if we wished to incorporate it into our usual register. . . . But the important point is that this form situates the agency of the ego, before its social determination, in a fictional direction."[99] The Imaginary, in short, misleads us. We are duped by what we see. We only *appear* to be all of a piece. The Imaginary is in this sense a thoroughly *un*productive imagination. Lacan's description of this stage is accordingly bleak and ominous: "The mirror stage is a drama whose internal thrust is precipitated from insufficiency to anticipation—and which manufactures for the subject, caught up in the lure of spatial identification, the succession of phantasies that extends from a fragmented body-image to a form of its totality that I shall call orthopaedic—and, lastly, to the assumption of the armour of an alienating identity, which will mark with its rigid structure the subject's entire mental development. Thus, to break out of the circle of the *Innenwelt* into the *Umwelt* generates the inexhaustible quadrature of the ego's verifications."[100]

It is clear from the above account that the schemata of the mirror stage are unequally weighted. A rift between insufficiency and anticipation splits the subject to begin with and governs its development. At the same time, the illusion of an Ideal-I is born, thus dividing the subject further and simultaneously dictating its armoured progression. The ego, in effect, is trapped in the "quadrature" drawn by its *corps morcelé*, its totalized mirror-image, its projected Ideal-I, and an external world which is everywhere Other.[101] Only the true subject, the subject of language, can resolve the illusion and put the individual on the right path. In psychoanalytic terms, the stage of the Imaginary is a hoax played on a newly hatched ego, a hoax structured by schemata which lead the subject astray and which can only be seen through by Lacanian psychoanalysis. Thus what the productive imagination has purportedly joined together is, according to Lacanian psychoanalytic "fact," a hoax. The Lacanian Symbolic, in turn, does not in fact rend anything asunder. The subject is already sundered. The unproductive Imaginary misleads us into thinking we are whole when we are not.

Two questions at least are apposite at this juncture. First, insofar as images are symbolic—one can *think* in images[102]—how can the Imaginary be categorically separated off from the Symbolic? Why is verbal language the exclusive preserve of the latter? Surely Lacan's blindness to the fact that images are symbolic must be called into question since dreams are imagic phenomena; images in fact are the very stuff of dreams, and dreams are recognized as the most common products of the unconscious.[103] There is a deep irony here that is apparent from many perspectives beginning with Freudian psychoanalysis itself. Although Freud's interpretation of dreams proceeds on the basis of free association, i.e., language, still Freud regards dreams themselves as thoroughly symbolic. Insofar as Lacan bases his return to Freud on precisely the early works of Freud, among them, most notably, *The Interpretation of Dreams,* it is remarkable that he deviates in such an extraordinary way from the original. What Lacan does with the *images* of a dream in fact appears the same as what he does with the *real* of the living body. Just as the living body is stripped of affectivity, of its felt reality, so also are dream images. Both are instead formulated by language, and duly scientized. In effect, both are deprived of their inherent powers.

The second, obviously related question—where does the imagination, or the Imaginary, end and the unconscious, or the Symbolic, begin?—is historical and quasi-ontological, but its answer too ultimately turns on a dispossession of power. Lacan's answer, that language marks off the unconscious from the Imaginary insofar as language *is* the unconscious, is no answer at all, for we would need then to ask, Where does language come from? How, in the very real sense of making articulatory gestures, did it begin? These questions recall Lacan's difficulty in answering his own historical, quasi-ontological questions about creation, about *origins*. The singular existence of the subject is "radically unassimilable to the signifier," he says. "Why is he [the subject] there?," he asks, "Where does he come from? What is he doing there? Why is he destined to disappear?" Lacan has no answer. As noted in the previous section, Lacan states that "nothing in the Symbolic explains creation." Yet just a sentence before that remark, he affirms that "creatures are unthinkable without a fundamental act of creation."[104] Surely we can ask, Is language any more thinkable than creatures "without a fundamental act of creation"? In fact, was not the fundamental act of creation in this instance clearly the act of a creature cognizant of itself as a sound-maker and quintessentially aware of its corporeal powers to articulate and to refer?[105]

Clearly there are gaps—logical and historical ones—in a psychoanalytic in which rules set down by schemata fail to accord with empirical evidence that theory tries to muster on its own behalf. Lacanian rules are not tethered to facts of experience—to what Kant identifies as the *sensible*. They are, in other words, tethered to "facts" necessitated by theory, and not only at the general level of infant-childhood stages as described above. For example, Lacan appropriates the term "body-image" in its psychoanalytic sense from Paul Schilder at the same time that he divests the concept of its tactile-kinesthetic dimensions. There is, in effect, nothing remotely suggestive of Schilder's memorable "woman with a feather in her hat" who spatially comports herself in ways appropriate to her increased stature, ways that have to do with tactility and kinesthesia.[106] Indeed, the seeds of language that Lacan sows in his optical schemata by way of a specular image sprout a visual detached from its tactile-kinesthetic moorings: the image that is "total" is hollow—in fact, not even. Since it is *planar,* i.e., two-dimensional, it has no insides whatsoever. Moreover in actuality, the image is not even total in Lacan's thoroughly optical sense. If we go by *actual experience,* we perceive that our mirrored body-image is a body-image that is not and never can be complete. We never see ourselves fully. We never *see* our three-dimensional form. To think that we do is to be misled by twentieth-century Western hypervisualism.[107] Were we to go by actual visual experience, attending without bias to what is there, we would know immediately in a similar way that animate form is something other than a static assemblage of parts—or rather slits, as Lacan describes the real body in deference to the overarching and dynamic register that is the Symbolic.

In telling contrast to Lacan's theoretically factual schemata, Kant attempted to tether his schemata to empirical evidence such that rules governing the imagination are anchored in realities of everyday experience. Lacan's rules assert that the Gestalt the subject sees in the mirror is in disaccord with "the turbulent movements that the subject feels are animating him"; they state that the Gestalt (the totalizing image in the mirror) symbolizes "the mental permanence of the *I,* at the same time that it prefigures its alienating destination"; they affirm that the Gestalt "is still pregnant with the correspondences that unite the *I* with the statue in which man projects himself, with the phantoms that dominate him, or with the automaton in which, in an ambiguous relation, the world of his own making tends to find completion."[108] Lacan's rules *affirm* this turbulence, this discordance, this alienation, and these various projections to exist, but his affirmation is a decree and not a description of what is actually there in clinical (or experimental) experience. Lacan's refraction of the Freudian unconscious into optico-linguistic schemata in truth forces the unconscious into a mold which, because it is not ready-made and immediately functional at birth, necessitates a forced structuring of infant life into stages that are first of all necessary to the infant's later capacity to speak and second, capable of bearing the weight of the privileged status of language. A pre-determined linguistic end thus defines the means. Lacan's optico-linguistic schemata are retroactive necessities. No wonder then that the rules have not been deduced from the sensible—from actual experience, from the living body, from animate form. No wonder too that nothing in the nature of an introspective methodology is sanctioned. Introspection would countermand rules that have been induced on the basis of necessity, namely, those ordained in the light of the pre-determined end that is language.

Radical differences notwithstanding, a Kantian-Lacanian commonality of sorts bears notice. The schemata of both the productive imagination and the unconscious are in actuality difficult to pin down, that is, to ground empirically in the sensible. In Kant's words, schematization "is an art concealed in the depths of the human soul, whose real modes of activity nature is hardly likely ever to allow us to discover, and to have open to our gaze."[109] Kant thus affirmed how much humans do not know about their abilities to understand the world. They organize their experiences, but in ways that are "concealed in the depths." Although he sketched out a number of schemata in relation to basic everyday experiences such as the experience of magnitude and of substance, he affirmed that the schemata of the productive imagination are shrouded in darkness—in depths as dark, we might say, as the schemata of the Lacanian unconscious, which, though "elucidated" by Lacan in his writings, remain uncommonly idiosyncratic and empirically opaque. Rules grounding the imagination and the unconscious, in effect, cannot be straightaway identified and readily described; they elude our experience. Surprisingly enough, Lacan himself actually remarks upon this fact and in just these terms when he avers, "There is a whole section of the real in our subjects which eludes us." What is more, he goes on to remark, cryptically or ironically, "[This elusive *real* is] something we deal with all the time."[110]

V. The Double Bind: A Final Word on Verification

Lacan's failure to ground his optico-linguistic schemata in the real brings to a head the question of verification and the missing body of evidence. Maligning introspection as a method that vitiates the science of psychology and as in fact furnishing no more than "abstractly isolated moments of the dialogue [of psychoanalysis]," which, were we to dwell on them, would "create an obstacle" in the path of the dialogue, Lacan distances himself absolutely from such "magical thought."[111] His abandonment of introspection leaves us two major alternatives. On the one hand, lacking any detailed case histories and the empirical assurances that would go with them, we may believe that Lacan has actually communed with his unconscious and that his texts are a transcription of that encounter. On the other hand, and again, lacking any detailed case histories and the empirical assurances that would go with them, we may believe that Lacanian psychoanalytic theory is not at all an elucidation of the unconscious but through and through a product of Lacan's own imagination, its pieces propped up here and there by empirical studies and observational reports (e.g., by research on pigeons and locusts, by others' observations of children, and so on),[112] but in essence and in fact not introspectively verifiable, certainly not in the classical first-hand psychoanalytic manner. From this latter perspective, Lacanian psychoanalytic theory is an imaginative invention fabricated from optical and linguistic flights of fancy. This is not to say that because it is a product of his imagination—or in more precise psychoanalytic terms, the product of *projection*—that there is no truth whatsoever to be found in it. It is rather to suggest that creative ideas, spiritual visions, artistic compositions, dreams, and delusional adventures—all products of the imagination—exist along a psychological continuum of spontaneous upsurges from the unconscious.[113] Accordingly, to say that Lacanian psychoanalytic theory is fabricated from optical and linguistic flights of fancy means only that *as theory,* we cannot regard it as cutting the scientific path it purports to cut, which in turn means that its schemata, at the purported sensible end of their tethering, cannot be regarded reliable. Indeed, to take Lacanian theory as a product of the imagination obliges one to look closely at how it is—or is not—tethered to "the real." The problem is precisely that Lacan himself obfuscates the real at the same time that in one instance at least, he avers that in the psychoanalytic situation, the real "eludes us." Insofar as obfuscation and elusiveness might be related, and insofar as the *real* is at times a kind of Lacanian pseudonym for the body, the problem of the Lacanian real can be viewed fundamentally a problem created by Lacan himself in his failure to take the living body into adequate account. This, of course, is why the living body haunts his psychoanalytic.

We might conclude that the Lacanian unconscious is indeed simply a form of the imagination. In other words, while Lacan would have us believe that he has gleaned the schemata directly from *the* unconscious and that the empirical supports he offers on behalf of his theory in references to pigeons or to a single study of a single infant, for example, are merely adjunctive to, and not foundationally supportive of, the truths he articulates (or purports to establish), Lacan has in large measure invented the schemata. Moreover we cannot consider the supportive evidence he offers to be adjunctive to his invention because the evidence itself is open to question. What the single study of a single infant offers is in fact the *conjectured* experience of the infant; what evidence from pigeons and other such animals offers is in fact an *extrapolation* of behavior from radically different nonhuman animal species with which, be it noted, Lacan acknowledges no continuities much less delineates continuities. In neither case, furthermore, does Lacan corroborate the adjunctive evidence by case histories. In short, the rules that Lacan says run the unconscious have no mode of justification apart from Lacanian psychoanalysis. In effect, unless one enters into Lacanian psychoanalysis, one cannot validate the rules Lacan specifies. To verify his findings, we ourselves, individual by individual, must enter into his psychoanalytic treatment. To enter into his psychoanalytic, however, is to be confronted with a psychoanalytic ideology that, like any ideology, has the potential power to brainwash us. Lacanian theory in particular articulates an all-powerful, absolute theoretical infrastructure. It asserts our psychic history—*what it was and what it ever shall be.* By giving ourselves over to that ideology, we theoretically discover our irreparably split selves. Were we not to give ourselves over to that ideology after the fact of coming face to face with it and finding it experientially unverifiable, how would our refusal be greeted? Would we not be charged with failing to see the light? With being obtuse? With being dishonest and of evading the truth? With being stubborn? Or incompetent? Or *incurable?* Faced with what appears to be Lacanian psychoanalytic dogma, we would be stymied for an answer and, indeed, our psychoanalytic fate would be mooted.

In sum, with respect to Lacanian psychoanalysis, we are damned if we do—we necessarily become the delusional but ultimately enlightened split subjects Lacan tells us we are and will be—and damned if we don't—we are written off as above: myopic in some way and/or beyond redemption. Either way, we are written up in the Lacanian record as in need of psychoanalytic help. But the dilemma we face is in actuality brought about by a psychoanalytic theory lacking a credible mode of verification. A replication that guarantees its own replication is not scientific; as indicated above, it is dogma. Moreover because Lacanian psychoanalytic theory hangs on language, and because language cannot take us back to origins, the real, that is, everything Lacan conceives to be corporeal and/or related to the corporeal, cannot but be compromised. The rules Lacan affirms he has found cannot be found by introspecting feelings, memories, or motivations. They are rules which, bolstered by conjectural and extrapolated "truths" from a sparse if diverse literature, in analytical practice become mathematically based principles that Lacan purports to find in linguistic condensations (metonyms) and displacements (metaphors) in the analysand's speech, and that, in the transference relationship between analyst and analysand, language naturally articulates. What purportedly universalizes Lacanian psychoanalytic theory is language, that is, language that is discourse addressed to an Other. But even here, in the speaking of language, there is no body to contend with for *language speaks the subject;* the true subject is the unconscious. Because language—the unconscious—speaks the subject, introspection is yet again clearly displaced; one cannot, after all, introspect language. Specifically, one can neither find nor not find the rules of the unconscious in language except insofar as one is induced to speak a certain language and except insofar as the language one is induced to speak is interpreted as verifying the rules of the language one is speaking. The unconscious, *as specified,* is thus manifest; and we, as Lacan has already told us, are its manifest destiny. We are, in effect, powerless subjects, an afterthought, as it were, of our tongues, themselves the faithfully wagging pawns of an invariant and all-powerful unconscious.

VI.   Desire, Lack, and the Other: the Sartrean Connection[114]

If "The whole history of ideas should be reviewed in the light of the power of social structures to generate symbols of their own," as the epigraph from anthropologist Mary Douglas suggests, so also the whole history of ideas should be reviewed in the light of the power of individuals to generate ideologies by elaborating symbols generated by the social structure in which they live, and, by extension, the evolutionary heritage from which they derive. We can exemplify this power of individuals by specifying fundamental aspects of Lacan's psychoanalytic makeover of Sartre's ontology. In doing so, we will attain a first vantage point on the power of the phallus in Lacan's psychoanalytic in particular and in the Western social psyche in general.

The confluence of the theoretical underpinnings of Lacan's work with Sartre's existential ontology is as substantively informative as it is unmistakable. Lacan follows Sartre in locating desire, specifically sexual desire, at the heart of human being. He follows Sartre in conceiving desire first and foremost as lack, lack of being. Where he does not follow Sartre is in the latter's existential disavowal of the unconscious, and in turn, in his actual descriptive account of desire. Lacan strips desire of its ontological moorings in freedom and embeds it instead in language. Though transplanted, its dialectic of subject and object remains, and in a fundamental sense; the Other, for Lacan as for Sartre, is a perpetual aspect of one's own being. Lacan, however, defines the Other in distinctly different terms. For Sartre, the Other is not a matter of signifiers at all but of being. The for-itself—human freedom, human consciousness—is always in flight toward a fullness of being, toward being-in-itself. This flight is never successful because, according to Sartre, we are what we are not and are not what we are. We are our past in the mode of not being it, for example, because we are already present; we are not our present because we are already in flight toward a future. "Fundamentally," Sartre says, "man is *the desire to be.*"[115] Lack defines desire because desire is never fulfilled. Lack thus defines being, hence the existential categories, *being* and *nothingness*. In sum, and in one of Sartre's most memorable phrases, "nothingness lies coiled in the heart of being—like a worm."[116]

For Lacan, desire and lack are not existentially but linguistically defined. Accordingly, the Other is given a decisively linguistic turn; the Other is first of all *language:* "The moment in which desire becomes human is also that in which the child is born into language," says Lacan.[117] With its entrance into the Symbolic, the child becomes human in precisely the sense specified earlier. He/she is irremediably split. Lack is born because the wholeness the child wants is never satisfied. In particular, the *"All"* that the child has known with the mother is ruptured.[118] The child's entrance into language marks that rupture. With language, the rift in the mirror becomes the permanent rift of desire, desire which is always desire of the Other. Because desire is tied to language, it is properly understood as "the effect of the signifier in the subject"; it is "the nodal point" at which sexuality is linked to the unconscious."[119] The Other is, in effect, the unconscious. At the moment a child is "born into language," Lacan states, "the desire of the little child has already become the desire of another, of an *alter ego* who dominates him and whose object of desire is henceforth his own affliction."[120] What Lacan means is that, through language, the subject enters into the discourse of the Other, i.e., the unconscious, and becomes *its* subject.[121] Hence, the subject does not speak the unconscious; the unconscious speaks the subject. "The Other," Lacan affirms, "is already there in every opening [e.g., in "dreams, slips of the tongue or pen, witticisms or symptom"], however fleeting it may be, of the unconscious."[122] The subject thus begins "in the locus of the Other"; the first signifier emerges from the unconscious.

The phrase "desire of the Other," it should be noted, has two senses (at least): desire is always the desire *of* an Other (than oneself); desire is always the desire *for* an Other (than oneself). The child's desire at the moment it enters into language is a lack of being—a "wanting-to-be"—in both senses: a desire for that being it knew before language, a desire for that "lost paradise" that was its fusion with its mother[123]—thus a desire for an Other; and a lack of being, a "wanting-to-be" the object of desire of the (m)other— thus the desire to be the desire of an Other.

Although remarkably transposed and radically compounded, the Sartrean ontological connection is so obvious that it is puzzling that commentators who clearly recognize certain ontological commitments within Lacan's psychoanalytic do not specify or exemplify dimensions of the Sartrean connection. Benvenuto and Kennedy, for example, remark that "The Lacanian view of the unconscious plays around with the philosophical problem of ontology, the study of 'being' and 'non-being'."[124] It is equally puzzling that other commentators who specifically point out a philosophical relationship between Lacan's psychoanalytic and Sartre's existential ontology do not pinpoint and discuss either the particulars or the depth of the correspondence.[125] Clearly, even from the above brief exposition, Lacan's "wanting-to-be," his principle that "man's desire is the desire of the

Other,"[126] coincides precisely with Sartre's claim that "fundamentally, the desire of man is to be." Clearly, Lacan's divided *psyche* is ontologically equivalent to Sartre's divided *being*. Neither is whole; neither can ever be whole. The fate of each is to be a lack.

Examination of this fundamental coincidence might appear abruptly foreclosed the moment one turns attention to the linguistic, i.e., signifying, foundation of Lacan's psychoanalytic. While for Sartre nothingness lies coiled like a worm in the heart of being, for Lacan nothingness lies like a decisively uncoiled worm in the heart of being. The privileged Lacanian signifier is the phallus. The phallus is lacking to the mother. It is also, however, lacking to the father. While the father *has* it, he *is* not it; the mother *is* it. But while the mother *is* it, she does not *have* it; the father *has* it. To appreciate fully this linguistic turn of psychoanalytic events, we must in fact quickly return to Sartre and examine the categories he says are fundamental to desire. In point of fact, with Lacan's dialectic of *being* and *having*, we are already back on common ground. The categories that Sartre says are fundamental to desire are precisely the ones which, more by proclamation than by explanation, Lacan takes over in describing desire and lack, namely, *being* (the phallus) and *having* (the phallus). The privileged object of desire aside for the moment, the only change Lacan makes in the categories—and it is a substantive change made at the very beginning—is an addition. In particular, the two existential categories—being and having—that Sartre discusses at length in the last sections of *Being and Nothingness* acquire differential sex labels in the theoretical writings of Lacan. It is relevant to point out that Lacan's psycho-sexual makeover of Sartre's ontological categories closely recall his makeover of linguist Ferdinand de Saussure's linguistic tree, which becomes the occasion for Lacanian ribaldry as well as theory. Saussure's picture of a tree with the word "tree" above it is transformed into a picture of two identical doors with the word "Ladies" written above one, the word "Gentleman" written above the other. According to Lacan, the "law of urinary segregation"[127] signals the fact that sexual difference is originally embedded in language. (The "law of urinary segregation" is in fact related to the "Law of the Father," i.e., the Symbolic order that structures all human relationships.) A preliminary summary of Sartre's categories will be instructive before examining the corporeal significance of Lacan's linguistification of sexual difference.

Sartre's notion of being and having derive from his determination of human being as being for-itself. The project—or *desire*—of the for-itself is always a choice, a concrete venture within a concrete situation; it is not a conceptual project—something on the order of airy musings.[128] This is why Sartre speaks of the project of the for-itself as being always the choice of a particular world, and why in consequence he can say that any particular choice is the creation of a particular world. But even with its creation by choice, the *being* of the created world is precisely what escapes the being of the for-itself. No matter that the latter has created it. The latter never can *be* the world it has created. Moreover it can never *have* that world. Whatever its desire, it can never realize it. Any attempt to appropriate that world conclusively meets with failure. For example, while it can be said that those objects that are a part of my life are me, those objects are also completely independent of me.[129] "Possession," says Sartre, "is a magical relation; I *am* these objects which I possess [e.g., my lamp, my pen, my clothing, my desk, my house], but outside, so to speak, facing myself; I create them as independent of me; what I possess is mine outside of me, outside all subjectivity, as an in-itself which escapes me at each instant and whose creation at each instant I perpetuate."[130] He goes on to affirm that "In the relation of possession the dominant term is the object possessed; without it I am nothing save a nothingness which possesses, nothing other than pure and simple possession, an incompleteness, an insufficiency, whose sufficiency and completion are there in that object."[131] It is understandable, then, why possession is "nothing save the *symbol* of the ideal of the for-itself or value." Sartre emphasizes this notion: "We can not insist too strongly on the fact that this [possessive] relation is *symbolic* and *ideal.*"[132] Immediately thereafter, he states, "My original desire of being my own foundation for myself is never satisfied through appropriation any more than Freud's patient satisfies his Oedipus complex when he dreams that a soldier kills the Czar (i.e., his father)."[133]

Clearly, *having,* for Sartre, is an appropriative act in which one is always left empty-handed. Like "the desire of being," *having* is a perpetually failed enterprise, a never satisfied lack at the heart of being. Not only this but appropriation has a destructive element. Sartre gives as example the buying of a bicycle, declaring that "no particular act of utilization really realizes the enjoyment of full possession."[134] Merely seeing and then touching the newly bought vehicle, for example, is insufficient; getting on the bicycle is what matters; then taking a ride is what matters; but again, taking more than a ride is what matters in that endless trips are part of its purchase. As Sartre puts it, the bicycle ends by *referring* me to more and more pedallings, to more and more trips which themselves *refer* to a myriad of behaviors so that "finally as one could foresee, handing over a bank-note is enough to make the bicycle belong to me, but my entire life is needed to realize this possession."[135] Because the appropriative enterprise always remains unfulfilled, there is "a violent urge," Sartre tells us, to *destroy* what we appropriate. We want to destroy the things we appropriate for the very reason that they are a constant symbol for us of a fullness of being that is beyond us, that is independent of us. Thus it is that appropriation is always symbolic: "In itself appropriation contains nothing concrete. It is not a real activity (such as eating, drinking, sleeping) which could serve in addition as a symbol for a particular desire. It exists . . . only as a symbol; it is its symbolism which gives it its meaning, its coherence, its existence. There can be found in it no positive enjoyment outside its symbolic value."[136]

The above exposition and quotations demonstrate the extraordinary conceptual linkage between Lacan's notion of having and Sartre's notion of having. There is indeed no difficulty in translating the foregoing into Lacanian terms: to appropriate is never actually to have because the Other is always at a remove, whether "Other" designates in one context the unconscious, in another context language, in a further context the analyst, or in still another context persons generally. The destructive element enters into Lacan's notion of having for the very reason it enters into Sartre's. Both Lacan and Sartre affirm: "I want to destroy the Other for the very reason that the Other is a constant symbol of my Otherness, of the lack that defines me." Aggressivity—"aggressive competitiveness"—in fact enters the Lacanian scene precisely in the mirror stage, i.e., in my original experience of my own *otherness.*[137]

Now the problem with Lacan's categories of being and having is that, unlike Sartre's, which derive from an ontological examination of being, Lacan's seem to descend *deus ex machina* into his psychoanalytic theory, and furthermore to attach themselves to pre-designated individuals with at most half a reason: males really do *have* a penis, a reason not without complications and implications with respect to the Lacanian *phallus* as will presently be shown in this chapter, and shown in further detail in chapter 10. Lacan's consistent difficulty in achieving the precision of Sartre's ontology appears in fact adducible precisely to his sexualization of being and having. On the one hand, and with theoretical tongue-in-cheek, one might say that because he "appropriates" the categories, they are never really his but indeed remain Other. His problem with them, then, is a matter of doing battle with them. But he cannot conquer them; he cannot set them forth clearly in his analytic. A candid appraisal of his difficult battle with them is given by Muller and Richardson when, in attempting to give an exposition of what it means for a subject to " 'have' the phallus-as-signifier (as opposed to 'being' the phallus for the Other) yet 'not have' the phallus-as-organ (because female)," they comment that if their previous interpretation is correct, then it "suggests a way to avoid dizziness through the following skid." "The following skid" is a passage from Lacan which reads: "Paradoxical as this formulation may seem, I am saying that it is in order to be the phallus, that is to say, the signifier of the desire of the Other, that a woman will reject an essential part of femininity, namely all her attributes in the masquerade. It is for that which she is not that she wishes to be desired as well as loved. But she finds the signifier of her own desire in the body of him to whom she addresses her demand for love."[138] A Sartrean translation of the passage would read: a female is what she has (is) not. A Sartrean amplification of the passage would read: a male has what he is (has) not. The aim of Lacanian psychoanalysis in Sartrean translation and amplification is to make the analysand in each case realize his or her always unfulfilled desire: either to be what he has or to have what she is. The Lacanian dialectic readily plays off the Sartrean original. But the essential difficulty remains. The categories are not clearly articulated; they remain indubitably borrowed from Sartre and consistently escape Lacan's grasp. If one asks why Lacan did not choose merely to utilize them—perhaps in some heuristic way—but instead insists on them as substantive aspects of his psychoanalytic, it is clear that his purpose is to articulate "the law of urinary segregation" as the law of ontological separation.

The law is enunciated in general form in the first of two paragraphs within Lacan's essay "La Métaphore du Sujet."[139] However arcane and variably decipherable the second paragraph, Lacan initially states unequivocally in the first paragraph that being and having proceed from language.[140] Glossing on this statement in the second paragraph, he first recalls the example of a metaphor that he gave in "The Agency of the Letter in the Unconscious or Reason Since Freud": "His sheaf was neither miserly nor spiteful. . . ."[141] The sheaf in question belongs to Booz, the character in whose name Victor Hugo wrote the poem, "Sleeping Booz." Clearly the metaphor has sexual overtones, and just as clearly it has ontological significance, for Lacan immediately goes on to state that the metaphor evokes the linkage that "unites the circumstance of having"—i.e., the circumstance of the sheaf being *his* sheaf—to "the refusal inscribed in his being"—i.e., Booz cannot *be* his sheaf—and that this linkage of having and being marks "the impasse of love." He who has (it) cannot be (it); she who is (it) cannot have (it).

With this fundamental dialectic suggestively in place, Lacan goes on to say that the metaphor's negation would in fact do no more than affirm it; that is, it would introduce a substitution, the substitution of Booz for "his sheaf," with the result that not the sheaf, but Booz himself "was neither miserly nor spiteful." The relationship of having to being is thus, through substitution, turned into a relationship of being to having. From this substitution, Lacan says, "arises [*faisait surgir*] the only object of which the fact of having necessitates the lack of being: the phallus."[142] The negation affirms the metaphor in that the sheaf refers to Booz all along, and this because it refers to the only object that can at the same time be something of which it can be said that to have it is not to be it and to be it is not to have it.[143] The negation of the metaphor simply reverses—or in Lacanian terms, "displaces"—its own signifiers, which, whatever their order, are not equal partners but differentially weighted, "the occulted signifier remaining present through its (metonymic) connexion with the rest of the chain."[144] The "formula" for metaphor, Lacan says, is simply *"[o]ne word for another."*[145] With the particular substitution of Booz for sheaf, however, the phallus necessarily comes into play because, as Lacan affirms, the distinction between being and having can only derive from the phallus, the privileged signifier or "signifier of signifiers." Thus from metaphor springs forth the dialectic of being and having. Being and having are in turn simply ways of speaking "the law of the signifier." Although Lacan distinguishes two different paths to the law of the signifier—metaphor (displacement) and metonymy (condensation)—both lead to Lacan's psychoanalytic Rome, namely, the phallus.

What Lacan wants us to recognize is the metaphoric nature of ontology, that is, the fact that ontology is really a matter of language. That this is the case becomes all the more evident when we consider his original—and less opaque—commentary on the same metaphor in his essay "The Agency of the Letter of the Unconscious." Patient and careful reflection upon his commentary makes evident that three possible meanings of sheaf—the actual sheaf, Booz's penis (or his "phallus"), and Booz in person—enter into a three-way metaphoric association. At the lowest level, metaphor is simply the result of an anthropomorphic reading, the sheaf itself being the instantiation of human qualities (miserly, spiteful). At a first psychoanalytic level, the sheaf is Booz's penis (phallus), which means of course that it is not attributes of the actual sheaf that are being spoken of but that it is Booz's penis (phallus) that is neither miserly nor spiteful. At a further psychoanalytic level, i.e., at a further level of substitution, it is Booz himself who is not miserly and spiteful because he, Booz, *is* not his penis (phallus); his penis (phallus) is something he *has*. Lacan in fact writes that if the sheaf replaces Booz, it replaces him "in the signifying chain at the very place where he was to be exalted." In other words, taken as metaphor at this third level, Booz is reduced to "less than nothing." True, he still *has* his sheaf even if he *is* not it, but he is "less than nothing" because he has been reduced "by the munificence of the sheaf which, coming from nature, knows neither our reserve nor our rejections, and even in its accumulation remains prodigal by our standards."[146] Clearly the sheaf at this level is full-blown Sartrean being-in-itself; it is the epitome of fullness of being, so full in fact that it can waste itself in munificent ejaculatives. Just as clearly, Booz is Sartre's being-for-itself; he is the epitome of a lack of being, so lacking that the munificence of the phallus is by his own ontologically impoverished standards prodigal. At this level, and as indicated earlier, Lacan assures us that the phallus arises because it is the only object of which it can be said that to have it is not to be it, and conversely, to be it is not to have it.

In admittedly rough terms, the legitimate question to ask is, does the Sartrean transplant take or can the same thing be said of a pickle? In more refined terms, are we to take being and having as metaphors, and thus say that the female is the phallus in the mode of not having it and that the male has the phallus in the mode of not being it? Or does such a way of putting the relationship between being and having sexualize the dialectic in a way that is unsubstantiated ontologically? Lacan designates the Sartrean onto-logical dialectic a metaphor; he analyzes it to its phallic core—as metaphor. All talk of being and having thereby becomes simply a metaphoric way of speaking of the object of desire. Females *are* what males desire; males *have* what females desire. From a Lacanian psychoanalytic perspective—and one could ask from a Lacanian perspective whether any other perspective in fact exists—life centers around the phallus. This singular and exclusive genitalic centralization is precisely what has exercised French feminists Luce Irigaray and Julia Kristeva. Irigaray in particular has written a polemic against Lacan's one-world order.[147]

One might say that the difficulty in accepting the Lacanian makeover arises because Lacan anchors being and having in something absolute and culturally pre-given, namely, the phallus. Clearly, the introduction of the phallus complicates Sartre's clearly stated existential analysis of being and having, not the least because it is difficult to conceive the phallus as something a female *is*. However handy such a conception might be as a psychoanalytic anchor point—both theoretical and practical—any psycho-analytic experience which would corroborate it is never even remotely suggested, and any self-initiated introspective rummagings on behalf of validating it fail to turn up any support. Even as exalted to the status of "privileged signifier," a neutered absolute, the phallus carries vestiges of its lowly male origin, a fact which returns us to the previously asked question, why a phallus and not a pickle? The answer, tentatively suggested here and to be later expounded in chapter 10 in full detail, is that Lacan can in fact not jettison the living body completely, for if sexual desire is at the very heart of Lacanian psychoanalysis, then however linguistically clothed, corporeal realities must enter the scene. The phallus is indeed the ultimate way in which the living body haunts Lacanian theory, and it is precisely a penis and not a pickle that gives substance to this haunting.

Now if Lacan's basic theorems are true—that "man's desire is the desire of the Other" (or in Sartrean terms that "fundamentally, man is the desire to be"), and that the phallus, that signifier of signifiers that circulates in language and that accordingly can be either hidden or revealed, symbolizes both the object of desire and the lack that forever marks desire and as such symbolizes both male and female desire and lack in terms of being and having—then a further fundamental question arises. This question arises on the basis of the fact that in such a scheme, all desire is taken as *sexual.* Indeed, as pointed out earlier, both Lacan and Sartre agree not only that desire is fundamental to human being but that desire is basically sexual. Does such a view jibe with the world of the infant-child? Or is such a view *sexualizing* the infant-child and everything in its world? If so, then the view is adultist since the infant itself has not been consulted but rather has had the view foisted upon it from on high. Psychoanalytically speaking, one might say that the characterization of the infant is likely a characterization deriving from the psychoanalyst's own problems; the first instance of projection in psychoanalysis could be said to be Freud's reduction of human psychological development—and of all psychological problems as seen in neurotics and psychotics—to infantile sexuality. Indeed, his unwavering insistence that all psychoanalysts uphold the psychoanalytic truth of infantile sexuality without reserve and without modification strongly suggests that an understanding of his own earlier psychological life was at stake.

Certainly Lacan's idea, and quite possibly Sartre's too, that sexual desire is the motor of all human activity—not only behaviors but thoughts, dreams, and so on—undoubtedly derives in good measure from Freud's theory of infantile sexuality. Lacan's phallus may in fact be regarded an attempt to give firm and lasting linguistic anchorage to what might otherwise be considered empirical weaknesses in Freud's theory. Indeed, Lacan's phallus holds together the whole. It is the *point de caption par excellence,*[148] the meeting point of all the forces and strains in the fabric of Freudian psychoanalysis. Freud's theory of infantile sexuality was, after all, strongly contested even in his own lifetime. It was at the heart of his dispute with Jung, for example, and a primary cause of their ultimate separation.[149] Freud's theory is also of course strongly contested by infant research over the last twenty years, as we shall see in more detail in the following chapter. Consultation of this literature shows that much of what is and has been consistently maintained as factual psychoanalytic information about infants is in fact dogma, beginning with the notion of a helpless, uncoordinated piece of protoplasm, and ending with the notion of an undifferentiated existence (with one's mother). We can see this dogma clearly revealed in

Lacan's psychoanalytic. The infant is a powerless, inept chunk of matter that is somehow in blissful union with its mother. In finer terms, it is differentially dissolved in language, and when it comes out of solution, it is as fragments of its former self, all the way from a *corps morcelé* to the bodily slits and holes which mark its sites of desire and lack. Close examination of the infant body in Lacanian theory will provide a justly rich context for examining the case against prevailing dogma. In undertaking this examination in the next chapter, we will set in relief an archetypal view of the infant that is sweepingly negative, that ignores the ontogenetic roots of power, and that owes its formidable strength to a privileging of verbal language. In the course of the examination we shall have the occasion to compare Lacan's archetypal infant with Jung's, and thereby set the stage for a further critical examination: that of the unconscious within the Lacanian psychoanalytic.

Before proceeding to these critical examinations, an interesting postscript should be added concerning the way in which Sartre not only influenced Lacan but may have even inspired him. The penultimate part of *Being and Nothingness* that culminates in a psycho-ontology is a propaedeutic for an existential psychoanalysis. After showing the similarities between empirical psychoanalysis and his envisioned existential psychoanalysis, Sartre moves on to discuss the fundamental differences between the two and begins to sketch the direction in which a bona fide existential analysis would lie. In contrast to empirical psychology, existential psychology would not "cause [the subject] to attain *consciousness* of what he is"; it would cause him "to attain *knowledge* of what he is"[150]—precisely as Lacan would over. Sartre goes on to declare that existential psychoanalysis "is a method destined to bring to light, in a strictly objective form, the subjective choice by which each living person makes himself a person."[151] Though Lacan would eschew the idea of subjective choice, he would most certainly agree to a strictly objective psychoanalytic. As Sartre specifically envisions it, this psychoanalytic would be a study of being and the relations of being. In particular, it would direct itself toward a comprehension of "being and the mode of being of the being confronting this being." Within an existential psychoanalysis, in other words, an understanding of the analyst as well as the analysand is mandatory. The investigator himself must be comprehended in the analytic, Sartre says, "inasmuch as he is himself a human reality."[152] This necessity is, of course, exactly what Lacan's psychoanalytic emphasizes, and not only over and over, but in controversially innovative ways in the context of the transference relationship. The role of the analyst is in fact *decisive* in Lacanian psychoanalysis.

The striking resemblances between Sartre's and Lacan's psychoanalytic formulations go even further. Sartre states that insofar as the whole aim of the psychoanalytic is "to discover being," and since "it is forbidden to stop before attaining this goal," what the psychoanalytic method must do is "to detach being from its symbolic expressions . . . [in order to] rediscover each time on the basis of a comparative study of acts and attitudes, *a symbol destined to decipher them*" (italics added).[153] Clearly Lacan's psychoanalytic theory radicalizes Sartre's; it pushes it to its furthest extreme. Lacan's signifier of signifiers, in all of its algebraic and graphic guises, deciphers *everything;* it confers ultimate meaning on all acts and attitudes. A similar key coincidence is moreover apparent with respect to the Sartrean subject who is reduced to docile compliant. Sartre writes not only that "The criterion of success [of the psychoanalytic] will be the number of facts which its hypothesis permits it to explain and to unify" but that "[t]o this criterion will be added in all cases *where it is possible,* the decisive testimony of the subject" (italics added).[154] Here again, there is a remarkable resemblance to Lacan's psychoanalytic. In Lacanian analysis, the subject is secondary to the theory which explains him/her. In this sense, the subject bows to established formulae, and his or her "decisive testimony" is in essence similarly superfluous—a mere postscript—to the scientific success of the psychoanalytic.

Put in future Lacanian perspective, Sartre's visions and affirmations are rich in significance. As a final validation of Lacan's idiosyncratic realization of Sartre's psychoanalytic, it should finally be noted that in completing his sketch of existential psychoanalysis, Sartre emphasizes the importance of its possibilities and actual realization. In the course of doing so, he remarks that, "This psychoanalysis has not yet found its Freud."[155] While surely Lacan did not take up existential psychoanalysis as Sartre envisioned it, just as surely he could not fail to read between the lines. Cloaking himself in Freudian robes, he takes up the Sartrean challenge as it were and makes it his own. Indeed, as Benvenuto and Kennedy point out, "One often has the impression that Lacan considered himself to be the only original analytic thinker since Freud."[156]

8

Corporeal Archetypes and Power: A Critical Examination of Lacan's Psychoanalytic of the Infant-Child

We have shown that babies perceive a distinction between self and the visual spatial environment before they have independent locomotion, that they are aware that others have 'points of view', that basic intersensory relations exist from birth, and that babies can make use of contextual cues and spatial landmarks to link successive visual events and to ensure coherence and continuity of experience. . . . An intensive period of research over the last 15 years has had the effect of changing the stereotype of the young baby from one of 'incompetence' to one of 'competence'.

George Butterworth[1]

Between eight and ten months William was making a good start at self-sufficiency. . . . His forefinger-thumb coordination had enabled him for some time to feed himself pieces of bread and other food skillfully; he could with great fascination open and close a doll's eye with his forefinger. At nine months he was a real explorer with curiosity and intrusiveness and a special mechanical bent. He made wheels spin and examined various protuberances—light switches, radio and phonograph knobs. He pulled out drawers, turned on radios, flushed toilets, and unscrewed tops of jars.

James A. Kleeman[2]

I.   Introduction

Though he would efface the living body, Lacan's thinking is modeled on it. His psychoanalytic rests upon a tripod base, each of whose three supports is a corporeal archetype: the corporeal archetype of the infant-child, of the unconscious, and of the phallus. The whole conceptual infrastructure of the Lacanian psychoanalytic can be shown to derive from the particular positive or negative value that Lacan gives the corporeal archetype and that he takes to be its full and only weighting. The value he accords the infant archetype is *wholly* negative; that of the unconscious *wholly* positive; and that of the phallus *wholly* positive. There are in fact no greys in Lacan's psychoanalytic. Everything is presented as being either black or white. Moreover everything that is positive exists (in Lacanian theory at least) in the full brightness of light that is language; everything that is negative is a deficiency. In theoretical practice, this means that whether negative or positive, the archetype is exaggerated out of all proportion. In consequence, archetypal meanings consistently exceed the bounds of the real. They float off instead into the heady realm of algebraic formulae, erudite allusions, linguistic analyses, and *linguisterie*.[3] This indeed explains in a further way why in the Lacanian psychoanalytic there is no recognition of affect or of preverbal phenomena and, above all, no descriptive accounts of experience—and, correlatively, no introspection. What we will consider in this and the following chapters is how, in addition to charging Lacan with the neglect of such phenomena, a case can also be made against Lacan's psychoanalytic from the inside, that is, from the perspective of the corporeal archetypes that structure his psychoanalytic but that lie hidden within it. What is at stake in this enterprise is an account of *the psychoanalytic roots of power,* which is true not only to the truths of animate form but to the possibilities of human experience. In this chapter and those following, we will accordingly examine the facts supporting the thesis proposed above, that the whole of Lacan's psychoanalytic is based on three corporeal archetypes. That infant-child, unconscious, and phallus are fundamentally corporeal archetypes in Lacan's psychoanalytic and not just assumed to be such will be both implicitly and explicitly demonstrated in the critical discussions of Lacan's work that follow.

II. The *Real* Infant Living Body

Lacan's presentation of the infant as powerless articulates a cultural translation of a corporeal archetype, but one that is, in addition, individually honed and highly idiosyncratic. Lacan's reworking of the archetype is tied to a conception of language as the absolute locus of power within the whole of human life. Insofar as infants do not speak, they are necessarily powerless. Because infants must be accounted for within the psychoanalytic, however, their "words" must be inserted for them. But in inscribing them in this way into his psychoanalytic, Lacan at the same time silences them. He bars them in the same way that he bars the living body and biology or nature. Infant, body, nature—all are in fact part of the same piece of cloth. They are all *outside* language. None is examined or studied in its own right because, categorically, none warrants examination. The infant, however, is different in the sense that it has specific psycho-archetypal ties. It is the harbinger of the future, and it is thoroughly dependent on others for sustenance, for example. What cultures (and individuals) do, as we have seen from earlier chapters, is to rework corporeal archetypes in ways congenial to their interests and values. The infant or child in our culture is typically *undervalued* because it does not speak; only adults speak. In turn, an infant or child does not *know;* only adults *know.* Until a child accedes to language, it has no value. Such, in brief, is the Lacanian infant.

The task of this chapter is to flesh out this infant in Lacan's own terms and at the same time subject that infant to detailed and critical appraisals in the light of empirical psychological studies. In order to pave the way for language in the sweeping way that Lacan conceives it, the roots of power in an ontogenetic sense must be effaced. The surest and quickest way to do this is to render animate form unintegrated, ineffective, and inconsequential from the beginning. Accordingly, Lacan describes the neonatal months of infancy as a time of "primordial Discord," which is "betrayed" by "signs of uneasiness and motor unco-ordination."[4] At an earlier point, he speaks of man's "organic insufficiency in his natural reality—in so far as any meaning can be given to the word 'nature'."[5] At a later point, he speaks of the infant's "fragmented body," which he associates with "a certain level of aggressive disintegration" in the adult individual who is undergoing psychoanalysis and who dreams of "disjointed limbs, or of those organs represented in exoscopy, growing wings and taking up arms for intestinal persecutions."[6] Lacan in fact views the infant up to six months of age as being still a fetus—even a monstrous one, given its "disjointedness." He believes the infant's anatomical incompleteness "confirms" as fact his "view" that in humans, there is a *"specific prematurity of birth."*[7]

Critics and proponents of Lacan alike dutifully record his pronouncements on the "fragmented and uncoordinated" infant from birth to six
months of age without the least critical notice. It is important to emphasize
this fact. Neither critic nor proponent apparently finds anything to contest
in Lacan's rendering of neonatal life as unintegrated, ineffective, and
inconsequential. Philospher Maurice Merleau-Ponty is among those who
accept Lacan's account of infancy without question. He accepts the notion
that infant life is a "confused reality"—"strongly felt," he says, "but
confused."[8] What is perplexing and in fact illogical is that given this state of
infant reality, why would anyone speak of the subsequent experience of the
infant seeing itself in the mirror as being an *alienating* experience,[9] both in
itself and as a prelude to an alienation by others. Personalizing this
experience of the infant, Merleau-Ponty describes it as an experience that
"tear[s] me away from my immediate inwardness."[10] The phrase "my immediate inwardness" appears honorific, a special and even treasured kind of
awareness; yet it is in fact, as Lacan tells us and as Merleau-Ponty affirms, a
thoroughly distressed organic state.[11] How then could one possibly find it
excruciating—as in a *tearing away*—to be alienated from it? On the
contrary, would it not seem a relief to the infant to get away from its
"immediate inwardness"?

Two further points, one pertaining to Lacan's "fragmented body" stage
of infancy and one pertaining to his "mirror" stage of infancy (both of them
glossed by Merleau-Ponty), will demonstrate in a further way why, in the
interests of a credible psychoanalytic of infancy, taking Lacan's infant on
faith is foolhardy and reviewing empirical findings on infancy is altogether
essential.

The first point has to do with the idea that "The child in no way
distinguishes at first between what is furnished by introception and what
comes from external perception."[12] When Merleau-Ponty makes this comment, he is in the process of examining the same material Lacan has
examined in forging his mirror stage of infant development. Like Lacan, he
too consults Henri Wallon's text, *Les origines du caractère chez l'enfant.*[13]
Like Lacan, he too, in summarizing Wallon's thesis, takes the infant's lack of
differentiation as fact. But, as suggested above, Merleau-Ponty also draws on
Lacan's own psychoanalytic in detailing and highlighting aspects of the
confused and undifferentiated state in which the infant lives. Now one of
the most interesting observational and experimental findings that confutes
the notion that an infant fails to distinguish between introception and
exteroception concerns a pair of Siamese twins ("Xiphophagus conjoint
twins") whose finger-sucking habits were studied prior to the twin's
separation at approximately four months. The twins were ventrally joined

but had entirely separate organs and nervous systems. In the several experiments performed, it was clear from movement patterns that each twin knew when its *own* fingers were being sucked and when "other" fingers were being sucked. In particular, there was bodily resistance to the attempted removal of the fingers if the twin's *own* fingers were being sucked; there was no resistance, but there was a straining forward of the head in pursuit of the withdrawing fingers if the fingers of the twin's sibling were being sucked.[14]

Granted that such studies were not available to Lacan or to Merleau-Ponty, the paradigm of an infant as a totally inept, nondiscriminating, uncoordinated, unintelligent piece of protoplasm is as erroneous as it is precipitous. The availability of studies notwithstanding, the paradigm in this instance rests on evidence that is flimsy in the extreme. The evidence—a single observation of an infant being held in front of a mirror for the first time by its father—is indeed hardly evidence at all. It is more fittingly characterized as *adultist,* perhaps even *sexist-adultist,* a rendering from on high of what it must be like to be both new to the world and capable of being surprised by the nature of things in the world. A body that meets with something new, that is caught off-guard, as it were, is perceived as simply not in command. *It has no power.* Thus we have initially a body whose "immediate inwardness" is "a confused reality." The thrust of the interpretation is that until a body learns the ropes, and until it is in fact capable of articulating those ropes, such a body is no-body at all.

The second point concerns the experience of the infant in front of the mirror: what is the infant experiencing? Again, Merleau-Ponty, like Lacan, takes up Wallon's notion. Although he faults Wallon's explanation as not wholly convincing, that is, although he emphasizes that it is the image and not the mirror that is of moment for the child since that image "is mysteriously inhabited by me," he passes over the significance of the infant's—and, as he reports, a chimpanzee's—reaching *behind* the mirror. What he overlooks is the fact that the infant and chimpanzee for as long as each has lived, has felt the three-dimensional *form* of its own body, a three-dimensional *moving* form at that. Moreover, any object each has ever handled has been three-dimensional. In effect, given its own bodily experiences and its experiences of things in the world, neither the infant nor the chimpanzee would readily understand how things in the world can be two-dimensional. The concept of the three-dimensionality of things is a concept modeled on the infant's and chimpanzee's own body. What infant and chimpanzee alike must understand if they are to understand mirrors "correctly," that is, *as human adults understand mirrors,* is that in seeing themselves in a mirror, they are seeing an illusion: what *looks* solid may not *be* solid; what one typically assumes three-dimensional may be two-dimensional. Infants and chimpanzees learn this in part precisely by reaching behind the mirror and discovering nothing there. The act of reaching is from this perspective an astute act, an intelligent act; *it is an attempt to validate the visible.*

There is an even further dimension of this corporeal intelligence. An infant's first experience of itself before a mirror and its attendant so-called "jubilation" (Lacan's term)[15] may well concern not the joy of seeing itself as a "totality" or as an "Ideal-I"[16] relative to a putatively impoverished, fragmented body, but the joy engendered in the sheer experience of discovering what it looks like, of discovering that it has a "face," for example. What has been a felt moving reality since birth, the site of untold gestures from grimaces to furrowings, pinchings, and so on, has a visual form. No wonder there is fascination: all this felt movement has a visual semblance. Had the infant words, it might exclaim, "I have a face!"[17] Indeed, an infant of eight months—the age at which an infant is assumed by Lacan to see itself in the mirror for the first time—in all likelihood never gave a thought to the possibility of seeing its own face. No doubt too, it literally never dreamed of or imagined its own face. Even if it had so dreamed—or so imagined—it would be astonished to see what it actually looked like precisely because it would experience that actuality for the first time. The Zen koan, what was your face before you were born?, is perhaps tangentially topical to the point at issue. There is no empirical answer either to the question of what one as an eight-month-old infant thinks it looks like, or to the question of what one's face was before one was born. If an eight-month-old infant could speak, and if it were asked, what do you look like?, it undoubtedly would answer, "I've not the slightest idea—beyond perhaps affirming that I have two eyes, a nose, and a mouth just like others; but understand, please, that these are things I can feel and move. *What I look like*—well, I've not the slightest idea!"

Given this initial sketch and critique of the Lacanian infant, it is apparent that Lacan turns the archetypal helplessness of infancy into an archetypal state of psychological distress, and then turns the infant's subsequent experience of its visual body into an "Ideal-I" that utterly swamps its archetypally distressed tactile-kinesthetic body. Let us compare this rendering of infant life with the research of infant-child psychiatrist Daniel Stern whose psychiatric clinical work is complemented by a developmental perspective informed by experimental research, including his own. As Stern himself observes, the two infants "constructed" by developmental psychologists on the one hand and infant-child psychiatrists on the other hand must be brought together: "the clinical infant breathes subjective life into the observed infant, while the observed infant points toward the general theories upon which one can build the inferred subjective life of the clinical infant."[18] In short, the clinical and observational compliment each other, the one providing features that the other lacks. In a section of his book, *The Interpersonal World of the Infant*—"Observing the Young Infant: A Revolution in Infancy Research"—Stern highlights many of the recent findings that have shed light on infant experience. For example, infants three days old recognize the smell of their own mother's milk; infants trained to correlate sucking with certain happenings control what they see or hear by maintaining a certain rate of sucking; infants can discriminate vertical symmetry from horizontal symmetry.[19] Stern gives four general principles of infant affect-cognition-perception which are or may be summarized as follows: 1. "Infants *seek* sensory stimulation" (italics added). 2. With respect to the kind of stimulation they seek, they have certain preferences that are innate. 3. Infants evaluate as well as discriminate, determining, for example, what features of the experience are invariant. 4. "Affective and cognitive processes cannot be readily separated."[20]

Studies that validate these principles validate certain other principles of perception, for example, the fact that infants are able to transfer information from one sensory modality to another. Infants perceive amodally or synaesthetically. In one well-known experiment involving two kinds of pacifiers, one with a spherical-shaped nipple and the other with a nubbed nipple, three-week-old infants who were initially blindfolded later distinguished visually the nipple they had previously sucked while blindfolded. Another experiment showed that infants perceive the correspondence between a sound heard and the articulatory gesture seen in conjunction with the sound. In other words, they correctly associate the sound with "the right face." Other studies concerned with affect have shown that infants only two days of age "would reliably imitate an adult model who either smiled, frowned, or showed a surprise face."[21] Still other studies have shown that twelve- to twenty-one-day-old infants protrude their tongues in imitation of an adult model.[22] The data strongly suggest that they even correct their imitations; that is, the infants' imitations involve "active intermodal matching."[23] Stern comments that the strongest inference that can be drawn from the latter studies is that there is an innate correspondence between what infants see and what they do. But he also points out that studies in affect raise the question of whether the infant is imitating or whether the response is "reflex-like."[24] Even were the infant's response deemed "reflex-like," however, it would still constitute an innate disposition and capacity for imitation.

Stern furthermore describes how physiognomic perception—affective aspects of tone, color, smell, and so on—and "vitality affects"—kinetic aspects of perceptual experience—play a central role in an infant's interpersonal world. Together and separately, these perceptual capacities call

Piagetian theory into question. Infant psychologist Andrew Meltzoff succinctly summarizes this questioning in proposing that we change our viewpoint on the infant: "Sensory-motor behaviour is not 'internalized' to give birth to the infant's representation capacities at eighteen to twenty-four months of age. Quite the contrary. The ability to act on the basis of abstract representations or descriptions of perceptually absent events needs to be considered as the starting point of infant development, not its culmination."[25] These perceptual capacities furthermore call upon us to acknowledge the fact that in the everyday world things appear to us in multi-sensory ways. In our experience of eating, for example, affective and kinetic aspects are intertwined. Acknowledging actual experience, we might realize how as adults we project backward onto infants what we have been taught and in so doing fragment not only ourselves but the world of the infant. I have noted elsewhere how this textbook-inspired fragmentation conflicts with the truths of experience.[26] In actual fact, unless we are intentionally and intently so concentrated, we do not perceive in discrete sensory modes. Whatever textbooks may tell us in separate chapters of visual perception, of auditory perception, of olfaction, of taste perception, and of tactility—and of the latter they tell us precious little, and of kinesthesia, they consistently tell us less than nothing—in the everyday world, we do not sense things in these categorically partitioned ways. What Stern, Meltzoff, and other psychologists have discovered and demonstrated by way of clinical and experimental evidence is that infants certainly do not.[27]

Studies of infant social experience complement studies of how infants perceive the world. In particular, the idea that infants fail to distinguish a human from any other item in their environment is belied by evidence that shows that, rather than experiencing a human form—whether face, voice, or breast, for example—simply as a physical stimulus, infants "experience persons as unique forms from the start."[28] Stern notes that the evidence that demonstrates the capacity of infants to discriminate humans from other things in the environment is multi-faceted; that is, the evidence comes from a wide variety of studies. One study shows, for example, that by one month of age, "infants . . . show appreciation of more global (nonfeatural) aspects of the human face such as animation, complexity, and even configuration." Another study shows that infants "gaze differently when scanning live faces than when viewing geometric forms. They are less captured by single featural elements and scan more fluidly during these first months." Interestingly enough, still another study shows that "When scanning live faces, newborns act differently than when scanning inanimate patterns. They move their arms and legs and open and close their hands and feet in smoother, more regulated, less jerky cycles of movement. They also emit more vocalizations."[29] In short, when it comes to social relations, infants do not start wholly from scratch as in Piaget's construct of infant perception; they bring their own innate social abilities with them into the world.

When Stern comes to consider "The Sense of a Core Self," specifically, the dimension of "Self versus Other," he directly addresses the important, basic question of whether infants live in an undifferentiated state of oneness with the mother until the end of their first year, or whether, on the contrary, they differentiate "a sense of self and other."[30] He states to begin with that "At the age of two to three months, infants begin to give the impression of being quite different persons"; at that age, they interact with others with an intact sense of themselves.[31] He goes on to challenge the opposing, long-entrenched received view that infants are wrapped in a state of oneness with the mother. He points out that "recent findings about infants . . . support the view that the infant's first order of business, in creating an interpersonal world, is to form the sense of a core self and core others. The evidence also supports the notion that this task is largely accomplished during the period between two and seven months."[32] What is furthermore enlightening is Stern's own view that the experience of fusion is actually *developmentally dependent* upon a prior sense of self: "First comes the formation of self and other, and only then is the sense of merger-like experiences possible."[33] On the basis of research findings, his own as well as those of others, Stern theorizes that the initial sense of self is formed on the basis of *self-agency, self-coherence, self-affectivity, self-history.* Together, these experiences of the self form the "core self." Although Stern does not explicitly say so, these experiences are all *corporeally* rooted. Self-agency specifies awareness of oneself not only as the author of actions— "your arm moves when you want it to"—but as a subject expecting certain consequences—"when you shut your eyes it gets dark."[34] Self-coherence specifies "a sense of being *a nonfragmented, physical whole with boundaries and a locus of integrated action, both while moving (behaving) and when still*" (italics added). Self-affectivity specifies an awareness of "inner qualities of feeling . . . that belong with other experiences of self." Finally, self-history means an awareness of oneself as enduring, continuous with one's own past, as well as an awareness of "regularities in the flow of events."[35]

A core self obviously constitutes an empirical challenge to both a *corps morcelé* first stage of infancy and a later "Ideal-I" mirror stage. It challenges as well those who claim that "I" and "not I" are undifferentiated by the infant in its first six months. Psychologist Jane Flax, for example, following the theoretical stand of object-relations psychiatrist D. W. Winnicott, states that "Neonates do not yet have a firm sense of their own body boundaries. I and not I are not yet fully differentiated, and inside and outside the self are only gradually distinguished."[36] If we interpret Flax's statement—and

Winnicott's claim—in sensory-kinetic terms, it becomes immediately clear why in a quite literal sense "neonates do not yet have a firm sense of their own body boundaries." Touch—like movement—is a self-reflexive sense. We adults have a firm sense of our own body boundaries not because of brain maturation but because of our *visual* experiences. In other words, we do not have a firm sense of our body boundaries in our everyday tactile contact with things in the world any more than an infant does. When we pick up a glass, we do not know *tactilely* in any precise sense where the glass ends and our hand begins. The space of each is "undifferentiated," i.e., unclear. The same may be said of the experience of bumping into another person. Where in the bump does my body begin and the other person's body end? "Object-relations" in the realm of tactility are by their very nature vague. Husserl long ago noted and examined the peculiar, dual touching/touched nature of tactility. Adults do not outgrow this tactile duality. They simply do not normally pay attention to it. Whether practically or sensuously oriented—steering a car or rubbing one's hand along a velvet cloth—the act of touching for adults is typically unilluminated by an awareness of the fundamentally undifferentiated self/other reflexive nature of touch.

For an infant, on the other hand, that nature *is* touch. Stern points out that four-and-one-half-month-old infants live each experience as it is in the moment, not as it is connected with other experiences.[37] Their tactile experience of their own body boundaries is accordingly not likely deficient in the least. It is simply in the beginning unilluminated by vision, i.e., by perceptions of themselves as visual bodies, and in turn not yet elaborated by visual experience. When we look to sensory-kinetic aspects of experience, we begin to understand not only why a neonate does not have any firm sense of its own body boundaries but how we, as adults, have come to build up our notion of our own body boundaries. Our sense of touch has not changed; *it* has not "improved"; and *we* have not become more discriminating in a tactile sense. On the contrary, we (in Western society at least) have only come to override what is there in tactile-kinesthetic perception by setting ourselves off as bodies—as predominantly visual objects—apart from the world. This visually based separation is undergirded by at least two fundamental developmental aspects. We come to individuate ourselves by *seeing* other bodies as separate from our own and by *kinetically establishing ourselves as separate from others,* that is, by standing and moving by ourselves. The mastery of these latter acts categorically sets us apart. In standing up and moving by ourselves, we claim a space entirely our own, a space independent of others. Our repertoire of 'I can's' is radically augmented as are our potential powers.

In light of all of the above evidence, it hardly needs emphasizing that the helplessness of an infant has been erroneously equated to a state of idiocy by many in Western culture, to a period of cognitively degraded being, and this because the infant does not yet speak, let alone stand upright by itself. Philosopher Daniel Dennett's remarks are particularly memorable in this respect. In answer to the question, what is it like to be a human infant?, he answers, "My killjoy answer would be that it isn't like very much. How do I know? I don't 'know,' of course, but my even more killjoy answer is that on my view of consciousness, it arises when there is work for it to do, and the preeminent work of consciousness is dependent on sophisticated language-using activities."[38] Held up to the likes of Stern's observed-clinical infant, Dennett's human infant merely typifies arrogant, adultist, and all too prevalent Western misconceptions of infancy. The Lacanian infant is paradigmatic of this Western-received disdain, but it is also, in addition, a work of the imagination *par excellence,* imagination in precisely the sense Lacan omits in speaking of the Imaginary. His rendition of infancy is based on the imagined marriage of infant neurology and adult psychopathological dreams, neither of which are related to studies of actual intact infants in the throes and pleasures of actual life. The four fundamental experiences which Stern identifies as core experiences of self are in contrast based on experimental and clinical data. They all are grounded in the living body. They all articulate, explicitly and implicitly, a recognition of what it is like to be an infant body. Indeed, as noted, at their deepest level, they are all constituted on the basis of tactile-kinesthetic experience. One finds nothing of such a body or of such experiences in Lacan's psychoanalytic. There is in fact no way Lacan's psychoanalytic infant could possibly be melded with or onto Stern's observed-clinical infant.

A particularly significant dimension of the disaccord turns on the perceptual experiences and acuity of infants, a topic about which Stern writes at length in *Diary of a Baby.* An infant's attention, Stern observes, is riveted by sensuous intensities. It may be magnetized by a patch of sunlight on a wall or by the place two contiguous forms come together—a forehead and tufts of hair, for instance. The sunlit patch is a perceptual experience unmediated by notions of illusion and reality; what is experienced is simply a vibrant play of light, a moving dynamic that changes shape and color. All infant perceptions are like this, Stern writes. "There are no 'dead', inanimate objects out there. There are only different forces at play."[39] The infant Stern is describing is six weeks old. S/he is capable of distinguishing what is near from what is far.[40] S/he is furthermore capable of transferring sense from one modality to another, as described earlier with respect to amodal perception. Not only is synaesthesia a basic faculty,[41] but so also is a sensitivity to movement. "All babies are extremely sensitive to things that move."[42] In this context, Stern describes an infant looking at the distinctive shape which is the bar on its crib, then at a second distinctive shape like the first. He describes these perceptions as harmonious, qualitative configurations, as spatial frequencies. He describes then how suddenly "a different note" sounds and how, "like a shooting star, it flashes past and quickly disappears."[43] He explains that when an infant is intensely engaged in something, its arm may suddenly swing forward, enter the visual field, and as quickly disappear. The "different note," Stern writes, is the beginning of another integration. The hand that the infant sees moving comes to be known as the same as both the one that it feels moving and the one that it intends to move. "Out of such perceptions, the infant constructs a unified world made up of many different kinds of events."[44]

What is emphatically set in relief in Stern's renderings of infant life is an infant's raptness and competencies in face of its creative task. The *powers* that an infant brings to the world are indeed prodigious. They are rooted in precisely those kinesthesias Husserl describes in terms of *if/then* relationships and '*I can's.*' Stern's observations in fact conclusively corroborate Husserl's descriptive analyses of fundamental human experiences. Stern writes, for example, that even a six-week old infant anticipates consequences. It begins "to form expectations of things to come."[45] Moreover, a bodily integrity, an intentional cohesiveness is apparent. A hungry infant, for example, caught up in the swelling and collapsing waves of its "hunger storm,"[46] is intentionally all of a piece. Stern's account of infant experience decisively invalidates the fragmented body Lacan ascribes to infants.

In sum, there is an obvious and substantial mismatch between Stern's descriptive account of infancy and Lacan's theoretical one. Indeed, the disjunction is so great it would seem that the two psychiatrists are describing two different species of creatures—one explorative, perceptive, a center of powers; the other thoroughly incapable and fundamentally inept. The body of the one is now active, now drowsy, now alert though non-active, now sleeping; the body of the other, initially floundering in its own "turbulent movement," is then jarred by the experience of seeing its ideal body in a mirror and in turn alienated from itself.

Before bringing the radical incompatibility between a Lacanian and Sternian infant into finer focus with respect to the question of power, it will be instructive first to take a brief but closer look at the received idea that infants live in an undifferentiated state, i.e., a state in which they are merged as one with their mothers. The belief that they are so merged obviously affects one's conception of the roots of power.

Many experimental psychologists support Stern's conclusion that infant capabilities have been undersold and that an initial mergence-with-mother does not mark the beginning months of infant life. Essays in *Advances in Infant Research,* for example, testify to the fact that infants are capable of an impressive array of sensory coordinations and that they experience themselves as separate from their environment. George Butterworth, for example, notes that research has shown that newborn infants "will look for the source of a sound . . . [and] perceive basic correspondences between something touched and something seen. . . . [T]hey perceive the distinction between 'self' and 'environment' before they are capable of independent locomotion, and they are aware that others have a different point of view and that an object can be the focus of shared visual attention. Babies of 8 months can be shown to make use of contextual cues and spatial landmarks to link successive visual events and thus ensure coherence and continuity of experience."[47] Butterworth's review and critique of Piagetian constructionist models, which aver that "there is no information *in the structure of sensory stimulation itself,*"[48] show how the models are confuted by an abundance of research evidence. Some of this research is based on the work of psychologist J. J. Gibson who showed that basic knowledge of space and objects is given in perception itself.[49] In a constructionist model, in contrast, knowledge of space and objects is precisely *constructed from scratch* by a know-nothing infant. In consequence, "the infant must be presented as having an extremely limited and chaotic awareness of self and others, precisely because it has not yet had the opportunity to construct the necessary concepts for social relations to be possible. The world is experienced as an extension of the infant's own actions and the baby is considered totally egocentric and unaware that there exist points of view other than its own."[50]

One of many interesting and innovative experiments which conclusively contest the letter this notion demonstrates that "babies from about 2 months reorientate their own line of gaze when an adult in the course of interacting with the baby changes the focus of her attention."[51] The baby's focus of attention, in other words, changes to follow the mother's. Butterworth summarizes findings (from work done in the 1970s and 80s) on the question of infant "egocentricity" by saying that "even very young infants . . . engage in pre-verbal communication through the intermediary of a world of permanent objects that they hold in common with adults."[52] Concluding a later section on intersensory coordination, and answering those who characterize the infant's sensory world as simply a welter of stimuli, Butterworth remarks that "the pre-existing relationships among the senses may be better characterized as harmonious and mutually supportive, as if the early sensory functioning of the infant acts to ensure coherence, rather than confusion of experience."[53] In summation of research findings as a whole, he states that "an intensive period of research over the last 15 years [approximately since the late 1960s] has had the effect of changing the stereotype of the young baby from one of 'incompetence' to one of 'competence'."[54]

III.   The Radical Incompatibility in Finer Focus: A Question of Power

Combined with Stern's writings, research such as that summarized above makes amply clear from yet a further perspective why Lacan must seal off the living body from his psychoanalytic. A living body is too alive with knowledge, with affect, with nonverbal understandings and capabilities. It is too alive to itself, to the world, and to a growing repertoire of powers. It must indeed be shut up. The peremptory silencing effectively clamps shut precisely that "structure of sensory stimulation" which is the correlate of animate form, the template of all our initial knowledge of the world, the source of our affective life, the core of our being, the core in each of the senses Stern enumerates: agency, coherence, affectivity, history. It silences tactile-kinesthetic life. The body the infant affectively feels, for example, is a body that is always there. It is not a fragmented part of its world any more than it is an illusion. Clearly, any time an infant—or an adult for that matter—cares to attend to its body, there it is. Not only this, but any time an infant—or an adult —*moves* and attends to the movement, in stretching, for example, or reaching, or positioning itself to do any act or engage in any activity, whether purposeful or playful, a sense of wholeness pervades. The movement *is* the body; the body *is* the movement. Where kinesthesia is peculiarly lacking in renditions of bodily life, so also is a veridical account of the body. Kinesthesia is not an arbitrary dimension of existence but its very core. Stern describes kinesthesia—proprioception—as "a pervasive reality of self-action whether the action is initiated by self or passively manipulated by another."[55] We are aware of movement, in other words, from our earliest days.[56] Our potentiality for movement and our capacity to sense movement is indeed what makes us *animate* forms. To ignore kinesthesia and tactility is in fact to ignore those very experiential dimensions of self that Stern identifies: the experience of moving and of knowing the consequences of moving; the experience of feeling oneself all-of-a-piece—"a locus of integrated action," even if lying still; the experience of feeling in an affective sense; the experience of being now, then, and in the future. These experiences are felt bodily realities. Lacan's *corps morcelé* bears no resemblance to these living realities because it bears no resemblance to animate form. In turn, it denies the ontogenetical roots of power. It conjures up a corporeal archetype bearing no resemblance to the core realities of infant life.

It bears no resemblance furthermore to those aspects of social living in which attachment and exploration are dominant. Attachment and exploration are central to primate social living in general. Because they are, it is not surprising that behaviors connected with them can be identified as intercorporeal primate patterns. Being held, being hugged, being curious, being reassured—all are behaviors that are emotionally bound up with the sensuous, an intercorporeal sensuous. Stern writes that "The ultimate magic of attachment is touch" and in fact points out in passing that this ultimate magic is part of our primate heritage.[57] Chest-to-chest contact with another, one's head resting upon the other's shoulder and neck, is a primate intercorporeal gesture of shared affection, of sympathetic relationship, of caring. It is indeed an intercorporeal primate archetype that defines a dimension of *power,* the power to express fondness and affection, the power to comfort. This dimension of power is rarely recognized in present-day Western culture. What is not a matter of control is not a matter of power. In this respect it is important to note that when everything to do with infant—not to mention adult—life is psychoanalytically funneled into sexuality, we lose sight of those fundamental powers of caring that are *not* sexual, that, in addition, are not social constructions or political inscriptions of one kind or another, and that, finally, are not tied in any way to a verbal language but are as much a part of our natural history as part of our ontogenetic heritage. Powers such as these are indeed part and parcel of an intercorporeal evolutionary semantics. They are fundamentally structured in intercorporeal archetypes—precisely as in hugging another body or in patting another's hand. These intercorporeal forms of attachment are clearly part of the world all of us have lived through (whether happily— "normally"—or not). They are part of our being the animate forms we are. Whether we actually had such positive experiences in our infancy and childhood is beside the point: *we know what they are.* We know that they have nothing to do with sexuality. They have rather to do with a positive sociality and caring, with an inter-camaraderie, with shared pleasures, including the sheer pleasure of being with another person whose company we treasure. Because such experiences fall outside the sexual but resonate with a sense of attachment, any psychoanalytic that bases the whole of its theory on the sexual not only misses "the ultimate magic of attachment" in touching another but fails utterly to understand it. As indicated in the prologue, "there are other ways to be human." Only in a phalli-crazed (and phallic-raised) society is preeminent value placed on control. Indeed, caring is an obstacle to control. While Ellis's study of rape discussed in chapter 5 makes the phallocratic relationship between control and lack of caring patently evident, chapter 10 will make explicit the bodily infrastructure of the relationship between control and power, a relationship that makes the archetype of the phallus not just possible but necessary.

Curiosity is a further testimonial both to infant powers and to the power of touch, movement, and affect. One need only consider how the felt wonderment "What is it like?" translates directly into powers of tactile-kinesthetic exploration and how what is later in a verbal world epitomized in "Why?" is in a preverbal world epitomized by sensory-kinetic investigations. Exploration demands attentiveness. The world is stopped as it were and all concentration is poured into the object of focus. The act of exploration by its very nature presupposes "a locus of integrated action." Moreover, in such an act there can be an appeal of movement and touch for their own sake—for the sheer pleasure in moving or making move, for the sheer pleasure of touching or being touched. In making such an appeal, movement and touch potentially call forth the fact and pleasure of our aliveness. What a sexually tethered psychoanalytic does is jettison such existential moments of aliveness—and jettison them wholly—in favor of symbolic moments of the sexual. In sexualizing life in this way, Lacan turns all thoughts, feelings, actions, and beliefs into that which can be read symbolically, that is, into signifiers. In effect, nothing is what it is. It is another thing. What is effaced in the process is bodily being—inner sensations of delight, for example, or of euphoria. When Stern writes of the inner sensations of delight that flood an infant, who, in exploring, finds a hidden toy, and when he describes these sensations as unfolding in time, as having a certain temporal dynamic (and, one could add, at times a certain spatial dynamic or center in the body) and thus a certain contour, he is describing the *real,* the real bodily felt experience of delight. When he further writes of a mother's capacity to attune herself spontaneously to her infant's delight such that she can mirror it in a vocalization, he is, again, speaking of an actual felt experience, one that in being understood and reflected back by another becomes an experience of fine-tuned intercorporeality. When this intercorporeal moment—or when moments like it—are read off symbolically in terms of infantile sexuality, the experience is dismembered. The infant's delight drains off into talk of desire, lack, absence, aggression, and, ultimately, the phallus.[58]

Lacan of course straightaway removes the possibility of delight or any other affect entering into his psychoanalytic, and in fact derisively denies the very idea that an infant is capable of affect at all since nothing is there in the way of a tactile subject to begin with. He does this in a particularly striking way in making a comparison between consciousness and affect. "Consciousness," he says, "is a feature as inadequate to ground the unconscious in its negation . . . as the affect is unsuited to play the role of the protopathic subject, since it is a service that has no holder" ("c'est un service qui n'y a pas de titulaire").[59] If affectivity resounds in emptiness, it is because, like consciousness, it is a groundless conceit; in the Lacanian psychoanalytic, affect, like consciousness, is by itself a baseless function. It is a "service" that may come with the package but that has no real place within it. In effect, both an affective and tactile corporeality and intercorporeality are categorically denied. An infant's social relations are cleansed of their felt dynamics, even as those dynamics might later enter expressively into the speech of psychoanalytic discourse. What is of moment is desire, sexual desire. Whatever interferes with articulating this "structural of the subject"[60] is decried in one way or another. Thus Lacan writes,

> What psychoanalysis shows us about desire in what might be called its most natural function, since on it depends the propagation of the species, is not only that it is subjected, in its agency, its appropriation, its normality, in short, to the accidents of the subject's history . . . but also that all this requires the co-operation of *structural elements,* which, in order to intervene, can do very well without these accidents, whose effects, so unharmonious, so unexpected, so difficult to reduce, certainly seem to leave to experience a remainder that drove Freud to admit that sexuality must bear the mark of some unnatural split[61] (italics added).

In light of the evidence Stern and others present, the remainder—or unnatural split—is the result of forcibly grafting sexuality onto the infant, and from head to toe. It is most certainly not the result of empirically discovering "a remainder" or "unnatural split" in an infant's behavior and experiences.[62]

IV. Infancy, Biology, and Intellectualization

By silencing the powers of the infant, Lacan effectively silences even more. To give the living body its due means, as suggested above, to give something to biology—to nature. The above quote makes clear that for Lacan this amounts to psychoanalytic heresy. Anything outside a language-derived and language-tied psychoanalytic is derided, even despised. Yet as David Macey comments with respect to Lacan's not infrequent usage of the biological, neurophysiological, and human sciences to bolster his theory, "Given Lacan's reputation for militant anti-biologism, there is something improbable, even uncomfortable, about this repeated appeal to developmental psychology, ethology and biology." Macey goes on to point out that when Lacan summons pigeons and locusts and asserts that research on these creatures "reveal[s] facts that prove the validity of [his] own insights into the importance of the mirror stage, . . . he places 'facts' within inverted commas, a mark, perhaps, of a certain epistemological embarrassment at having to fall back upon an empirical argument. The same illustration [appeal to research on pigeons and locusts] appears in [Lacan's] 'Some Reflections on the Ego', where the facts are shorn of their inverted commas and where Lacan adds that 'psychoanalytic experience substantiates in the most striking way the speculations of philosophy'."[63]

Lacan's distancing of himself from the biological except as it serves his needs coincides with his denigration of infancy. Just as there is certainly nothing *intellectual* about infants, there is nothing *intellectual* about biology. Indeed, biology appears to be as fetal a science as an infant is a fetal human specimen. Both are immature; both are intellectually distressed. What is truly intellectual comes from language. Nothing in Lacan's work suggests even remotely that language has any ties to biology. Except for his own general admission that symbols can tell us nothing about creation (recall the quotations and discussion in chapter 7), there is nothing in his work that comes close to acknowledging a provenience of language, neither in terms of its phylogenetic nor ontogenetic beginnings. This gap in his psychoanalytic is of considerable significance precisely insofar as Lacanian psychoanalytic theory hangs on language and insofar as, ontogenetically speaking and however preprogrammed in deep cortical structures, language is an acquired skill. This point and the issue it raises will be more specifically addressed below. What is of moment here is Lacan's failure to take into account the fact that language did not arrive *deus ex machina* on the human scene. Language was invented, and not in the last several thousand years, but at some time in our more distant phylogenetic past. Any theory— psychological or otherwise—which hangs the whole of human nature on language must deal with this fact, and this because although all humans are hominids, not all hominids are humans.[64] At some point in our evolutionary past, our nonhuman hominid ancestors created a verbal language. Accordingly, language itself has a past; it developed over time. Even supposing the Lacanian script were correct—that we are spoken by language—at what moment or period of time did this language that speaks us begin? More importantly *how* did it begin? Did a bevy of tongues one day begin wagging in concert? Clearly, there is a point at which psychoanalytic theory must make historical sense; it must accord with and even complement what we know about humans as evolved creatures. This is precisely what Lacan himself claims for his own theory; namely, that it accommodates the facts.

Lacan's preeminent concern with intellect and his idea that intellect has nothing whatsoever to do with biology are both equally transparent. At one point in a lecture when he indirectly, even coyly, suggests that what he is saying about analysis—that associations lead to "free speech"—is an indubitable truth, he asks whether the method used in analysis is "a progress towards truth." He immediately tempers this immodest claim by saying, "I can already hear the apprentices murmuring that I intellectualize analysis: though I am in the very act, I believe, of preserving the unsayable aspect of it."[65] The fact is that Lacan *is* intellectualizing analysis precisely in trying to preserve what he declares is unsayable about it, and this for the very reason that what he specifies as unsayable is a concocted scenario that has no descriptive anchor points, either in documentations from self-analysis or from case histories. The unsayable is validated only by what Lacan affirms becomes manifest in Lacanian analysis. The core problem as it relates to the silencing of the living infant body and of biology lies precisely here. Because regression is of such moment in Lacanian analysis, that is, because a return to the impoverished and alienated world of the infant is in fact the *sine qua non* of the psychoanalytic, the analysand must finally be led to articulate by way of the Symbolic both the infant-Imaginary and the infant-Real, i.e., the infant mirror stage and its earlier, original stage of a *corps morcelé*. Lacan's delineation of infant life is therefore critical. If his delineation is validated by his clinical techniques, then we should expect that delineation to be reflected—if even only obliquely reflected—both in the work of other psychoanalysts and in the research of developmental psychologists. As we have seen, however, it is not so reflected. On the contrary, infant life as rendered experimentally and clinically gives not the slightest evidence of a *corps morcelé*, a jarring mirror stage, or an entrance into the Symbolic where demand is transformed into desire. Lacanian psychoanalytic theory and clinical techniques are themselves intellectualizations. Lacanian theory is in fact a radically cerebralized version of that distancing and distanced verbal world that Stern describes in describing a child's entry into language.[66] Because it is so thoroughly cerebralized, its clinical *modus operandi* cannot even be said to be "all in the head"—or even "all in the tongue." Only the most tenuous threads of Lacan's imaginative "stories"[67] link the Symbolic to the Imaginary and the Imaginary to the Real. In consequence, anything remotely resembling a living infant body and biological life—what might be called "the natural and ontogenetic history of infants"—is at an intellectualized remove.

Intellectualization is actually elsewhere both explicitly and positively linked by Lacan to his psychoanalytic. In inveighing against "autonomous ego" psychoanalysts, Lacan writes that "I would like to say, to all those who are listening to me, how they can recognize bad psychoanalysts; this is by the word they use to deprecate all technical or theoretical research that carries forward the Freudian experience along its authentic lines. That word is 'intellectualization'—execrable to all those who, living in fear of being tried and found wanting by the wine of truth, spit on the bread of men, although their slaver can no longer have any effect other than that of leavening."[68] Clearly Lacan believes his analytic to be a monument to intellectual truth, at the pinnacle of a truly ratiocinative psychoanalytic. The "nucleus of our being,"[69] as Lacan tells us, lies not at all by way of nature; our bodies are simply an amalgam of so many material parts. The biological teaches us next to nothing. Any psychoanalytic worth its salt will accordingly proceed not by way of the body but by way of language. Lacan's priorities distinctly recall those of another renowned Frenchman whose three-hundred-year-old meditations are far from forgotten. Spelling out the connection will prove informative.

At one point in their annotated commentary on Lacan, Muller and Richardson write, "Good Frenchman that he is, Lacan begins all over again with Descartes."[70] They are speaking of Lacan's concern—in a 1957 lecture—to delineate the subject, specifically to delineate the psychoanalytic place of the unconscious vis-à-vis the subject. But a Cartesian point of departure can be found much earlier in Lacan's psychoanalytic; it can be found at the very core of his work. In Descartes's Second Meditation where Descartes moves from his discovery "I am, I exist" to the question "What is this I?," he initially prompts himself to answer "a rational animal." But he just as quickly dismisses this answer on the grounds that he would have to inquire "what an animal is, what rationality is, and in this way one question would lead me down the slope to other harder ones, and I do not now have the time to waste on subtleties of this kind." He proposes instead to concentrate "on what came into my thoughts spontaneously and quite naturally whenever I used to consider what I was." Remarkably enough, Descartes's "spontaneous and natural" thoughts are of the body. But remarkably enough too, they are of a body far removed from an experienced living body. The body Descartes immediately thinks of has four dominant features. It is a visual entity; it is a mechanical thing; it is a possession; and it is—by Descartes's own words—a corpse. In particular, Descartes writes that "the first thought to come to mind was that I had a face, hands, arms and the whole mechanical structure of limbs which can be seen in a corpse, and which I called the body."[71] This rendition of the body is strikingly similar to Lacan's. It is a body once removed from livingness. The body parts Descartes lists, for example, begin with the face, a part which one has and knows as such only by reference to the bodies of others. In a tactile-kinesthetic sense, i.e., in a felt bodily sense, one does not have a face *as such*.[72] The body Descartes describes is furthermore an assemblage of material parts, a mechanical rig by which one gets about in the world. There is no vestige of animate form in such a rig. In this sense the body which comes to Descartes's mind is an impoverished biological specimen. There is indeed nothing "bio" about it at all since all its attributes are related to a corpse.

The way in which Lacan writes off the body as a biological entity corresponds to the way in which Descartes writes off the body as a dead piece of machinery. Animate form is in each case reduced to nothing. The threads that bind the whole of animate life together never surface in such renditions of the body. Indeed, for Lacan, lines are clearly drawn by way of language between humans and other creatures. Humans are thereby averred wholly different, that is, not different in degree but in kind, even in cases where Lacan finds it necessary to cite biological evidence to support his psychoanalytic claims—for example, when he declares a chimpanzee's behavior in front of a mirror to be "indifferent" in contrast to the behavior of a human infant.[73] Granted that this exemplary "fact" was included in a 1948 essay and that the bulk of research on chimpanzee behavior has been gathered and published only in the past twenty-five to thirty-five years—in other words, granted that Lacan had only limited evidence available— Lacan reports the finding in a conclusive, near arrogant manner, at the same time giving no indication of his source. Furthermore, along with affirming a separation in kind, Lacan jettisons instinct as something pertaining only to animal life. Humans have *drives,* and Lacan is at pains to separate these clearly from instincts. The latter could hardly pertain to language; they could hardly carry an intellectual impress. The child in fact renounces "instinctual satisfaction" with its birth into language.[74] It gives up living in the Imaginary with its "imagos" of itself and others when it begins to speak. "The laws of recollection and symbolic recognition are . . . different in essence and manifestation from the laws of imaginary reminiscence, that is to say, from the *echo of feeling or instinctual imprint*"[75] (italics added). Lacan moreover forthrightly insists that Freud's *Treib* carries with it "the advent of the signifier."[76] This is why he can distinguish "objectivity" from "objectality" (Lacan's terms)—the Symbolic from the Imaginary—with respect to psychiatrist Karl Abraham's theory of object love. Objectality is tainted by its "affective substance," whereas objectivity remains pure and clean.[77] Lacan refers derisively to Abraham's theory as "an ectoplasmic conception of the object," which conceivably can mean not only an external or superficial rather than internal or penetrating conception of the object, but also an understanding based on something like a spiritual reading of the object, i.e., an affective reading based on the object's "emanations."[78]

A remark from Lacan himself nicely summarizes the foregoing discussion of the living infant body and its connection to biology and to intellect, that is, to what Lacan denigrates and to what he correlatively pedestals and admires. At the same time, the remark opens a vantage point upon archetypes, both Jungian and corporeal. Lacan writes that "Psychoanalysis is nourished by the observation of children and by the infantilism of the observations."[79] Putting aside the propriety of Lacan's denigrating the very observations that are lacking to the validation of his own work, we are face to face with the question of power, not the power of psychoanalysis to be nourished—or not nourished—or the power of psychoanalysis to make faulty observations—or correct ones—but the power of adults to denigrate infancy and childhood. Any psychoanalysis that would nourish itself by observing children apparently regresses to infancy in the process. Lacan's derogatory tone leaves no room for doubt. An infant and child know nothing. They are helpless or near so; they are dependent. An infant, after all, is filled with an "original distress resulting from [the] intra-organic and relational discordance [of its] first six months, [a time during which it] bears the signs, neurological and humoral, of a physiological natal prematuration."[80]

V.   The Archetype of the Infant

According to Lacan, we are had not by instincts but by *drives* which are heralded by bodily slits—by openings like the eyes, the vagina, the anus, and so on—but which have nothing whatsoever to do with anything biological. Instincts, with their associated kind of knowledge—*connaissance*—are peculiar to biology—to nonhuman animal life; drives, with their associated kind of knowledge—*savoir*—are unique to humans. Our power to come to terms with our drives is determined by language and psychoanalysis. In Lacan's view, then, adults alone can accede to power. Infants, being by definition without language, are powerless. What the clinical and experimental psychological work reviewed above demands of us, however, is precisely an acknowledgment of infant powers, that developing repertoire of 'I can's' and if/then awarenesses that are at the heart of infant life. Those 'I can's' are not mere physical accomplishments nor rote mechanical processes. An awareness of if/then relationships consistently informs their development. Stern's description of how a four-and-one-half-month-old infant learns to recognize regularities succinctly exemplifies just such an awareness: in beginning to construct "the world of people, including himself . . . ["Joey" first recognizes] those happenings that are always the same: for example, when he wants to move his arm and does, he always feels the feedback from his muscles. Things that always go together and do not change are called *invariants*."[81] Stern actually describes invariant if/then relationships as invariant "consequences of action,"[82] but the meaning of the two descriptive terminologies is clearly coincident. The same if/then relationship between moving and sensing obtains when an infant shifts its vision and finds that with such shifting a new vista comes into view as when it moves its arms and is kinesthetically aware of the movement and of the new position of its arm. Similarly, when an infant puts a round knob in a round hole and a square knob in a square hole, it recognizes certain invariants. Its capacity to recognize invariants is a power. At the same time, its acts are themselves a validation of certain powers. An infant's developing repertoire of 'I can's' is thus not merely a list of performatives, a compendium of rote capacities. An infant's acts are knowledgeable and it is because they are knowledgeable that they are experienced as powers to do certain things. In the course of its first two years especially, an infant develops prodigious capabilities.

Now according to Freudian theory, if one does not navigate effectively through the stages of infantile sexuality, one does not mature. As we have seen, however, the very term—*infantile sexuality,* suggests more than what a growing infant actually delivers, and this on the basis of research that shows a myriad of new and ongoing nonsexual involvements that occupy it. In this sense, the term *sexuality* is again of particular concern. All of the 'I can's' and all of the multitude of if/then relationships in which those 'I can's' originate—e.g., if I hold my lips just so and expel air just so, then I can make *this* sound—are completely ignored. Indeed, given the documented accomplishments of infants, one has the impression that psychoanalysts such as Lacan have never encountered or observed a growing infant.[83] They have neither watched the astounding and complex insights of infants into the ways of the world take shape nor have they observed an infant exploring a new toy, learning to walk, or discovering itself as a sound-maker. The Lacanian infant grows up in a nonlinguistic vacuum, and on that account is deemed "fetal." When it comes to see itself in a mirror, it comes to recognize how fetal it has been through a retrospective awareness of itself as a fragmented blob; its awareness of itself as an organically distressed piece of goods retroactively comes to light. When it comes to speak, it is catapulted into the realm of the signifier and recognizes how serious the schism is that marks its being. The acquisition of language is thus the beginning of the Other and of death, both of them inscribed in "the negativity of [the child's] discourse."[84] It is sobering in the context of this negative archetype of the infant-child to consider an empirically grounded view of a child's acquisition of language.

Language alters an infant's life. The acquisition of language does not come without a cost. Stern's description of the effects of language brings out in a thoughtful way what not merely recedes, but what is broken into, with the acquisition of language. Most dramatically from an experiential point of view, language fractures global self-experience, experience in which perception is multisensory and mixed with affect. Stern gives as example an infant "perceiving-feeling" a patch of yellow sunlight on a wall. The global experience is a mixture of the intensity, the warmth, the shape, the brightness, and the pleasure of the patch. In particular, the infant is not aware of the patch as *a visual experience*. That awareness enters when someone comes in and remarks, "Oh, *look* at the *yellow* sun*light*."[85] As Stern goes on to say, the language version "becomes the official version, and the amodal version goes underground, resurfacing only when conditions suppress or outweigh the dominance of the linguistic version."[86] (He of course notes too that at times language beautifully captures the quintessential aspects of experience such that there is no discord.) Moreover, language can be and is ineffective in rendering the character and quality of feelings. There is much that is experienced that is nameless because it cannot be labelled. Furthermore, there is a "slippage between experience and words" in that experiences of self having to do with the sense of coherence and continuity "fall into a category something like your heartbeat or regular breathing."[87] They are experiences of self that are not ordinarily put into words. In those situations where they are verbally expressed, something quite extraordinary happens according to Stern. In fact, his description gives pause for thought. He states that "Periodically some transient sense of this experience is revealed, for some inexplicable reason or via psychopathology, with the breathtaking effect of sudden realization that your existential and verbal selves can be light years apart, *that the self is unavoidably divided by language*"[88] (italics added).

In one sense Stern's observation validates Lacanian psychoanalytic theory: language *is* Other. It introduces "an *other* world" to which one must give allegiance in order to be part of the society in which one lives. In another sense, however, the self-divided-by-language is wholly contrary to a

Lacanian psychoanalytic and this because at its core, the self is, and has been, a quite different self in precisely the ways Stern has described. Its core is indeed an *existential* self, a preeminently bodily felt presence that carries with it a sense of coherence and continuity, of aliveness. While language distances us from this existential experience of ourselves, and from inner personal experience generally, it cannot conclusively efface it. Moreover, it in no way prevents us from acknowledging and appreciating intercorporeally shared experiences that are nonlinguistic. Gazing into another's eyes at the same moment the other gazes into our own eyes is one such experience, as Stern points out. Neither does language in any way prevent us from acknowledging and appreciating simple corporeal experiences—the sense of another person's "vitality affects"—his or her way of walking, for example.[89] Stern in fact parenthetically comments that "it is little wonder we need art so badly to bridge these [language-induced] gaps in ourselves."[90] Such existential experiences, like those too in which we speak of *communing* with nature, very often stand out as heightened moments of being. At these moments, we are more than metaphorically in touch with an other; we resonate corporeally in and with their presence.

In sum, to whatever degree the self is divided by its entrance into a linguistic world, language cannot wholly efface animate form and the tactile-kinesthetic life that is its core. We ourselves modulate their fading by our attentiveness or lack thereof. As Stern's reflective observations show, however, language brings with it momentous changes; it introduces new ways of being with others. Well aware of the psychoanalytic impact of language, he writes that "prior to this linguistic ability, infants are confined to reflect the impress of reality. They can now transcend that, for good or ill."[91] In fact, as he later points out, "Paradoxically, while language vastly extends our grasp on reality, it can also provide the mechanism for the distortion of reality as experienced . . . language can force apart interpersonal self-experience as lived and as verbally represented."[92]

What Stern has observed in infants and children with respect to the introduction of language is patently at odds with Lacan's theoretical explications of an infant-child's entrance into the register of the Symbolic. Again, this is because the young Sternian subject, at least the normal and even neurotic one, remains unswallowed up by language. However much it speaks, it remains most fundamentally a living body. Stern, it should be noted, does not speak of the body in this way, but implicit in what he writes is the fact that the "amodal" self, like other aspects of the self, is fundamentally a felt body, an animate form that is first of all a tactile-kinesthetic locus of sensation and activity.

It is significant that in his short, exploratory chapter "The Psychology of the Child Archetype," Jung makes no mention of a distressed, ineffective, uncoordinated creature. But it is significant too that in this essay, originally written in 1940, thus *before* Lacan's 1948 essay on aggression in which Lacan speaks of the infant's "visual Gestalt of his own body . . . [being] invested with all the original distress resulting from . . . [its] intra-organic and relational discordance during . . . [its] first six months," Jung speaks of an "original distress"—in italics even—but it is not an original physiological or relational infant distress; it is an *"original psychic distress,* namely, a state of *unconsciousness."*[93] Jung so describes the state of those who preceded the "conquerors of darkness" of primeval times, conquerors who brought light and whose original darings are recorded in legends. He likens the child to just such conquerors because the child too is a "bringer of light," an "enlarger of consciousness."[94] The child archetype is in fact connected with three quite positive qualities: the child is the harbinger of a new beginning; it is invincible; it signifies wholeness through a creative union of opposites. It is the harbinger of new beginnings in that "it is the most precious fruit of Mother Nature herself, the most pregnant with the future, signifying a higher stage of self-realization." In this sense, it "paves the way for a future change of personality."[95] For just such reasons, Jung points out, child gods figure frequently in legends as saviors. The child is invincible in that it is "equipped with all the powers of nature and instinct."[96] In this respect its strength is unequalled. It can do nothing other than realize itself. All of its being is thrown into self-realization. Being born from "the womb of the unconscious," it is begotten "out of living Nature herself," and it is because its origins are what they are that the child is "an incarnation of *the inability to do otherwise."*[97] The child as a unity of opposites draws on its issuance from male and female, but also, in a related way, from its symbolic wholeness, that is, from its representing a unity of both conscious and unconscious psychic aspects of personality. Jung writes that "Just as every individual derives from masculine and feminine genes, . . . so in the psyche it is only the conscious mind, in a man, that has the masculine sign, while the unconscious is by nature feminine. The reverse is true in the case of a woman."[98] (If Jung is right that the archetype of the unconscious is *either* male or female, depending upon one's sex, then we have an explanation of why the standard corporeal archetype of the unconscious as female is the formidable and implacable thorn it is in Lacanian theory. See chapter 9.)

In sum, if as psychic archetype the infant-child is a unity of opposites whose powers are tied to Nature and to instinct, then it is evident first of all that the psychic archetype is fundamentally a corporeal archetype. Second, it is evident that the impoverished picture of infancy that Lacan presents is an extreme individual reworking of the negative pole of the archetype. His radical reworking reduces the inherent powers of the original to absolute zero, thereby rendering the infant-child utterly helpless, lacking in any and all resources, and destitute of meaning. Third, it is evident that as a unity of opposites, the archetypal infant-child necessarily conjoins innate capacities with helplessness, invincibility with dependency, natural talents with cultural ineptness. Rather than recognizing this gamut of corporeal attributes, a certain typical twentieth-century Western way of thinking that consistently pedestals language consistently suppresses attributes at the positive end of the continuum: to be an infant is to be by definition without language, hence impoverished and deficient. It is this literal cultural translation of the archetype that is exaggerated out of all proportion in Lacan's psychoanalytic.

9

Corporeal Archetypes and Power: Lacan's Psychoanalytic of the Unconscious

Let us look at the facts. The reality of the unconscious is sexual reality . . . At every opportunity, Freud defended his formula . . .

Jacques Lacan[1]

The oldest and best meaning of the word 'unconscious' is the descriptive one; we call a psychical process unconscious whose existence we are obliged to assume—for some such reason as that we infer it from its effects—but of which we know nothing. In that case we have the same relation to it as we have to a psychical process in another person, except that it is in fact one of our own.

Sigmund Freud[2]

For Freud, . . . the unconscious is of an exclusively personal nature, although he was aware of its archaic and mythological thought-forms. A more or less superficial layer of the unconscious is undoubtedly personal. I call it the personal unconscious. *But this personal unconscious rests upon a deeper layer, which does not derive from personal experience and is not a personal acquisition but is inborn. This deeper layer I call the* collective unconscious.

Carl G. Jung[3]

I. Introduction

It is precisely in the context of discussing the natural and instinctive dimensions of self that Jung obliquely adumbrates the notion of *corporeal* archetypes. A lengthy quote is apposite in order not only to demonstrate the closeness of Jung's thought to corporeal archetypes but also to bring out both the theoretical coherency of his thoughts regarding the relationship among archetypal symbols, the unconscious, and the body, and the startling correspondence of Lacan's psychoanalytic with his (Jung's) theoretical orderings. The quote is akin to a condensed essay on what Lacan came later to redefine chronologically and to reformulate axiologically as three "registers": the real, the imaginary, and the symbolic. Jung wrote that

"Fantasies" are the natural expressions of the life of the unconscious. But since the unconscious is the psyche of all the body's autonomous functional complexes, its "fantasies" have an aetiological significance that is not to be despised. From the psychopathology of the individuation process we know that the formation of symbols is frequently associated with physical disorders of a psychic origin, which in some cases are felt as decidedly "real." In medicine, fantasies are *real things* with which the psychotherapist has to reckon very seriously indeed. He cannot therefore deprive of all justification those primitive phantasms whose content is so real that it is projected upon the external world. In the last analysis the human body, too, is built of the stuff of the world, the very stuff wherein fantasies become visible; indeed, without it they could not be experienced at all. Without this stuff they would be like a sort of abstract crystalline lattice in a solution where the crystallization process had not yet started.

The symbols of the self arise in the depths of the body and they express its materiality every bit as much as the structure of the perceiving consciousness. The symbol is thus a living body, *corpus et anima;* hence the "child" is such an apt formula for the symbol. The uniqueness of the psyche can never enter wholly into reality, it can only be realized approximately, though it still remains the absolute basis of all consciousness. The deeper "layers" of the psyche lose their individual uniqueness as they retreat farther and farther into darkness. "Lower down," that is to say as they approach the autonomous functional systems, they become increasingly collective until they are universalized and extinguished in the body's materiality, i.e., in chemical substances. The body's carbon is simply carbon. Hence "at bottom" the psyche is simply "world." In this sense I hold Kerenyi to be absolutely right when he says that in the symbol the *world itself* is speaking. The more archaic and "deeper," that is the more *physiological*, the symbol is, the more collective and universal, the more "material" it is. The more abstract, differentiated, and specific it is, the more its nature approximates to conscious uniqueness and individuality, the more it sloughs off its universal character. Having finally attained full consciousness, it runs the risk of becoming a mere allegory which nowhere oversteps the bounds of conscious comprehension, and is then exposed to all sorts of attempts at rationalistic and therefore inadequate explanation.[4]

As the quotation makes clear, Jung's idea that Nature is the 'rock bottom' of our being is not in the least incompatible with the notion of an unconscious. On the contrary, the two are tied together. Our psychical being enters into the materiality of our nature just as the materiality of our nature enters into our psychical being. As we descend into the deeper layers of the latter, we move away from our unique individuality, our individual consciousnesses, and descend into the collective aspects of our being. Ultimately, we reach that microscopic materiality of which we are all composed. The psyche, Jung says, is thus coterminous with the world. Hence its symbols are *corpus et anima,* and when it speaks, "the world itself is speaking."

Jung's enfoldment of nature—of the cosmo-biological—within the psyche obviously runs counter to Lacan's categorically divisionary scheme. Psychosomatic illnesses are *real* for Jung in a way quite unlike they are *real* for Lacan. Rather than looking down on this *real* as from on high, Jung conceives the real to be the ultimate foundation of the psyche. Everything of the nature of the self starts in earnest with the living body, including symbolic thought. Where symbols of the self are otherwise generated, they float off, Jung says, into the abstract and are anchored in an *individuality.* They are, in other words, products of a unique consciousness; they are not universal; they come not from the depths but from the shallows. From a Jungian point of view, Lacan's psychoanalytic clearly derives from the shallows: it originates near "full consciousness." From an extended Jungian point of view, his psychoanalytic is clearly anchored in corporeal archetypes that have been shaped not only by a highly acculturated but a highly idiosyncratic individuality.

II.   The Unconscious as Corporeal Archetype

As an actual psychic phenomenon, the unconscious is a powerful force conceived in basically opposed ways by Freud and Jung. It is the site of repressed feelings and desires and of unresolved conflicts for Freud. Though Freud says the processes of the unconscious are timeless,[5] the unconscious is preeminently connected to the past. It is a repository of experiences too painful to recognize; it is the site of feelings that, most especially in our infancy-childhood past, were too disturbing actually to feel and live through, and that reverberate still in the form of dreams, slips of the tongue, jokes, physical symptoms, and so on. With respect to psychic disturbances, the hidden past of the unconscious creates the present, and the present

recreates the hidden past. In effect, the unconscious generates something of
a vicious circle. Freudian analysis—"the talking cure"—known for its
interminability even by Freud,[6] might well be interminable because there is
no way out of the circle: the past gathers as fast as the present, that is, as fast
as talk-making displaces it. For Jung, the unconscious is the site of creativity.
Jung connects the unconscious with the imagination: in dreams, in artistic
creations, in fantasies, the unconscious produces symbols that show us a
way forward, away from our painful and stalemated modes of being and
onto more healthful paths. The unconscious is thus future-oriented as well
as creative. For both Freud and Jung, however, the unconscious is above all a
dark and unknown site, a psychic aspect of our being that remains hidden
except for those moments in which, whether through gaps in consciousness
as in dreaming or through lapses in consciousness as in slips of the tongue
or, with specific reference to Jung, through an active engagement of the
imagination as in fantasizing, we are brought face to face with its products.
Hence in spite of their differences about the unconscious as an actual
psychic phenomenon, Jung and Freud are in basic accord: as a topographical
unit of the psyche, the unconscious is the dark and unknown side of our
being.

   This conception of the unconscious is the classical conception. The idea
of the unconscious as hidden and obscure, a dark and abstruse psychic
terrain needing effort to understand and indeed patience should one
attempt to fathom its mysterious powers, is not an arbitrary view of the
unconscious. While it might at first seem merely a conception more or less
demanded by the word itself, it is not the only conception possible. Granted
the terminological opposition of consciousness and unconsciousness, and
granted too the literal and metaphorical opposition of light and dark, there
is a more intricate and fundamental way of comprehending the unconscious
as the dark and unknown side of our being, namely, as a corporeal
archetype. This corporeal archetype derives in the most primitive sense from
the side of our being that is precisely *our bodily inside*. This *side* of our bodies
is consistently hidden from view; it is never brought up to the light of day.
We catch glimpses of it now and then in various kinds of exudings, sounds,
and the like, but its complex of features and actual workings are not part of
our everyday experience. Moreover, there is an autonomy about this bodily
inside that defies our control. Given both its hidden and autonomous
nature, the analogical link of our bodily inside with the unconscious is
virtually transparent. What is called the underside of consciousness is akin
to the underside of our bodies. What Lacan does is suppress this corporeal
archetype, turning the unconscious instead into a thoroughly revealed

structure. But as we shall see, the corporeal archetype of the unconscious haunts his psychoanalytic. It haunts it along the lines of "the problem of female sexuality," precisely in the direction of being "in the form of a hole." In his attempts to deal with this problem, he is forced to compromise the very psychoanalytic he has labored to produce.

Lacan's radical privileging of language makes the *real* of the body ultimately all the more insistent. The *real* body, animate form, will not jump compliantly through the linguistic and algebraic hoops and loops Lacan invokes and, in effect, docilely consign itself to oblivion.[7] On the contrary, it will install itself—albeit hidden—in the analytic itself. It will in fact hide beneath the psychoanalytic anatomy that Lacan conjures into existence, an anatomy whose developmental stages, as we have seen, are each triggered by an original happening that sends a hypothetical infant on its fateful way into a standardized psychoanalytic history. The psychoanalytic anatomy that Lacan conjures into existence is a psychoanalytic body that in fact *masquerades* as the real body. It will be our purpose here to show the results of this masquerade, namely, the ways in which Lacan's unconscious comes to be modeled on the female body—up to a point, in fact a decisive turning point since the modeling proves at odds with Lacanian doctrine and will not work—and correlatively, the ways in which Lacan's female body, in turn, is initially linked to the unconscious. This connection of the unconscious with the female body and the female body with the unconscious is rooted in archetypal understandings of both the unconscious and female sexuality. We will begin first by considering Lacan's view of the unconscious, a view which in spite of its seeming concordance with the classical conception, diverges from it absolutely.

Lacan speaks of the unconscious as "that zone of shades." He speaks of bringing patients "up to the light of day." He speaks of "that ultimately unknown centre." He speaks of "the veils of the unconscious."[8] He furthermore makes clear that *in the light of this darkness,* i.e., in the light of Lacanian psychoanalysis, consciousness loses something of "its privilege." In and through our speech, "something other demands to be realized."[9] The unconscious is thus not only catapulted into prominence; *it* becomes privileged, and in an extraordinary way. No longer is it a figure on the background of consciousness; it is descriptive of the subject him/herself. It constitutes the subject as indeterminate and inherently split.[10] Accordingly, Lacan comes ultimately to speak not of the unconscious of a subject—thus in essence of a person *having* an unconscious (or an unconscious side)—but of the subject of the unconscious—thus in essence of a person *being* the (spokesperson of his/her) unconscious. It is this thoroughly commanding unconscious that peeps through—that "manifests itself"[11]—in gaps in consciousness.

To be noted first is that Lacan anchors these gaps in consciousness in the body, specifically, in bodily holes and slits—notably, the eyes, the vagina, the penile slit, the anus, the urethra. It is through these holes and slits that the unconscious speaks in the form of desire. The aim of Lacanian psychoanalysis is to cause the "refused thoughts" of the unconscious to appear through its bodily slits. His psychoanalytic does this by making the subject—the analysand—regress, or, in other words, accede to paranoiac speech. In this procedure, Lacan avers that he is following to the letter Freud's thesis regarding the unconscious. Freud wrote that "Unconscious processes only become cognizable by us under the conditions of dreaming and of neurosis—that is to say, when processes of the higher, Pcs. [Perceptual consciousness], system are set back to an earlier stage by being lowered (by regression)."[12] In his psychoanalytic technique, regression is Lacan's rule of thumb. Unconscious processes are indeed "set back" to the earliest possible stage in Lacanian psychoanalysis—but not directly with respect to infancy; rather, with respect to psychopathology, i.e., psychosis.

Now it is not simply that Lacan has an interest in, not to say an infatuation with, psychosis or that commentators regularly find psychopathological features in Lacan's work, one of them, for example, labelling Lacan's self-proclaimed "psychotic rigor" a "scientific delirium," and persuasively demonstrating that "the principle of incoherence" informs his psychoanalytic.[13] But it is not either a question of Lacan's own estimate of his work, as when he forthrightly links its rigorous character to psychosis: "Psychosis is an essay in rigor. In this sense, I would say that I'm a psychotic. I'm a psychotic for the single reason that I've always tried to be rigorous."[14] What is of moment is that regression is intentionally cultivated in his psychoanalytic. As suggested earlier, Lacanian psychoanalysis *induces* paranoia. Lacan writes that "The analytic maieutic" (Lacanian psychoanalysis) approaches the problem of the ego (a psychic manifestation that blocks "the realization of the subject") [in] a round-about [way] that amounts in fact to inducing in the subject a controlled paranoia."[15] Lacan, be it noted, immediately assures us that the paranoiac "mechanism" he is using is "highly systematized, filtered, as it were, and properly checked."[16] Our apprehensions placated, we might presumably come to accept the idea that through induced paranoia, analysands have first-hand experiences of paranoiac states, paranoiac alienation, and ultimately paranoiac knowledge. When analysands come into a paranoiac state and attain to its attendant knowledge, however, it is clear that they regress back toward the original slit—or hole—from which they came. Lacan himself states unequivocally that in analytic regression "the whole past opens up right down to early infancy."[17] Thus, to regress is to regress both psychopathologically and in a physico-chronological manner, indeed to go back toward the original slit—or hole—from which we came. It begins to dawn on us, then, how, in Lacan's psychoanalytic, the unconscious is tied to the living body, but how all the same, the living body remains hidden, that is, *unexpressed,* in the analytic: in regressing, we not only travel a psychopathological path to the unconscious, into a darkness which has heretofore precluded an awareness of our true *psychoanalytic anatomy* and which now becomes light by dint of our speech; we travel backward toward the dark hole from which we came, a darkness unilluminated by speech but implicit both in Lacan's psychoanalytic anatomy and in the preverbal stages of infancy that order his psychoanalytic. Unconscious and female body thus coincide: in our actual birth as in Lacanian psychoanalysis, we are "brought up to the light of day." We journey in reverse directions with respect to each, but in each case we journey in relation to an "ultimately unknown centre." Thus, as the subject of the unconscious progressively speaks its regresosive speech, it at the same time leads us back along a path that stops short of our actual expulsion from darkness—the moment at which we came out of the dark hole[18]—but that unequivocally features along the way those other holes that once connected us directly and indirectly to a female body.

The extraordinary coincidence of the unconscious and the female body in Lacan's psychoanalytic obviously rests fundamentally on a synecdochic rendering of the female: her genitals, which are dark and unknown. It is just this rendering that allows an archetypal figuration of the female and in turn links female with unconscious. On the surface, then, the synecdochic female archetype constitutes an initial and perfect fit with the archetypal rendition of the unconscious. Before following up this initial linkage, we will first take up the notion of the unconscious as a "zone of shades" in the light of Lacan's actual psychoanalytic, and this in order to show specifically how radically and absolutely Lacan's conception of the unconscious diverges from the classical conception.

III. The "Zone of Shades" and Its Transformation

The Lacanian unconscious is in actuality an unconscious in name only. Lacan's poetic metaphors notwithstanding, the unconscious is not some bottomless dark psychic pit which, with or without psychoanalytic treatment, will always remain residually dark. What the Lacanian psychoanalytic weaves together is a complicated tapestry, the whole of which is held together ideologically by Lacan's aim to make psychoanalysis into a science, but what at a finer level can be seen as a reversal of the classical archetypal figuration of the unconscious as dark and unknown. That reversal is evident in the installation of a psychoanalytic program that will not simply elucidate the unconscious in theoretical and everyday terms, i.e., in a Freudian manner, but will elucidate the unconscious *in fact*. Indeed, a Lacanian unconscious appears only terminologically related to a Freudian unconscious since it is a consummately specified psyche to every letter of its laws. In consequence, it is a completely transformed unconscious; it is a *completely lit* unconscious, thus no longer properly an *un*conscious.

We can see this transformation from a variety of perspectives. Superficially, it is documented by Lacan's habit of baldly affirming "facts," as, for instance, when in writing of the moment the infant enters the mirror stage, he states that "We cannot fail to appreciate the affective value which the gestalt of the vision of the whole body-image may assume when we consider *the fact* that it appears against a background of organic disturbance and discord"[19] (italics added). Numerous instances of such "facts" abound in Lacan's writings, giving credence to a thoroughly transparent psyche from birth onwards. While it is important to highlight Lacan's factual pronouncements with respect to the transformation of the unconscious, they are trivial in comparison to the method by which Lacan carries out his psychoanalysis. The same may be said of his tendency to invoke the writings of others (often with no proper citations) as authorities for his own claims, as when he invokes the physiology and behaviors of pigeons and locusts to support his claim of a mirror stage in human infants.[20] Lacan's factual invocations are furthermore trivial in comparison with the program that is Lacanian psychoanalysis itself. Put in the perspective of that program, the transformation of the unconscious is indeed virtually transparent. It is decisively and readily evident in four major dimensions of Lacanian psychoanalysis: interpretation, training, cure and/or goal, and invariants. We will briefly exemplify each of these dimensions, showing how, as threads in the Lacanian tapestry, they not only are interrelated but strongly reinforce one another.

Lacan insists that interpretation bow to his "doctrine of the signifier,"[21] thus assuring that what is understood in psychoanalysis is uniform and invariant. Interpretation, says Lacan, "cannot be bent to any meaning. It designates only a single series of signifiers."[22] In other words, interpretation is controlled and to be controlled; only by going through Lacanian training can one come to know the correct "place of interpretation" in psychoanalysis.[23] That place appears to be a matter of *translation,* i.e., interpretation is a matter of carrying meaning from one language to another such that missing elements are made to appear in the discourse of the subject: "In order to decipher the diachrony of unconscious repetitions, interpretation must introduce into the synchrony of the signifiers that compose it something that suddenly makes translation possible—precisely what is made possible by the function of the Other in the concealment of the code, it being in relation to that Other that the missing element appears."[24] Interpretation is thus reigned in. It cannot possibly stray from a Lacanian course. It adheres to a set program. It is admissible only within the framework of the Other who controls the subject's discourse.

The result is that there is a single schema that legitimizes interpretation and that is locked in by language. The single schema is based on the rules laid down by the Lacanian stages of infant development. By these rules, Lacan safeguards his psychoanalytic precisely from the kind of criticisms addressed to Freud, notably by Jung, who would not accede to Freud's fundamental precept of infantile sexuality. With the jettisoning of infantile sexuality, the whole Freudian edifice falls. Just so with Lacan's psychoanalytic—save for the fact of his buttressing the Freudian fortress with language, thus ostensibly anchoring the latter to a universal and in turn mapping its laws.

It is notable that the question of the correct use of interpretation links up with the question of Lacan's psychoanalytic teaching in that, when it comes to giving an example of a correct interpretation, Lacan relies not on his own clinical work, but on the case history of another psychoanalyst, who, he shows, did *not* make the correct interpretation. He excuses his use of the work of another on the grounds of confidentiality:

> I cannot make use of my own analyses to demonstrate the level to which interpretation reaches, when interpretation, proving to be coextensive with the history [of an analysand], cannot be communicated in the communicating milieu in which many of my analyses take place without risking an infringement of anonymity.

He goes on, however, to say: "For I have succeeded on such occasions in saying enough about a case without saying too much, that is to say to cite my example, without anyone, except the person in question, recognizing it."[25]

Lacan's demurral and his seeming retraction are telling. He appears to want to keep the actual training of analysts a company secret. Correct interpretations may be learned first-hand by going through Lacanian psychoanalysis. To set forth the analytic technique, including the use of interpretation, by presenting case histories or even moments of case histories would be to risk "the purity" of the technique. It is not surprising, then, that Lacan avers that "The aim of my teaching has been and still is the training of analysts."[26] His dedication to the training of analysts precisely insures the purity of his psychoanalytic technique. Recalling his notion of "transmission through recurrence," one might say that his dedication is to limitless but controlled dissemination—passing on his unmixed psychoanalytic genes *in perpetuum*. What his technique promises in each transmissively recurrent case is an illumination of the unconscious—not a temporary flash of insight and certainly not an extended period of stability, but a fundamental encounter with the Other that, psychically speaking, explains everything. Lacan goes so far as to say that "There is only one kind of psycho-analysis—the training analysis—which means a psycho-analysis that has looped this loop to its end."[27]

Lacanian psychoanalysis, by such affirmations, in fact promises a *cure*. Analysis is not *interminable* but terminable. The psychoanalytic program is a set program in that all the pieces of the psychic puzzle are in place. All the analysand must do is give voice to each piece. This is to say that all of us have basically the same malady since the same treatment cures us all. Lacan himself defines the termination of analysis in various ways, most forthrightly when he states that "The subject . . . begins the analysis by talking about himself without talking to you, or by talking to you without talking about himself. When he can talk to you about himself, the analysis is over."[28] Clearly, Lacanian psychoanalysis hangs on the use one makes of one's capacity for language. When that capacity results in what Lacan calls "full speech," or "free speech," or "true speech,"[29] it leads to illumination. The unconscious at this point is clearly lit up; the analysand is "brought into the light." There is no more darkness, no more unknown. In the end, Lacanian psychoanalysis does away with the unconscious. The analysand knows him/herself as an irremediably split subject whose desires have never been satisfied and will never be satisfied. Clearly, the unconscious is no longer *un*conscious. If "the unconscious is that chapter in my history that is marked by a blank,"[30] then once I articulate it in full speech and see the desire/lack that is the mark of my being, I am cured. This very notion is subsumed in Lacan's statement regarding the demand of the analysand: "Of course, his demand is deployed on the field of an implicit demand, that for which he is there: the demand to cure him, to reveal him to himself, to introduce him to psychoanalysis, *to help him to qualify as an analyst*"[31] (italics added).

We see then that the path carved out by Lacan's psychoanalytic is a path that in a quite particular sense already exists. Language is already there; it "pre-exists the infantile subject,"[32] or as Lacan otherwise puts it, "the gap of the unconscious is *pre-ontological.*"[33] We are born into language, into a world of the Other, which was there before we came onto the scene and which will be there after we have departed. Actually, by Lacanian lights, we are "born again" when we are born into language, and it is this, our second birth, that is our most significant birth since, according to Lacan, it is with this birth that we become human.[34] Insofar as our second birth takes precedence over our first, we can appreciate from yet another perspective how the psychoanalytic body of slits and holes effaces the *real* living body. By the same token we can also appreciate how Lacan's psychoanalytic manages to stop short of the dark hole from which we came and why our birth ties with a female body are ignored. The ties are ignored because a birth-giving female body is not a sexual body; it is a *reproductive* body, and a reproductive body, being part of the Lacanian *real* (or *Real*), is not substantively recognized in the psychoanalytic. *This* hole is, in other words, an altogether different dark-and-unknown bodily hole from the ones on Lacan's psychoanalytic anatomy chart. *This* hole is a mere piece of human anatomy with no particular psychoanalytic significance—a surprising and actually puzzling construal considering that the hole in question is the very source of Otherness.

Although the word is not his, *invariants* are the ultimate meaning of Lacan's graphs, formulae, doctrines, schemata, and so on. These graphical-mathematical features of his writings specify absolute features of the unconscious. Lacan's demurral concerning case histories is actually a testimonial to his concern with invariants. What is actually experienced by the subject is not of interest to him, for what is actually experienced by the subject is "subjective," and the subject "goes well beyond what is experienced 'subjectively'."[35] Put somewhat more cryptically, "the truth of his [the subject's] history is not all contained in his script."[36] What concerns Lacan is the unconscious rather than the subject because the unconscious is the blank in the script. *It* is what is marked by invariant structures. Accordingly, Lacan is interested in what commentators Muller and Richardson at one point speak of as "the larger text"[37] wherein psychoanalysis takes place and where the Other in its original Lacanian guise is brought forth, that is, where the structuring of the unconscious as a language comes into its own. Case histories would merely deflect from the identification of the invariant structures of this language. Lacan clearly articulates his fundamental concern with invariants when he states that "there is no progress for the subject other than through the integration which he arrives at from his position in the universal: technically through the projection of his past into a discourse in the process of becoming."[38] The subject's "position in the universal" is of course open to question since it is bought at the price of an ungrounded generalization. David Macey makes the critical point more broadly when he notes that "theses propounded with respect to language within psychoanalysis are projected on to [sic] language as such."[39] The criticism aside, Lacan's intent is clearly to anchor the unconscious in invariants and in so doing systematize psychoanalysis to the point that it produces the same effect in all its subjects. All articulate the unconscious, in minimally different ways perhaps, but all according to the same program of regression and all reaching the same conclusions. In effect, no matter to whom it might belong, the unconscious is no longer *un*conscious. By the same token, of course, no matter to whom it might belong, the unconscious is no longer a personal unconscious; it is a collective unconscious.

A final point needs to be made. Commentators of Lacan speak in various ways of "the subject's radical *inability* to know itself,"[40] or in other terms, of Lacan's "attempts to destabilize a metaphysics of the *cogito* and its epistemological underpinnings,"[41] and in turn, of his contribution to a critique of traditional Western philosophy that thinks in terms of an all-lucid *cogito*. Lacan's attempts are actually patterned on Freud's charge that "To most people who have been educated in philosophy the idea of anything psychical which is not also conscious is so inconceivable that it seems to them absurd and refutable simply by logic."[42] Theoretically, from Socrates onward, and most specifically from Descartes, Western philosophers have had the notion that, with patient philosophical spadework, humans can indeed come to know themselves. But the notion of a full-blown luminosity, a consciousness that is wholly and directly accessible is not typical of *all* Western philosophers. It is erroneous, for instance, to attribute to William James or Edmund Husserl a belief in just such a lucidity.[43] Husserl writes, for example, that the realm of the "mind" is rather like geology and paleontology: inferences must be made on the basis of what is observable. Accordingly, Husserl would undoubtedly insist that to begin to fathom our unconscious, we need to pay attention to "observables"—as Freud first directed us: to spontaneous productions such as dreams, slips of the tongue, and so on. We need, in other words, to pay attention to what slips into our experience unbidden, *what is out of our control,* and from there, infer (as Husserl's analogy suggests) unconscious processes. In following the Lacanian prescription, on the other hand, we are precisely not *inferring;* we are unequivocally bringing to light—to full light—the subject of the unconscious. In consequence, the idea that Lacan has destabilized Western metaphysics by showing that the subject is unable to know itself is grossly misleading, for in light of his psychoanalytic, the subject precisely *does* come to know her/himself. How otherwise would the subject arrive at "speaking *of himself* to you"? How otherwise could Lacan speak of analysis as that progress toward integration, integration being that moment in analysis which marks its termination and which the subject "arrives at from his position in the universal"?

In sum, the Lacanian unconscious is not irremediably *dark* and *unknown*. It is accessible to the light of Lacanian psychoanalysis. Hence in spite of references to veils, shades, and the like, the Lacanian psychoanalytic is not consistent with the classical view of the unconscious. Furthermore, as pointed out, Lacanian psychoanalysis is not interminable. Lacan's program of analysis is conclusive; his formulae and graphs leave no hidden aspect of the unconscious unturned. This is in large part because at a theoretical level, his sportive psychoanalytic body holds his psychoanalytic in place. Language notwithstanding, everything is ultimately laid out on the orifices and ontological stages of an idiosyncratic psychoanalytic anatomy. Indeed, Lacan has *mapped* the unconscious onto his psychoanalytic body. That body is the backbone of his theory. Where Freud gave us a much more general program of three bodily stages—oral, anal, and genital—specific conceptual anchorages in terms of the Oedipal complex and castration anxiety, Lacan gives us a concrete and detailed picture of the unconscious in action. We learn how it works; we learn its *structure*. Indeed, we become privy to its codes by acknowledging their psycho-anatomical provenience. In short, the unconscious is brought out of the dark—except for one nagging problem: female sexuality. It will not come out and into the light. To understand how this problem arises is to show first how Lacan in the beginning models female sexuality along the lines of the unconscious as a dark and unknown continent.[44] It is precisely as the unconscious comes to be lit, and female sexuality is *not* lit along with it, that an inconsistency develops in Lacan's psychoanalytic toward the end of his life and "the problem of female sexuality" comes brusquely and unequivocally to the fore.

IV. Female Sexuality and the Unconscious: An Initial Analysis of Psychoanalysis

Within the Lacanian psychoanalytic, a tenuous relation obtains between subject and unconscious. The patient is not cured until s/he voices the desire of the Other that speaks her/him. This voice speaks through bodily slits; the subject thus comes to be characterized as a *sujet troué*.[45] While in English, "being riddled with holes" suggests not only slits but deficient knowledge—thus perhaps even further that we are beset by the riddle of our holes, and in effect, that our bodies call our being into question—we must remember that in Lacan's psychoanalytic, the *unconscious,* not the living body, speaks desire; being dumb to language, the living body is pure symptomatology. It is thus the unconscious and not the living body that riddles us. The unconscious is the gap *par excellence.* But by the same token, because the body makes its appearance in Lacan's psychoanalytic in the form of holes, and because the unconscious speaks desire and all desire is sexual, the gap *par excellence* is also "[the] gaping cunt," to use Lacan's own expression.[46] In less abusive terms, the fact that "the orifice constitutes the essential structure of the subject,"[47] and the fact that sexuality is "strictly consubstantial with the dimension of the unconscious,"[48] make the confluence of female body and unconscious unmistakable. We could say, then, that according to the Lacanian psychoanalytic, we are all, male and female alike, preeminently riddled in the form of a female. We could furthermore say—in English at least, but also as Freud first affirmed in speaking of "the riddle of the nature of femininity"[49]—that psychoanalysis is riddled in face of all that is feminine.[50] Psychoanalysis is at pains to understand what it is to be female.

Macey writes that the impenetrability of female sexuality is a constituent element of psychoanalytic discourse. He demonstrates persuasively and in particular how "the Lacanian literature on the dark continent" fails utterly to shed light on the subject.[51] Now if Macey is correct in saying that "the riddle of the nature of femininity" is a built-in of psychoanalytic discourse, then psychoanalysis itself is in need of analysis. A critical examination of Lacan's radical psychoanalytic with respect to female sexuality and the unconscious will carry out in an initial way just such an analysis.

In her provocative essay, "Veiling Over Desire," Mary Ann Doane presents an analysis of how close-ups of the female are used in film to signify the tension between truth and appearance, that is, the tension between vision as a guarantor of knowledge and vision as deceiver, and, by extension, the tension between the female as object of desire and object of horror. In this context, Doane discusses Lacan's usurpation of the veil for the phallus, pointing out how this appropriation serves a patriarchal psychoanalytic, for even though it is possible to argue "that the phallus is not a masculine category, that it is a signifier and not equivalent to the penis . . . and therefore that we are not confronted with a situation in which the psychoanalyst snatches the veil from the woman in order to conceal his own private parts,"[52] still, truth comes to be associated with the phallus.

Lacan's usurpation of the veil by the phallus of course creates interesting filmic possibilities which have yet to be realized. Doane does not consider these or even point out that such possibilities exist because her specific interest is in what *female* veiling accomplishes. The film she analyzes is mainly the early Dietrich film *The Scarlet Empress*. She shows how the closer the camera comes to Dietrich's veil-covered face, the more the veil and the screen itself become objects of desire. In other words, the grain of the film and the grain of the veil over Dietrich's face meld together. Even as the granular filmic image takes precedence over Dietrich's face, Doane argues, Dietrich's face still supports the image's seductive appeal to the spectator. The point to be emphasized and elaborated here, however, is that the veil is snatched from the female by Lacan in the same way that it is snatched from the female by the camera close-up. The act of course in each case robs the female herself of her seductive appeal. More importantly, however, in each case the act is an appropriative act: someone—the film director or Lacan—takes the veil and uses it for his own ends. But although Lacan would snatch the veil because, as he affirms in his essay, "The Meaning of the Phallus," "the phallus can only play its role as veiled, that is, as in itself the sign of the latency with which everything signifiable is struck as soon as it is raised to the function of the signifier,"[53] still, he cannot, by his appropriative act, come to understand woman herself. Though he has unveiled her, he is no further ahead. On the contrary, he speaks over and over of an inability to comprehend woman and claims that women themselves do not know themselves.[54] Clearly, though he has snatched her veil from her, sufficiently to expose a "gaping cunt," he is not the more enlightened; he does not *know* her better. He is as speechless—dumb—as he was before the unveiling.

Now were one not bound by the Lacanian fact that the unconscious is structured in language and is therefore potentially transparent, one would say that, given this situation, the unconscious remains on the side of the female—veiled or not. It coincides with the female body, veiled or not, precisely because it *remains* "ultimately the unknown centre" in the same way that a gaping cunt, in spite of its gaping, remains "ultimately the unknown centre." In other terms, that Lacan cannot articulate the female does not point to a deficiency in the female; it points to the classical nature of the unconscious: to be ultimately dark—just like a woman. In effect, the original psychoanalytic archetypal bond remains.

In Lacan's psychoanalytic, the veil is related to masquerade. It is significant then that Lacan aligns masquerade with the feminine, and in a quite pointed way, as we shall see.[55] Masquerade is play at the symbolic level, Lacan says. It is different from *display* because display is part of animal coupling, and masquerade has nothing to do with such behavior. In addition, Lacan insists, "display is usually seen on the side of the male."[56] What Lacan says he will show (in a chapter titled "From Love to the Libido" in *The Four Fundamental Concepts of Psycho-Analysis,* published in 1973) is how masquerade—play at the symbolic rather than imaginary level—characterizes sexuality. However, because he has just previously linked masquerade with "the feminine sexual attitude" by referring to "one female psycho-analyst [who] has pin-pointed the feminine sexual attitude . . . [by] the term *masquerade*," Lacan clearly intimates that in what follows, the feminine sexual attitude in particular will be elucidated. Thus, when he says that "It is on this basis [of masquerade] that it now remains to us to show that sexuality as such comes into play, exercises its proper activity, through the mediation—paradoxical as that may seem—of the partial drives,"[57] one expects that what he will set forth in a focal way is how the "feminine sexual attitude," or more broadly, feminine sexuality, is related to "partial drives." His subsequent discussion of partial drives, however, shows the connection to be tenuous, even opaque. Commentator Jacqueline Rose, summarizing this passage in Lacan, writes both that "Sexuality belongs for Lacan in the realm of masquerade," and that "For Lacan, masquerade is the very definition of 'femininity' precisely because it is constructed with reference to a male sign."[58] She does not go on to elucidate the relationship of "the feminine sexual attitude" to partial drives in what follows nor does she elucidate the relationship either among masquerade, femininity, and male sign or among sexuality, masquerade, and femininity. She speaks immediately and instead of "the question of frigidity," and "the difficulty laid over the body by desire."[59] At a further point, she states that relative to Lacan's earlier works, "we could say that woman no longer masquerades, she *defaults*," and in support of the claim she quotes Lacan: " 'the *jouissance* of the woman does not go without saying, that is, without the saying of truth,' " adding that for the man, again quoting Lacan, " 'his *jouissance* suffices which is precisely why he understands nothing.' "[60]

What Rose is at pains to justify is the claim that Lacan is sympathetic to "feminine sexuality," in other words, that he does not bind feminine sexuality to a phallic psychoanalytic. Thus of his later work she writes, "Lacan talked of woman's 'anti-phallic' nature leaving her open to that 'which of the unconscious cannot be spoken'."[61] From this perspective, the relationship between masquerade and femininity does indeed appear to be a matter of something like default. A woman cannot be accused of dissembling if she really does not know, i.e., cannot speak of that which she is asked to speak. Before examining the question of what a woman knows or of what a woman can in fact speak, and this through an examination of *jouissance,* we need first to clarify further the Lacanian notion of masquerade.

At least in its first or early Lacanian guise, masquerade is akin to the unconscious. Masquerade *is* in fact exactly what the unconscious *does.* It dresses up in other people's clothing, so to speak; it disguises itself. The appearance it puts forth is a cover for something else. Personifying, we can say that it *knows* but dissembles. An unconscious that knows and a subject who does not know together present a problem for psychoanalysis that Sartre discusses at length. But it is not that problem that is of immediate concern here. What is of immediate concern here is that through the workings of the unconscious, feelings, affects, motivations, inclinations, and so on—what in Lacanian terms might be called "the *real* bodily-real"—are hidden. Thus, if masquerade characterizes "the feminine sexual attitude," if it is the quintessence of female sexuality, then females similarly hide "the *real* bodily-real." Like the unconscious, the appearance they put forth is a cover for something else, and although they *know,* they dissemble. In sum, it seems that females are as much like the unconscious as the unconscious is like females. Should we ask the question which came first, Lacan would likely tell us that "paradoxical as [it] may seem," sexual difference, being a matter of the phallus and not of biology (a Lacanian claim we will discuss in the next section), female and unconscious arrived together at the same time, namely, with the child's entrance into language.

Now in "The Signification of the Phallus," an essay presented in 1958, Lacan states what appears to be the original version of the paradox of how sexuality comes into play. He declares,

> Paradoxical as this formulation might seem, I would say that it is in order to be the phallus, that is to say, the signifier of the desire of the Other, that the woman will reject an essential part of her femininity, notably all its attributes through masquerade. It is for what she is not that she expects to be desired as well as loved.[62]

The precipitous and illogical jump from rejecting "an *essential* part of her femininity" to "*all* of its attributes" aside, Lacan three sentences later and in the same paragraph makes an equally jarring jump, this time in subject matter. He affirms that frigidity "is well tolerated in women and that impotence, on the other hand, is much harder on men." Male promiscuity —male desire for another and yet another woman—is in these terms given full field: a man's "desire for the phallus will throw up its signifier in the form of a persistent divergence towards 'another woman' who can signify this phallus under various guises, whether as a virgin or a prostitute. The result is a centrifugal tendency of the genital drive in the sexual life of the man which makes impotence much harder for him to bear."[63] Female infidelity, though given equal verbal assent, is reined in on the grounds that a female prefers staying attached to "the man whose attributes she cherishes."[64] Thus we find that while not specifically indicated as such, in a Lacanian psychoanalytic, there are centrifugal and centripetal genital drives.

But if this is so, then centrifugal and centripetal genital drives should be differentially articulated by the unconscious. All we are told, however, is that females *are* the phallus, or at least they *are* it to the extent that they desire to be desired and loved, and that precisely insofar as they "expect" males to desire and love them, they must masquerade as something they are not, namely, phalluses.

The notion is paradoxical indeed. And it is all the more paradoxical in light of the veil, for if females are to masquerade as something they are not in order to be desired and loved, then they need to keep the veil. If it is snatched from them, they can hardly show themselves as other than what they *really* are. Macey has admirably traced one dimension of this *real* as it appears in the Lacanian psychoanalytic. It is distinguished, he says, by a resolute female genital determinism—as the phrase "a gaping cunt" suggests. In brief, Lacan reduces female sexuality to a hole, a gross and unsightly gap in being that, genital determinism aside, is startling reminiscent of the obscene hole of which Sartre speaks and which we discussed in chapter 6. As for the *other* possible site of female desire, we might note that Lacan dismisses the clitoris with a condescending and sarcastic flourish: "I will not oblige women to measure by the yardstick of castration the charming sheath they do not elevate to the signifier."[65] Neither does Lacan's further traditional male justification of male infidelity and female fidelity help in providing further elucidation. All that is clear is that a woman must hide herself if she wants a man. By this act of dissembling, she mimics the unconscious. Like the unconscious, she disguises herself as something else, i.e., a phallus, in order to get her message of desire across. Masquerade is thus fundamentally descriptive of both females and the unconscious. When Lacan remarks of the unconscious that "the need to disappear seems to be inherent in it,"[66] he could as easily be speaking of females, whether with respect to being what they really are or speaking what they really know.

Lacan's abrupt changeover from a discussion of "the feminine sexual attitude" to frigidity and impotence, or, in Rose's terms, from "femininity" to *jouissance,* marks a substantive change of topic. We are catapulted into bed, so to speak. It is at this site that Lacan resolutely and decisively raises the question of female knowledge. As indicated above, we are told by Rose that here, that is, with respect to *jouissance,* it is not a matter of females' dissembling but of their defaulting. To clarify the points at issue, let us spell them out in terms of the unconscious.

The unconscious hides its knowledge—just like a woman: if it does know, it is not telling, at least not directly nor to our immediate understanding; if it does not know, it is really dumb in this particular matter—literally speechless. But this speechlessness becomes problematic in a particular way in the Lacanian psychoanalytic. How indeed can such speechlessness be? That is, how can the unconscious not give us the answer here? How can language fail us? In particular, how is it that the signifier of signifiers, which presumably has dominion *über alles,* does not live up to its own billing? That in this instance it falls short of omnipotence indicates a seeming flaw in the Lacanian psychoanalytic, indeed, a quite fundamental one. In addition, the speechlessness of Lacan's unconscious sets up a queer tension in his psychoanalytic. His initial and classical archetypal identification of females with the unconscious aside, the speechlessness of his unconscious raises the possibility that to be female is to be whole in the sense that perhaps, just perhaps, *jouissance* is utterly satisfying, which would mean that female sexual desire is not constituted by lack, that it is not definable in terms of partial drives, and that in addition, it is not sited on partial objects. Irigaray's affirmations on behalf of "this sex which is not one"[67] coincide with this analysis. In both cases, it is clear that female sexuality cannot be genitally reduced, for the sex which is not one is not reducible and wholeness is precisely wholeness. Epistemologically, this means that perhaps females actually know what they are "supposed-to-know"[68]—a phrase Lacan uses with respect both to the analysand's view of the psychoanalyst as having all the answers and to the psychoanalytic transference relationship—but that they choose not to tell; it means that perhaps like the astute and skillful Lacanian analyst who is "supposed-to-know" but who remains silent, they too choose to remain silent.

Now Lacan's synecdochic rendering of females is at bottom not only what fundamentally propels him initially to bind females to the unconscious; it is also what propels him later to adopt a particular archetype of females, a rendition of females as preeminently she-devils who will not tell what they know. The rendering is near fatal to the psychoanalytic because it does not just disrupt or create a trivial impasse; it calls into question both the structural lucidity and the systematic ordering of the Lacanian unconscious. In order to extricate his psychoanalytic from impending theoretical disaster, *it is necessary, logically necessary, for Lacan to claim that females do not know.* The claim, of course, comes at a price—of which more presently—but it saves the psychoanalytic. It saves the psychoanalytic by protecting its laws of the unconscious from female contradiction. Moreover it protects male power. By denying any such thing as *female* knowledge, Lacan can affirm that there is nothing to elucidate. Hence males as well as psychoanalysis retain power, which means that what is female can be kept strictly under patriarchal control—and subject to derision.[69] Because Lacan's denial of female knowledge is more a matter of proclamation than demonstration, however, the possibility of enlightenment is not altogether closed off, and a certain vulnerability remains. The undercurrent of threat, envy, and resentment that runs along in Lacan's psychoanalytic, an undercurrent commentators have noted but not traced to its source, is explained by the above analysis. The undercurrent will be further elucidated below in the discussion of male and female realities of *jouissance*. For now, it is clear that the possibility of females having knowledge, *carnal* knowledge, and of their having desire *and* satisfaction, cuts at the heart of Lacan's psychoanalytic: language is at a loss for words. To save the psychoanalytic, females must be cast out. To put the matter concisely and at the same time in far-reaching terms, insofar as female *jouissance,* by refusing incorporation into the Lacanian psychoanalytic, threatens to undo the whole of the psychoanalytic in the form of the unconscious as Lacan himself has formulated it, female sexuality must be dismissed; females must be "beyond the phallus,"[70] and indeed, Lacan characterizes them just so.

We should note that if we apply Lacan's rule about a subject who is "supposed to know"—"as soon as there is a subject that is supposed to know, there is transference"[71]—then we obviously find that a transference relationship obtains in the situation under discussion, Lacan, and perhaps a number of other males as well, putting onto females all of their demands, whether for knowledge or other benefits. It was remarked above that perhaps the putatively epistemologically deficient—or linguistically recalcitrant—female, like the astute and skillful Lacanian analyst of whom Lacan writes, actually *chooses* to remain silent, and by her silence, attempts in the very manner of a Lacanian psychoanalyst to bring Lacan and other males into the light of their own darkness. In addition to applying this Lacanian rule concerning the subject who is "supposed to know," we can further- more apply the Lacanian principle specifying the aim and end of Lacan- ian psychoanalysis. If the aim of the latter is finally to bring the analy- sand to "full" speech—by which term Lacan means that the analysand "can talk to you about himself"—then by analogy, the silence of the recalci- trant female is for psychoanalytical effect. In brief, if Lacan adheres to the tenets of his own doctrine, his begging of females—his "begging them on [his] knees," to use his very own words[72]—is of critical psychoanalyti- cal significance. If his own speech is indeed *full,* then he must actually be "talking to us about himself" with respect to his frustration with, and re- sentment about, female silence—and with respect to his own unfulfilled desire.

In the context of his rule that any time there is a subject who is "supposed to know," there is a transference, Lacan's painting of females into an epistemological corner is noteworthy in a further respect. When he introduces the notion of "a *jouissance* proper to her, to this 'her' which does not exist and which signifies nothing," Lacan specifically avers that of this *jouissance*, "she herself may know nothing, except that she experiences it—that much she does know. She knows it of course when it happens. It does not happen to all of them."[73] He then attempts to link the suggested female epistemological deficiency to the silence of female mystics, but only after having quickly and conclusively decided that, after all, "woman knows nothing of this *jouissance*," even though "we've been begging them . . . begging them on our knees to try to tell us about it, well, not a word! We have never managed to get anything out of them."[74]

Coupled with the failure of language to tell all, the possibility that female knowledge is not being shared and is indeed *other than what it is 'supposed to be'* forcefully shifts what was formerly a comfortable imbalance of male/female power, the male terms of Lacan's discourse obviously anchoring power on the side of the male to begin with. Not only this, but the two vulnerable facets of Lacan's psychoanalytic together suggest that, for all his claims to mapping *the* unconscious, Lacan is in fact mapping *his* unconscious. In particular, and in addition to the earlier discussion in chapter 7 of the possibility of a linkage between Lacan's psychoanalytic and Kant's productive imagination, we could say that from a Jungian point of view Lacan, in his initial archetypal joining of feminine and unconscious, is letting his anima come forward into the light. The possibility warrants serious if brief examination and comment.

V. Lacan's "Female Sexuality" in Jungian Perspective: A Further Analysis of Psychoanalysis

However great the threat of essentialism in Jung's anima/animus formulations, it is nonetheless surprising that those critical of Lacan, especially feminists assaying the relevance of his psychoanalytic to feminism, have not turned to Jung as offering another possible perspective on the psyche—and on Lacan himself. The suggestion that Lacan's psychoanalytic is an expression of his anima is strongly supported first of all by the evidence brought forward linking the unconscious to the feminine. Lacan's psychoanalytic of the unconscious is indeed interlaced with just those qualities Jung associates with a certain face of the anima (see below). At the same time, however, the interlacing is colored over by Lacan's own predilections. But then Jung shows and insists emphatically that an archetype is never given pure and simple—like a reflex, for example. Archetypes are *formal* images, which is to say archetypes are *"forms without content,* representing merely the possibility of a certain type of perception and action."[75] Archetypes are in effect not "inherited *ideas* but . . . inherited *possibilities* of ideas."[76] They are given specific content by the individual. Thus, like any archetype, the anima is variously elaborated—influenced—by one's personal unconscious as well as by one's culture.[77] Jung further contends that any archetype has a positive and negative side and that these dual sides must be taken into account. The descriptive analyses in chapter 6—in conjunction with Sartre's characterization of females' being "in the form of a hole" and with the further characterization of males being in the form of a phallus/copula[78]—support his contention. In what follows, we will critically explore Lacan's female sexuality in Jungian terms as above. We will in this way carry one step further the analysis of psychoanalysis.

Coincident with critical commentators' estimations of Lacan's attitude toward females, Lacan's anima surfaces as a pernicious one: the feminine divides; the feminine is Other; the feminine is evasive; the feminine obfuscates—in each case, just like the unconscious. All the same, Lacan's anima is subject by Jungian standards to Lacan's animus, which demands intelligence, formulae, power. The result is an *imposed* order. With his graphs, laws, vectors, and the like, Lacan maps a recalcitrant, uncooperative, deficient Other in indelible and all-knowing script, that is, in universal terms. There is nothing outside of his control. From a Jungian perspective, however, the Other onto which all the ordering is imposed is a product of his own unconscious, *a projection* of his own repressed feminine side. Projection, in a Jungian sense, is an activation of unconscious contents.[79] Jung describes projection at one point as "a process of *dissimilation* (v. *assimilation*), by which a subjective content becomes alienated from the subject and is, so to speak, embodied in the object." He goes on to say that "The subject gets rid of painful, incompatible contents by projecting them, as also of positive values which . . . are inaccessible to him."[80] Macey's remark quoted earlier—that Lacan's "theses propounded with respect to language within psychoanalysis are projected on to [sic] language as such"[81]—is pertinent in just this further context, and not simply in the paraphrastic sense that "theses propounded with respect to feminine sexuality within psychoanalysis are projected on to females as such." If we take Macey's claim to be correct, then projection is as much at the base of Lacan's propoundings about female sexuality as it is at the base of Lacan's propoundings about language. The equivalence is significant insofar as both language and sexuality are the primary stuff of the Lacanian unconscious. Moreover, if we put Macey's claim about Lacan in Jungian perspective, it strongly suggests that in imposing order upon language with his vectors, algorithms, metaphoric and metonymic analyses, graphs, and so on, Lacan is not simply *expressing* his animus; he is projecting his animus no less than he is projecting his anima. With this possibility, an interesting psychoanalysis of Lacan's psychoanalytic begins to emerge. A clear tension is evident between feminine and masculine elements. If Lacan's renderings of language and of feminine sexuality are each projections, and if they are in one way and another definitive of the unconscious, then not only is the unconscious itself a Lacanian projection, but the subject of the Lacanian psychoanalytic is also. The subject is split, alienated, full of holes, bifurcated, a damaged piece of goods, and so on, not because of a universal defect in the *human* psyche but because of a serious imbalance between masculine and feminine elements in Lacan's. This critical point is consonant with the earlier analyses of chapter 7; that is, it is supported by the fact that Lacan's psychoanalytic is undocumented by actual clinical histories and particularly by *self-analysis*.

Lest it be thought that an uncritical case is being made for Jung's analytic psychology, it should be noted that Jung's rendition of females is certainly not free of projection, derision—or violence. The most astonishing, utterly deplorable, and insensately violent instance occurs in the context of his discussion of a woman's animus: "No matter how friendly and obliging a woman's Eros [anima] may be, no logic on earth can shake her if she is ridden by the animus. Often the man has the feeling—and he is not altogether wrong—that only seduction or a beating or rape would have the necessary power of persuasion."[82]

VI.   Lacan's Female Sexuality and the Question of *Jouissance*

Lacan's necessity to speak of *jouissance* in a final attempt to explicate female sexuality is telling, not the least because, however much it falls outside of language and however obliquely he treats it, he must concern himself with actual experience. Irigaray implicitly recognizes this deviant necessity in describing the polymorphously perverse nature of female sexuality, and quite obviously turns her attention too to *jouissance* in her critical rejoinder. But the central psychoanalytic problem of the relationship between unconscious and female remains. On the one hand, there is the unanalyzed unconscious and a synecdochally rendered female sexuality; the two are archetypally conjoined in psychoanalysis in the form of a dark hole. On the other hand, in the course of Lacanian psychoanalysis, the unconscious should reveal female sexuality by the very rules Lacan has laid out. The doctrine of signifiers, the laws of the unconscious, ought to bring female sexuality to light, but they clearly do not, for Lacan is, and indeed remains, puzzled. In preface to his essays, "God and the *Jouissance* of The Woman" and "A Love Letter," essays which are regarded "the two central chapters of Lacan's Seminar XX, 'Encore',"[83] Jacqueline Rose and Juliet Mitchell write that "Encore marks a turning point in Lacan's work, both at the conceptual level and in terms of its polemic. It represents Lacan's most direct attempt to take up the question of feminine sexuality." They go on to state that

> It is the central tenet of these chapters that "The Woman" does not exist, in that phallic sexuality assigns her to a position of fantasy. . . . Against this fantasy [of Oneness], Lacan sets the concept of *jouissance*. *Jouissance* is used here to refer to that moment of sexuality which is always in excess, something over and above the phallic term which is the mark of sexual identity. The question Lacan explicitly asks is that of woman's relation to *jouissance*. It is a question *which can easily lapse into a mystification of woman as the site of truth*[84] (italics added).

Even with such an introduction, one might well ask, "Just what *is* the problem with female sexuality?" It is nowhere clearly or explicitly stated by Lacan. There is only the suggestion, as Rose and Mitchell point out, that feminine sexuality is an *excess* with respect to phallic determination. But surely Lacan himself is giving us a sufficient clue when he shifts gears from talk about masquerade, veils, holes, and inconsequential sheaths to *jouissance*. Stated most simply, female sexuality is a problem because females are capable of one orgasm after the other. They are capable of an *excess* of sexual pleasure or delight or joy, and such an excess is not accounted for in the Lacanian psychoanalytic libidinal economy. The psychoanalytic problem for Lacan is thus precisely to explain how and why this excess does not fit into his psychoanalytic. Let us consider the incompatibility in more exacting terms.

In Lacan's as in Freud's psychoanalytic (as in Sartre's psychoanalytic of things in the world), females are in the form of a hole. How can this hole, this nothing, be the source of a seriate joy? Though central, this is not the only question that female sexuality poses for the Lacanian psychoanalytic. There are others that are utterly incongruent with the psychoanalytic staples we have come to associate with Lacan as well as with Freud. In particular, how can *penis envy* exist in creatures whose capacity for *jouissance* so far exceeds that of creatures who have a penis that Lacan must speak of their *jouissance* as "supplementary"? How can *castration anxiety* enter into a female psychoanalytic since, even with their "charming sheath [that] they do not elevate to the signifier," females are still capable of multiple orgasms at one sitting, so to speak. There is a fundamental logical error in the psychoanalytic. Lacanian psychoanalytic theory, and psychoanalytic theory in general, must be festooned with *ad hoc* features and explanatory disgressions. The result is a lack of parsimony. The whole does not by itself hold together. It is fairly obvious that penis envy and castration anxiety do not enter constitutionally into a female psychoanalytic unless aided on the one hand by male projection, and on the other hand by a phallic-oriented culture that all the way from infancy to adulthood warps the female psyche at the same time that it laments its not coming into view. In short, penis envy and castration anxiety are made to enter the psychoanalytic scene by way of projection and a cultural privileging of the penis. Indeed, how could they possibly enter into an account of female sexuality unless everything sexual is made to rise and fall with the presence or absence of a penis?

Now strikingly enough, and as suggested by the foregoing analyses, the more the Lacanian unconscious comes into the light, the more enigmatic female sexuality becomes. We see this as a clear development in Lacan's psychoanalytic. Correlatively, the more Lacan's unconscious comes into the light, the more an emphatically and authoritative male sexuality comes into the light. This relationship will be elaborated in detail in the following chapter. Suffice to note that it is not surprisingly a question of the phallus. What is of moment here is how strongly the developing relationships—of female and male sexuality to the unconscious—recall Jung's description of the psyche as a continuum that reaches from the depths to the shallows, from archaic or collective unconscious understandings to singular and individual conscious ones that risk overstepping "the bounds of conscious comprehension." What progressively comes to the fore in Lacan's developing psychoanalytic is an enigmatic female and a fully transparent male, *aspects of Lacan's own psyche,* along with "all sorts of attempts at rationalistic and therefore inadequate explanation," and this because Lacan does indeed want nowhere to exceed "the bounds of conscious comprehension." What is furthermore of significance here is that the wonder side of female sexuality is suppressed; Lacan in fact razes that corporeal archetype. Like Sartre, he denigrates what is female. It clearly threatens his preeminently phallic psychoanalytic just as it threatens Sartre's patriarchal ontology. If we work backwards through Lacan's writings from *jouissance,* the term by which Lacan finally recognizes feminine sexuality in a distinctly positive but at the same time derisive way, we in fact find female sexuality consistently stifled. Whatever it might be, it is too saddled with male psychoanalytic baggage for it to be recognized. Females are made to carry that baggage and carry it they do insofar as they are indoctrinated into male culture. Of course, they lack a penis. In truth, they have no sexual "organ." That is precisely the conundrum for a psychoanalytic that sees nothing but a hole when it considers females. There is no way such a psychoanalytic can accommodate much less understand female sexuality, for it cannot fathom how *nothing* can possibly be the site of an excess of sensuous pleasures, one and then another, over and over again. How can a veritable gap in being bring a seriate *jouissance?*

If we reflect patiently and deeply upon this question, we come to realize that there is a different answer from the one offered by Lacan or by those feminists who turn toward a kind of speaking or a kind of writing that would celebrate the female body and thus secure females a place in the "libidinal economy," a socio-political foothold that would subvert phallocentric theory and practice. We realize first that the question is raised within a *male* "libidinal economy" and that it is precisely its euphemistic self-reference—"libidinal economy"—that must be translated into corporeal terms. Our attention must then be directed to the question of how and why this seriate *jouissance* can be a problem for those who do not experience it. To reflect upon the initial question is thus to realize that, psychoanalytically speaking at least, and psychoanalytically speaking within the present Western psychoanalytic frame of reference, there is a disjunction between male and female sexuality that, for some males at least, creates a disaccord, a friction. We might label the source of this friction *"jouissance* envy," but not without attempting to explain its provenience and arrive at a minimal understanding of it.

It is significant that when Lacan turns to his elucidation of female *jouissance,* he almost immediately deflects us from any corporeal investigations by turning our attention to *statuary.* He begins speaking of mystics and saints, then of courtly love and God.[85] He never comes close to speaking of the living body. Yet *jouissance* is clearly an experience of the felt body, in particular, the tactile-kinesthetic body. It is perhaps *the* consummate tactile-kinesthetic experience for human adults. How, then, can one possibly hope to explain females psychoanalytically by explaining "the problem of feminine sexuality" in terms of *jouissance* if one omits reference to the living body, and not simply the *female* living body, but the *male* living body for whom it is preeminently a problem? It bears emphasizing that the problem must be considered at its source.

Since within Lacan's psychoanalytic the living body is not a proper referent point, it is not surprising that tactile-kinesthetic experience never enters into the psychoanalytic of *jouissance.* The body in fact exits as quickly as it is introduced: "There is a *jouissance,"* Lacan writes, "since we are dealing with *jouissance,* a *jouissance* of the body which is, if the expression be allowed, *beyond the phallus.*"[86] It suffices for Lacan to claim that female *jouissance* is "beyond the phallus" in order for him to disavow the body, for whatever is beyond the phallus is beyond articulation, whatever is beyond articulation being of neither psychoanalytic significance nor interest. Lacan, in effect, upholds his psychoanalytic; he has not solved "the problem of feminine sexuality"—defined as *jouissance*—he has merely put it, i.e., female *jouissance,* outside its borders. By this move, the female living body, in addition to its categorical displacement within the psychoanalytic, is substantively removed from the scene of the unconscious. If we ask why Lacan finds it necessary to take such measures, we do not have far to look, for living bodies themselves, male living bodies, give us at least three answers.

On the way to considering these answers, we should note that if female sexuality is "beyond the phallus," then the phallus does not reach that far; female sexuality is an unknown density beyond language. But how, given Lacan's psychoanalytic, can there be an unconscious that does not speak the subject? Recall that with Sartre, Lacan believes that psychoanalysis is not a matter of *conscience* but of *connaissance.*[87] If the unconscious is a repository of knowledge and if a psychoanalytic cannot get that repository to open its doors and speak what is within, *or* if it cannot even get the conscious subject herself to speak, then the unconscious remains an unknown, dark pit. Like a female body, it remains a mystery. This corporeal archetype of a feminine mystique becomes a femme fatale to Lacan's psychoanalytic. His unconscious rises up to the light of day only to fall back for lack of support; it cannot rise to the occasion of female sexuality. Of course, for Lacan, it is not his unconscious that fails; it is females. They, not his psychoanalytic, are at fault. In banishing female sexuality to a realm beyond the phallus, he does not seem concerned that only some "unconsciousnesses" can be brought up to the light and that others remain a "zone of shades."

In fact, considering the whole of humanity, a sizable zone remains, and with it, the classical conception of the unconscious. Lacanian pronouncements about females' being beyond the phallus notwithstanding, the corporeal archetype of the female as model of the unconscious prevails. Something wonderful and foreboding cannot be reduced to the law of the signifier; the phallus is powerless to condense it in an algebraic formula. Ultimately unknown centres, like beings in the form of a hole, exceed Lacan's psychoanalytic. All such sites of wonder and foreboding are potentially sites of new discoveries and new beginnings, sources of creativity that are outside the endless chain of signifiers. The *real* bodily-real that grounds our thinking is hence not subdued. On the contrary, all of our powers originate on the grounds of that corporeal *real.*

10

Corporeal Archetypes and Power: The Phallus as Archetype

Nowadays, no one thinks of the phallus simply as a penis.

Serge Leclaire[1]

Feminism proclaims that the personal is political. Psychoanalysis—both vulgar and Lacanian—replies that the political is pathological.

David Macey[2]

I. The Three Answers

A. *Autonomy*

Eugene Monick, in his book *Phallos: Sacred Image of the Masculine,* speaks of being at the mercy of an organ that one cannot control. "Autonomy with regard to phallos is grounded in the experienced reality of men that they cannot, however much they might wish otherwise, make phallos obey the ego. Phallos has a mind of its own."[3] We should note before examining Monick's informative study further that one might expect to find autonomy linked to *personification of the penis,* the former being considered conceptually related to, if not the necessary antecedent or causal condition of, the latter. But that "Phallos has a mind of its own" does not inevitably figure either conceptually or conditionally in discussions of personification. Semiotician Daniel Rancour-Laferrière, for example, explains the personification of the penis on the basis of inclusive fitness, a sociobiological concept that fundamentally seeks to explain all behavior on the basis of benefit to one's constellation of genes. Hence, according to Rancour-Laferrière, "if ancestral hominids started treating it [the penis] as a person—talking to it, worshipping it, making special efforts to protect it, etc.—then the genes of penis possessors and their relatives may have benefitted. . . . What I am suggesting, of course, is the threat of castration."[4] The problem with such explanations is precisely that they ignore the experience of human males that Monick describes, the experience of being out of control: a penis behaves in an autonomous manner. Indeed, one could rightly say that a penis behaves like "a little person," i.e., a child. Not only this but it has certain well-known characteristics of "little persons." It is undependable, capricious, it does things that are "naughty," it is balky, it does not do what it is told, it wants its own way, it demands attention, it can make a scene, and so on. That personification of the penis is not seen in these terms is a measure of how theory can outrun corporeal matters of fact and how embarrassing questions of autonomy can be suitably silenced.

Describing the autonomy of phallos further, Monick writes that "Enspirited phallos is the reason teenaged boys are embarrassed in public as individuals. . . . Boys never know when phallos will appear. They are not sure of themselves as erectile beings."[5] Elaborating still further at a later point he says that

> Spirit comes to a man "bidden or unbidden." . . . Always it comes autonomously. Unbidden, it rises from the unconscious without warning, quickly making penis into phallos . . . taking one by surprise. Bidden, it is prepared for, as in masculine initiation calculated to poultice spirit from the unconscious and make erections ready to inseminate. When it arrives, bidden or unbidden, phallos makes males feel alive, in awe of the transformation within themselves. They are fascinated by the appearance of power in what had been flaccid. . . . [P]hallic spirit . . . is profoundly and archetypally masculine.[6]

Clearly, Monick, a Jungian analyst and former Episcopalian minister, is describing a fundamental, deeply meaningful experience of human males.

Multiple and interrelated aspects of phallic autonomy bear on the Lacanian psychoanalytic. All of them are tethered to the question of control, in particular, to the question, how does one control something that is unpredictable? Lacan's psychoanalytic furnishes an immediate answer: what an ego cannot control, Lacanian psychoanalytic theory can. The theory removes the wayward and recalcitrant member from the register of the real (Real) and places it in the register of the Symbolic. In effect, one takes what is wayward out of circulation. (It is ironic that "circulation" is the term Lacan uses to describe the way in which the phallus permeates all of language and anchors all signifiers: it *circulates* in language.) To remove phallos from everyday realities is already to affirm who is in charge. Removal from the everyday is compensated not only by placement in the Symbolic but by placement in the position of highest authority in the Symbolic. So situated, phallos is no longer unpredictable. It is no longer part of the real (Real) in any way. Its status as pure signifier protects it from any taint of unpredictability. It is, in effect, *always erect*. However there in name only, i.e., however Symbolic, it is on permanent display. We thus begin to see from the very start how the Lacanian phallus is an elaboration of a corporeal archetype.

When Monick writes that phallos cannot be made to obey the ego, he implicitly raises the question of what it is that phallos does obey. It is reasonable to assume that a Freudian would answer straightforwardly that in its ups and downs, phallos is obeying the id. The vagaries of penile erection are, in effect, tied to the unconscious—just as Monick says they are. It is difficult to imagine Lacan answering such a question, but assuming he would, his answer would necessarily be similar. After all, if language—the conscious—speaks the subject, the vagaries of phallos are controlled by the unconscious as well, i.e., by the Other. The Freudian and Lacanian "registers" may be different, but their answer is virtually the same: unpredictable upsurges of phallos are unpredictable gaps in consciousness. Unquestionably, then, phallos is outside of a male's control. Unquestionably then too, beyond the idea that phallos is subservient to the unconscious is the idea that phallos *behaves* like the unconscious. Monick supports such a notion when, aligning phallos with the mystical as well as the psychological, he writes that "Physical phallos has become a religious and psychological symbol because it decides on its own, independent of its owner's ego decision, when and with whom it wants to spring into action. It is thus an appropriate metaphor for the unconscious itself, and specifically the masculine mode of the unconscious."[7]

Given the plausibility of the association, one might wonder how Lacan could possibly avoid encompassing phallos within his psychoanalytic, specifically with reference to the psychoanalytic anatomy he proposes and to that anatomy's ties with the unconscious. Why, in other words, is his psychoanatomical body limited to holes? Moreover, why is it a thoroughly static body, a body devoid of movement? Wonder readily fades the moment we consider the consummately definitive psychoanalytic Lacan offers. There is no place for unpredictability in this psychoanalytic because to admit the unpredictable would be to admit an irrational element and irrational elements, as noted in chapter 7, are deplored; such elements are tied to flesh and blood bodies. This explains why only bodily holes and slits appear as sites of the unconscious in Lacan's psychoanalytic, i.e., as sites of desire, and why Lacan of necessity devises a quite circumspect psychoanalytic anatomy to accommodate them. Their openings and closings are not really the stuff of living flesh. They are rather places on a body that can be "linguistified," that is, turned into speech. In more exacting Lacanian terms, the openings and closings of bodily holes and slits are consonant with the "pulsative" nature of the unconscious, *its* openings and closings. The question of control in effect never arises with respect to these openings and closings because these acts are part of the inherent need of the unconscious to speak and to disappear.[8] But of course the same could be said of phallos. Indeed, psychoanalytically speaking, the likening of phallos's behavior with the behavior of the unconscious cannot be dismissed. It is clearly supported by male experience. Monick in fact links the autonomous behavior of both phallos and the unconscious to a mysterious dimension, which obviously suggests, and in a strong way, that there is something inarticulable about phallos and the unconscious, something that resists full analytic exposure precisely because each is autonomous. Assaying the similarity in behavior from Monick's Jungian perspective, we could say that what Lacan would bring to full light in the name of the unconscious turns out, at the level of living bodies, to be something that cannot be tamed, something that obstinately remains outside of a male's control, something that *is* an irrational element, and that in consequence is as beyond Lacan's psychoanalytic as "feminine sexuality." In short, if we accept Monick's analytical claims as genuine, that is, as supported by experiential and corporeal matters of fact, then not only female *jouissance* is beyond the phallus; *phallos is beyond the phallus*. If this is so, then the very Signifier of signifiers is, improbably enough, beyond language. It can no longer enjoy its Symbolic privilege. Something fundamental to its elevated position has outstripped its authority and rendered it powerless.

Clearly, animate form is not only of utmost significance to psychoanalytic theory; an appreciation of it is mandatory. What is said of the unconscious must jibe with animate form, which is to say that whatever is said of the unconscious must accord with what living bodies experience in the course of their everyday lives. Although Freud did not speak of animate form nor dwell on the body, certainly he made clear in his first writings on dreams, slips of the tongue, and so on, that whatever was said of the unconscious must accord with actual experience. What is said of the unconscious must furthermore jibe with *differences* in animate form. As is evident from analyses and discussions in previous chapters, Lacan mapped the unconscious, specified how it works, and specified as well what it is one must do in order to be disabused of false egoistic notions of desire and cured of one's psychoanalytic ills—all without substantive reference to living bodies; Lacan's own writings tell us unequivocally that his psychoanalytic technique brings the unconscious into the light. This being so, a critical analysis of his work might charge that the unconscious that Lacan brings into the light is a masculine mode of the unconscious. However removed from "physical phallos," his linguistic totem is nonetheless modeled on a male body. Accordingly, however much Lacan might claim that the Signifier of signifiers is neuter, Lacan's phallus is not neuter at all. There is indeed nothing to say against its being a purely male-privileged signifier. Lacan himself has difficulty justifying exactly why *it* is privileged.[9] This is why we have legitimately questioned, why a phallus and not a pickle? Put in this perspective, Lacan's phallus brings both animate form and the experience of unpredictability into decisive focus.

The removal of the phallus from the real (Real) and its placement within the Symbolic obviously cancel out its ups and downs. This overlooked fact is by itself portentous. Archetypes are culturally and individually reworked in just such ways, that is, by ignoring or suppressing certain aspects of life in favor of others. If we add to this overlooked fact the possibility that as a purely linguistic "emanation" of the unconscious—a *male* unconscious— the Lacanian phallus expresses male desire, we come to an appreciation of how Lacan archetypally expresses a male thematic at the same time that he avoids entirely the question of male autonomy. Lacan's phallus expresses male desire, the desire for Lacan's power, a *continuing* experience of power—in Monick's words, "what makes males feel alive." As Symbolic, of course, a perpetually erect penis is never spent but is always there unabridged, as it were, ever ready for insertion anywhere in Lacan's psychoanalytic. Accordingly, its power is never in question because it is never diminished. In effect, viewed within Lacan's psychoanalytic, the phallus *satisfies fully* the desire of males to control what otherwise "has a mind of its own." The desire is fulfilled by the linguistic instantiation of a priapic condition. Not only is there no lack attached to this condition, but satisfaction of desire is secure against all suffering. Ensconced within the Symbolic order, priapism is beyond pain.

Just as the linguistic totem that reigns supreme brings the unpredictable under control, so also does a linguistic priapism that is beyond pain. In fact, between totem and priapism, unpredictability is negated. These dual aspects of control are fundamental to the establishment and maintenance of patriarchal power. By erecting a linguistic totem and then by ensuring that it can never be toppled, Lacan preserves the phallus in perpetuity. The archetype of penile display is thus reworked into a Symbolic ultimate. This supreme Signifier furthermore performs a fundamental and altogether formidable service: it keeps castration fears at bay. Castration fears are kept at bay because they are encompassed within the Symbolic along with the phallus, and, like the phallus, they undergo transformation. To appreciate this fact we must introduce some new Lacanian themes.

B.   Castration

As was amply demonstrated in earlier chapters as well as by the phallos/ phallus relationship, the living body haunts and continues to haunt Lacan's psychoanalytic. To see this is to see why castration is a major Lacanian theme that is tied to death. According to Lacan, death enters the scene with a child's first words, which signify both a lack of being within itself, a *finitude* of its own being, and a symbolic replacement, that is, the death of the other since the other is replaced by words.[10] But as Monick describes, and as we shall begin to specify and to appreciate more exactingly, the connection between castration and death is secured at an earlier level, specifically, at the *bodily* level by the death of phallos, i.e., by detumescence.

Insofar as the privileged Lacanian signifier is phallos in linguistic disguise, it of course never undergoes detumescence. Indeed, if it did, why would the privileged signifier *not* be a pickle? His earlier references to sexual copulation and turgidity notwithstanding, the reasons Lacan offers on behalf of his choice of the phallus as privileged signifier are reasons that, as one commentator observed, "are never made explicit . . . [but are] dictated by the requirements of Lacanian systematization."[11] On the one hand, "requirements of thought,"[12] as Lacan himself once put it in setting forth the reason for a principle in his psychoanalytic, hardly suffice as reasons. Moreover, to *insist* that the Law (or name) of the Father ordains the phallus as privileged signifier is reasonable only insofar as one affirms the phallus to have bodily ties. Fathers, after all, are male. Nonetheless, at the level of the Symbolic Lacan literally insists on his ordination of the phallus as Signifier of signifiers. In speaking of the symbolic order, he baldly states that the order is "founded on the existence of the name of the father," and then writes, "I insist—the symbolic order must be conceived as something superposed."[13] On the other hand, to speak of the phallus in *real* terms, in terms of sexual copulation and turgidity, and then to say that "these propositions [concerning sexual copulation and turgidity] merely conceal the fact that [the phallus] can play its role only when veiled, that is to say, as itself a sign of the latency with which any signifiable is struck, when it is raised . . . to the function of signifier,"[14] is to cancel out the very *real* reasons offered for selecting the phallus in the first place. Moreover it is to warn that any references to the living body will veil over ("conceal") the fact that the phallus must be veiled in order to be and to remain a sign of latency. In brief, what Lacan wants us to believe is that we are led astray if we think that we can understand the phallus by any reference to living bodies, for the "true" phallus is hidden—veiled—within the Symbolic. Needless to say we are not told where the veil comes from, what it is made of, or who puts it on, but it is clear by Lacanian fiat that for the phallus to play its role, it must drape itself so that it can "function" as a pure signifier, as a latency of meaning.[15] Latency in consequence is always present, *but only because veiling conceals any change in actual status*. Translated into archetypal language, what veiling accomplishes is a reworking of penile display into an under-cover perpetual erection. Such a move is protective. There is no sign of detumescence; there is no death of phallos; and there is no castration. The Signifier of signifiers is absolutely removed from any such possible change in status and from the fears that go with it. Insofar as castration figures centrally in psychoanalysis, however, then to that degree the living body must figure just as centrally, Symbolic register or no Symbolic register. The notion of castration is, after all, *male* in origin. Lacan's claim that castration is just a metaphor for whatever the subject lacks and that the goal of analysis is symbolic castration[16]—the analysand's realization that s/he can never fulfill her/his desires, whatever they may be—is scarcely credible short of a living body, in particular a male living body. Clearly, no matter how privileged, the phallus retains ties with animate form. It cannot be thoroughly uprooted and conclusively carried off to a linguistic Olympus with a wave of the tongue. While conceivably a pure psyche might have desires, it could hardly envisage anything on the order of castration—even as metaphor—short of experiencing the ups and downs of a penis. In sum, the primacy of the phallus in the Lacanian psychoanalytic must in part be understood in terms of the bodily immediacy of male fears of castration.

Psychiatrist Robert Stoller writes that "With practically no exception, all the males we shall ever see, in our practices or anywhere else, fear castration." He adds that "This fear, with its underlying threat to one's identity as a male, creates a lot of what we judge in males to be masculinity, and when severe it contributes to that detour on the road to masculinity we call perversion."[17] Monick concurs, putting the fact in perhaps even stronger terms. He writes "As phallos enters a situation, an apprehension . . . takes place that could not take place without phallos. That is the horror of castration. It has always been so."[18] In light of this absolute avowal of castration fears, the question arises not only of the specific nature of these fears but of what dissipates them. What might keep, or help keep, castration fears at bay? A substantive bodily answer is not difficult to come by. Castration fears pale in face of a mighty phallus. Veiling aside, so long as the phallus is venerated, carried on high, whether literally, ritualistically, artistically, or linguistically, there is no chance of castration. So long as it is

worshipped, it is protected, not only from assault, *but from its own death.*
Roman festivals in which maidens paraded with a raised phallus are
testimonials to such veneration. These acts of worship are neither unique
nor idiosyncratic; they are or have been practiced in one form or another
cross-culturally. They are part of what has already been identified as
elaborations of penile display, elaborations that are documented in ethologi-
cal, biological, anthropological, and semiotic studies. Cultural elaborations
of penile display are also amply documented by illustrations in Monick's
book, illustrations of a Terracotta lamp sculpture, for example, of a marble
pillar with the head of Hermes at the top, of Priapus weighing himself (i.e.,
weighing his priapic penis), of figures such as the Cerne giant, and so on.[19]
Of course, to look upon females as castrated obviously elevates the phallus
even further, indeed beyond the need for totemic and/or priapic practices;
concentrated attention on females as castrated heightens beyond measure
the glories of the phallus. Penis envy thickens the plot even further,
ensuring an undying veneration insofar as, psychoanalytically speaking, it is
repeated generation after generation. The repetition guarantees its being
conceived not merely a psychopathological malady but a malady endemic to
females.

(In spite of the long Western tradition of regarding females as deficient,
to be fair, attention should parenthetically be called to the fact that penis
envy is, and has been, a controversial issue within psychoanalytic theory. As
psychoanalysts William Grossman and Walter Stewart recently remarked,
"because the theory lacks clarity, its clinical application does not always
produce the results expected." Their own clinical findings are of particular
theoretical as well as practical interest insofar as they strongly suggest that
interpreting psychological problems concerning "a sense of identity,
narcissistic sensitivity, and problems of aggression," on the basis of penis
envy has *"an organizing effect, but not a therapeutic one."*[20] This finding
explains why psychoanalyses anchored in penis envy are "stalemated"; they
provide a ready-made psychoanalytic anatomy on which one can hang one's
feelings, but they do not go to the heart of the problem. As Grossman and
Stewart indicate, penis envy "must be treated—like the manifest content of
a dream or a screen memory—as a mental product. It is not to be regarded
as an ultimate, irreducible, and even genetically necessary truth, impenetrable to further analysis." In short, penis envy is not the female "'bedrock'"
[a reference to Freud; see below] it is so frequently taken to be.)[21]

Now given that venerational festivals, monumental enshrinements, and
the like no longer take place in Western societies, and given the psychoanalytically affirmed absolute pervasiveness of castration anxiety, it is not
surprising that other means are sought and found to keep castration fears at

bay. In a sense, classical psychoanalytical theory, Lacan's included, appears to have taken the place of the festivals, enshrinements, and so on. No matter that the view of females that the theory presents is actually both a projection and the result of a phallic-raised culture. Insofar as envy of something is a way of venerating it—prizing it, treasuring it, and so on—then females as well as a totemic-priapic phallus can keep male castration anxieties at bay. The pivotal factor, as Lacan's psychoanalytic demonstrates, and as Monick's writings attest, is the distinction between penis and *the* phallus (or between penis and phallos). From the perspective of actual experience, it is evident that what merely dangles in front of one's eyes is powerless either to distract in a threatening way or to compel attention. What grows in size and stature and is firmly elevated has the potential power to do both. It has the power of an archetype. It can distract in an unsettling way, intimidate, and actually do violent harm. It can captivate or compel to worship. In this latter guise it is awesome in just the sense Monick describes when he speaks of phallos's autonomy and its fascinating power. When Lacan creates his linguistic totem, he draws on just this awesome dimension of phallos. He puts his phallus on par with ancient monuments and sculptures and with the phallic statuary carried overhead by Roman maidens in their processionals. His phallic pillar is a totem that can be inserted anywhere precisely because it is *the* omnipotent, every-ready signifier whose rule is law. To understand how castration anxiety can possibly enter into this psychoanalytic is to under-stand how Lacan's seemingly indomitable phallic autonomy is vulnerable. Specifically, it is to understand how vulnerability lies in the experiential difference between penis and phallos, i.e., in castration fears or the death of phallos. Lacan's indomitable phallic autonomy is, in other words, threat-ened by the uncontrollable lives and deaths of phallos. The challenge, then, is to understand how a male may be beset by fears that, having risen once, twice, or a hundred times, phallos may not rise again—in short, how a penis is not a phallus. As his commentators show many times over, Lacan was painfully unsuccessful in establishing this distinction so essential to his psychoanalytic.[22]

One quite obvious and simple distinction between a penis and a phallus, which Lacan did not call attention to and which his mentor Freud *did* call attention to and used inventively, focuses directly on the body. Freud remarked upon the distinction in a little known essay titled "The Acquisi-tion of Power Over Fire."[23] In this essay, he pointedly calls attention to the fact that a man cannot pee if he is on fire, i.e., in the heat of desire, and similarly, he cannot be on fire if he is peeing. Freud uses the myth of Prometheus as a starting point, mentioning that the fire Prometheus stole from the gods was hidden in a hollow rod—a fennel stalk. He states that if this myth is interpreted as a dream is interpreted, then the rod signifies the penis, and what "man harbours in his penis-tube . . . is the means of extinguishing fire," i.e., what he harbors is "the water of his stream of urine."[24]

Freud's initial concern is to answer the question of why stealing fire, a benefit to humanity, should mythically be considered a crime, and goes on to opine "that primitive man could not but regard fire as something analogous to the passion of love. . . . The warmth radiated by fire evokes the same kind of glow as accompanies the state of sexual excitation, and the form and motion of the flame suggest the phallus in action."[25] He even goes on to point out historical and locutionary correspondences between phallus and flames. Moreover he speaks of the symbolic similarity between fire and passion in terms of "daily consumption and renewal," that is, of how "the appetite of love" is "gratified daily . . . [and] daily renewed." In this same context he recalls the phoenix, "the bird which, as often as it is consumed by fire, emerges again rejuvenated," and states that "probably the earliest significance of the phoenix was that of the revivified penis after its state of flaccidity."[26] His point here is that libidinal desires are never sated once and for all but are continually revived. The myth of the hydra of Lerna furnishes a further instance of renewal after destruction. The myth of a water-serpent that is subdued by fire can be interpreted, Freud says, by reversing the manifest content of the myth. Accordingly, Heracles does not subdue water with fire but extinguishes fire with water. After linking the cultural acquisition of fire with myth and even with "physiological fact"—by quoting Heine's lines, " 'With that which serves a man to piss he creates his own kind' "—Freud finally states: "The male sexual organ has two functions, whose association is to many a man a source of annoyance."[27] The two acts, he says, are "as incompatible as fire and water. . . . When the penis passes into that condition of excitation which has caused it to be compared with a bird and whilst those sensations are being experienced which suggest the heat of fire, urination is impossible. Conversely, when the penis is fulfilling its function of evacuating urine . . . all connection with its genital function appears to be extinguished."[28]

A basic point of Freud's essay is to show that the acquisition of power over fire was connected with an awareness of the body, for Freud writes that "we may suppose that primitive man, who had to try to grasp the external world *with the help of his own bodily sensations and states,* did not fail to observe and apply *the analogies* presented to him by the behaviour of fire" (italics added). "Primitive man," in other words, was aware that the two penile functions are incompatible.[29] Accordingly, how did he gain power over fire? He learned not to pee on it, or, as Freud put it at the beginning of his essay, he "renounce[d] the homosexually-tinged desire to extinguish it by a stream of urine."[30] Now regardless of whether or not the desire is "homosexually-tinged," the point is that fire and water do not mix. If we take this point as a point of departure for understanding at a deeper level the distinction between penis and phallos/*the* phallus, then we need only articulate the psychoanalytic conclusion to be drawn from the fact that fire and water do not mix. To have absolute control over "the flame of love"—to have the power of phallos/*the* phallus—is equivalent to perpetual erection, which is to say *never having to pee,* for having to pee quenches the flames of passion and compromises autonomy. Power over passion in consequence means leaving the everyday body behind.

This is of course just what Lacan does when he creates his linguistic totem. The power of the flame, to use the terms of Freud's essay, is thereby made unquenchable and never dies. But even leaving the body behind, Lacan still preserves a sense of the real as he makes way for the power of the flame because according to his psychoanalytic, the Symbolic and the biological (in the form of sexual reproduction) *both* subserve the unconscious. Castration enters into Lacan's psychoanalytic precisely in the context of their junction. Castration is lack, and lack, Lacan tells us, has a twofold nature: at the Symbolic level, it is the subject him/herself who is castrated in that "the subject depends on the signifier and . . . [the] signifier is first of all in the field of the Other"; at the level of the real, it is the subject her/himself who is castrated in that in reproducing her/himself, "the living being loses . . . that part of himself *qua* living being . . . [and as such falls] under the blow of individual death."[31] With respect to Lacan's double characterization of lack, Roustang remarks, "The inevitable conclusion is that sexuality is simply inadmissible in Lacanian doctrine, which is why it is reduced to a hole. . . . If, in Lacanian doctrine, sexuality is defined exclusively with reference to lack, it is quite simply because sexuality is lacking in Lacan's doctrine."[32] Perhaps Roustang is correct and no more can be expected of Lacan's psychoanalytic. In any event, the price of Lacanian doctrine is high: insofar as "the subject, through his relations with the signifier, is a subject-with-holes,"[33] the subject is consistently and endlessly castrated, whether at the moment s/he opens her/his mouth to speak—in the register of the Symbolic—or the moment s/he begins having reproductive relations—in the register of the real.[34] What must be clarified, however, is the chronology of the two Lacanian registers with respect to castration and the consequences of that chronology.

Castration enters the scene well before sexual maturity, that is, well before losing part of oneself in reproduction. The moment of original castration is captured in psychoanalysis when the analysand recognizes that he or she is irremediably split, that is, *castrated,* and has been so since early childhood. With each and every desire, he or she has been (is, and always will be) a "want-to-be"—a want of being, or a lack—that is never satisfied. The original of the lack is the yearning for a "lost paradise" or primary fusion with the mother that is symbolized by the phallus. The phallus is the signifier of that "All" at the same time that it is the signifier of its lack. As Muller and Richardson put it, the subject is "cut off from its 'copula' simply by reason of its finitude, . . . [and] thereby suffers a primordial castration."[35] The moment the child is born into desire—i.e., into language, the moment at which desire becomes human—the child is simultaneously born into castration or death, this moment being recognized as such only in retrospect in Lacanian psychoanalysis.

Castration is thus not to be feared as something that might happen to one; it is something that has *already* happened to one. Not only this, but castration is spread around, so to speak. It has already happened to *everybody,* male and female alike. At one point, however, Lacan does actually tie castration to the death of phallos, i.e., to detumescence, and in a way that validates Monick's characterization. Phallos both "explodes" and "dies" according to Monick; and it "return[s] to life . . . after defeat and death."[36] In Monick's psychoanalytic, castration and death are linked through the body. There is an end of male identity in castration.[37] In a cryptic passage in a sentence that effectively devalues the bodily-real at the same moment it invokes it, Lacan writes that "in so far as he [a male] is deprived of what he gives, [he] finds it difficult to see himself in the retreat in which he is substituted for the being of the very man whose attributes she cherishes."[38] In a detumescent state, in other words, a male finds it difficult to accept his (spent) penis as a replacement of the "attributes" that a female cherishes. Lacan thus suggests that a male does not recognize himself without his phallus. His identity is at stake "in the retreat." For Lacan, then, detumescence is a state in which the penis is clearly distinguished from the phallus, but it is a state in which a male is uneasy for he has lost his power. Lacan thus sides with the "being" whose attributes are cherished over the body that is in a state of retreat. He ignores the "difficulty," i.e., castration anxiety or fear that phallos may not rise again, in favor of a priapic phallus that is invincible. In briefer terms, for Lacan, a penis is not *the* phallus any more than it is phallos since it is never *transformed,* whether into an instrument of power as described in chapter 6 or as Monick describes. The result is that Lacan's phallus is forever on fire and never disappears to pee.

Lacan's passing slur on fellow psychiatrist Gregory Zilboorg should be singled out and appreciated in this context. Zilboorg published an article, "Masculine and Feminine: Some Biological and Cultural Aspects," in 1944 in which he criticized Freud for "not com[ing] out of the confines of the age-long theory of feminine inferiority,"[39] feminine inferiority being, of course, psychoanalytically connected with a female's castrated state. Not only does Zilboorg emphasize "the androcentric bias under which psychoanalysis has apparently been laboring,"[40] but he adduces the lack of clarity in psychoanalytic theory with regard to female sexuality to "the concept of the superiority of man in the psychoanalytic theory of sexual development—a concept of values."[41] He goes on to propose a radically different psychoanalytic based on Lester F. Ward's sociological study *Pure Sociology: A Treatise on the Origin and Spontaneous Development of Society*. Taking as point of departure one of Ward's conclusions, that biologically speaking, " *'life begins with the female organism and is carried on a long distance by means of females alone'* " (Zilboorg's italics), Zilboorg goes on to affirm that "those who stress man's primordial striving toward paternity overlook the fact that paternity in the true sense of the word must be actually a later cultural development. At one time man's relationship to woman must have been purely hedonistic."[42] More striking still is his claim that "The facts and observations available—clinical, ethnological, and sociological—support more the assumption that the primordial deed, the primal rape—which may be considered the original sin in the true sense of the word—was an act devoid of any genital, object-libidinous strivings. It was a phallic, sadistic act."[43]

As Zilboorg develops his theory of how males have resented what might be called the biological standing and powers of females and how they have come in turn to assert superiority over females, he wends his way toward Ward's thesis that " 'the male sex was at first and for a long period, and [is] still throughout many of the lower orders of being . . . devoted exclusively to the function for which it was created, *viz.*, that of fertilization.' "[44] In taking up Ward's idea that " 'among millions of humble creatures the male is simply and solely a fertilizer,' "[45] Zilboorg comments that a male in proto-society not only had "no conception of paternity, no more than a rooster or a dog, . . . [but was] relentless in his egoistic demands."[46] In brief, Zilboorg claims that from a psychoanalytic viewpoint, a male in proto-society was an "unconscious fertilizer," a brutal and demanding one. Psychoanalytic theory, he says, suffers from an androcentric cultural lag. He states that "it is the deep ambivalence toward the primordial mother and the first wife-slave that, if properly understood, sheds considerable light on that which psychoanalysts have become accustomed to call the feminine castration complex."[47]

Certainly it is evident from Lacan's psychoanalytic that he would want to safeguard the phallus from such blasphemous encroachments. In the opening pages of his 1953 report to the Rome Congress in which he speaks of his "secession" from the *Société psychanalytique de Paris,* Lacan sarcastically castigates "the impenetrable M. Zilboorg" (for insisting "that no secession should be acceptable except on the basis of a scientific dispute").[48] Lacan's descriptive epithet for Zilboorg not only speaks for itself, but speaks in multiple tongues of Lacanian desire and Zilboorg's (Lacan-given) psycho-anatomical body.

However eternal the fire of the Lacanian phallus and firm its refusal to pee, the question of autonomy will not so easily disappear. A lack of control means that where the flesh is willing, the spirit may not be, or conversely, where the spirit is willing, the flesh may be weak. Monick documents just such an experience of phallos manqué in recounting the experience of one of his patients. The experience recalls St. Augustine's centuries-old lament about the plight of males.[49] Such descriptions are in fact rare. Males are typically silent about this experience, as silent in fact as Lacan says females are silent about *jouissance.*[50] It is significant with respect to this silence that female analysts—Lacanian or otherwise—do not deride males for keeping secrets, any more than they deride males for their lack of control. Differential sex practices of psychoanalysts aside, that phallos manqué mocks desire cannot be questioned. Where phallos should be is not something rather than nothing but psychoanalytically, simply nothing. Such a turn of events understandably undermines male power. Indeed, what reasons are there for thinking phallos will rise again? In light of this existential uncertainty, it is particularly odd that within Lacan's psychoanalytic, impotence, while mentioned, is nowhere given weight. Words that speak of phallos manqué, of desire trumped by the flesh, are lacking. Lacan's talk about impotence in terms of the "centrifugal tendency" of a male's genital drive, and of both the consequently greater burden males bear if they cannot fulfill this inborn tendency and the consequently "more important" nature of repression in males, says more about Lacan's view of how males differ from females— more about what might be termed "the inborn plight of males" and in turn why male promiscuity is justified—than it does about impotence. Indeed, while Lacan pointedly remarks that infidelity is not peculiar to males, his definition of female infidelity reduces to female *jouissance.* Though not outrightly stated, Lacan implies that in her excess of orgasms, it is as if a female is having sex with one man after the next. That is *her* infidelity.[51] Actually, what is at issue has nothing at all to do with impotence—as much a Lacanian side-tracking device as frigidity. The issue has to do with the vagaries of male life and the fears that go with it. It has to do with the unpredictability of a living male body and the necessity of compensating for that "deficiency" by claiming excessive power, thereby equalizing if not outstripping female *jouissance,* at the same time muffling if not silencing castration anxieties.

C. *Excess*

Monick does not discuss how the unpredictability of phallos makes males feel, that is, what being out of control does to their psyche, but certainly it is something with which males have to come to terms, in the same way that *any body* has to come to terms with the animate form it is. In particular, *any body* has to make sense of itself. The ambiguity of penile erection (discussed in chapter 5) shows the need for males to make sense of their bodies all the more pressing since it is precisely a question of semantics as well as of autonomy—or lack thereof. By categorically removing the living body from his psychoanalytic, and substantively removing it as well from the scene of female *jouissance,* Lacan avoids any such confrontation. Yet though a male's lack of control never surfaces as a topic in Lacan's psychoanalytic nor makes an appearance in any Lacanian equation or formula, it is as fundamental to human sexuality as *jouissance.* What the omission suggests is that there is a fundamental inequality that Lacan's psychoanalytic cannot accommodate, an inequality in male and female sexual power. It suggests that Lacan cannot admit being at the mercy of something *male.* He can admit only to being at the mercy of females with respect to knowledge about female *jouissance,* and to a certain ignorance about females. He can admit to an inability to grasp female sexuality because he can dismiss it from his psychoanalytic. He cannot, however, dismiss male sexuality since it is the *bedrock* of his psychoanalytic, in the same way that it was the *bedrock* (see below) of Freud's psychoanalytic. At its extreme, the inequality in male and female power might well precipitate a desire to trade the uncertainties of male sexual life for what is seen as an always potential and readily available female *jouissance.* The disparaging and resentful tone of Lacan's explanation of female *jouissance* appears consonant with this desire: one resents what another has if it *exceeds* what one has. Even further, Lacan may find the literal ups and downs of penile life—its joys and deaths—inferior to the continuous joys of female sexuality. Here again, what another has *in excess* may be too much to bear with equanimity. But there is even more reason to think that Lacan's resentful tone toward females harbors within it a desire to trade.

Lacan avoids any consideration of a male equivalent of female *jouissance,* the latter considered precisely as an *excess.* Yet clearly, there is just such an equivalent. What is in excess on the male side are sperm. With each orgasm, males release between 120 and 600 million sperm that, with the exception of one, are perfectly superfluous to insemination, in the same way that, from the viewpoint of some males at least, one orgasm after the other is superfluous, though not to insemination. If we ask what one orgasm after the other is superfluous to, there is hardly an answer, except from those who think that what is enough for them is enough for everyone else. The very idea of *excess* gives the secret away. It betrays a kind of thinking in which some individuals may feel themselves shortchanged with respect to other individuals. They find they have gotten the wrong end of the bargain. What *they* have in excess is not a pleasure to them at all. It is simply in the cards they have been dealt, cards from which they derive no immediate benefit whatsoever. Accordingly, to envy female *jouissance* is in the same cards; when some males see how the cards have been stacked, they see that they have been stacked against them. In short, envy is the result of having the wrong excess. Surely it is not extravagant to read Lacan's derision of females in just this way.

There is a further dimension to the question of why Lacan can admit only to a lack of control of female *jouissance* and thereby avoid confronting and possibly exposing a male lack of power within his psychoanalytic. This answer too is tied to excess. It is in fact a way of dealing with excess by turning the erstwhile useless cards one has been dealt to one's own advantage. What before counted as a negative is now a positive: an excess of sperm is turned into an excess of male virility. *This* excess is positively valorized. We have in fact already seen it validated within the Lacanian psychoanalytic. It is a question of a centrifugal—as opposed to a centripetal—sexual drive. As Lacan tells it, "[a male's] own desire for the phallus will throw up its signifier in the form of a persistent divergence towards 'another woman' who can signify this phallus under various guises, whether as a virgin or a prostitute."[52] As noted earlier, because of this "centrifugal tendency," impotence is "much harder for [a male] to bear [than frigidity for a female]."[53] "Persistent divergence towards 'another woman' " is taken as a given, a built-in of male sexuality. It is an *inherent* excess. This is precisely why its inexpressibility via impotence can be affirmed so hard to bear, in particular, as Lacan declares, "much harder . . . to bear." Lacan is here weighing impotence in men against frigidity in women. But the two hardly seem comparable to begin with. Impotence has to do with a failure to achieve or to maintain an erect penis. It has to do with *autonomy*. In the section of their book on impotence ("erectile dysfunction"), Masters, Johnson, and Kolodny write that "Isolated episodes of not having erections (or of losing an erection at an inopportune time) are so common that they are nearly a universal occurrence among men."[54] Clearly impotence has to do with being at the mercy of something out of one's control. Frigidity has to do with an inability to have an orgasm, with a disinterest in sex, or with a lack of sexual arousal.[55] It does not have to do with an *organ* dysfunction which leaves one at the mercy of something outside one's control. In consequence, frigidity does not deprive females of autonomy; it deprives them of sexual pleasure. In contrast, what impotence first and foremost deprives a male of is complete autonomy over his own bodily behaviors, *complete autonomy over what he can and cannot do; indeed, it deprives him of what he would typically regard one of his most basic 'I can's'.* For Lacan, this lack of complete autonomy translates into a deprivation of excess, not of sperm—an obviously inherent male excess—but of females. What his "centrifugal tendency" adds up to is thus a vindication of having one woman after the next. Having one woman after the next is like having one orgasm after the next. It *equalizes* male and female sexuality. Her *more* becomes his *more,* only his more is with other partners. The hands dealt are thus equalized. What is hers naturally becomes in Lacan's psychoanalytic what is his naturally. Fittingly enough, his *more* is even language-like: it is like "one word for another," which phrase, Lacan tells us, "is the formula for the metaphor." Metaphor is not a matter of a fundamental connection of any kind (of resemblance); it is a matter of pure *substitution.* Accordingly, says Lacan, "if you are a poet you will produce for your own delight *a continuous stream* [of metaphors]"[56] (italics added). Clearly, if you are a man, you will do the same: "persistent divergence[s] towards 'another woman'" equivalently produce delight in "a continuous stream." Moreover women, like language, are always there. They are the primary objects of exchange, says Lacan, "just like speech."[57] Indeed, as Macey points out, unlike Lévi-Strauss, Lacan does not qualify his outright equation of women to words.[58]

Considered in the light of the above discussion, we come to a deeper appreciation of why *the* phallus is hoisted to a place of supreme prominence. What is it that vanquishes unpredictability if not *the* phallus, first of all by being removed from the real and catapulted to pure Symbolhood, to a pure power of meaning having dominion over all other signifiers? What expresses a "centrifugal tendency" of one female after the next if not *the* phallus? Mixed registers aside, what releases an excess of sperm in the throes of *jouissance* if not *the* phallus? *The* phallus itself is excessive in ways that are positively valorized over females and *their* excess. It is excessive in being All-controlling; it has "supplementary" tendencies; its spermatic productions are prodigal in the extreme, as even Lacan recognized in speaking of Booz's sheath.[59] By removing female *jouissance* from his psychoanalytic, Lacan in truth establishes an unrivalled excess. Freud said that only infantile genital organization is phallic. He wrote that in contrast to "the final genital organization of the adult," in infant genital organization "only one genital, namely the male one, comes into account."[60] With Lacan, however, genital organization is phallic from beginning to end. His divergence from Freud is unequivocal. He turns Freud's statement about infancy—that at this period of life there is "not a primacy of the genitals, but a primacy of the *phallus*"—into a statement about the whole of human life, *all* human life. Lacan thereby solves Freud's "riddle of the nature of femininity." He declares "the nature of femininity" to be outside the psychoanalytic and blocks the door by phallic excess.

It should be added that precisely negative aspects proverbially attributed to females are aspects actually belonging to phallos. What is proverbially unpredictable but a woman? A woman changes her mind. She is fickle. She is every inch *la donna è mobile*. Is a woman really unpredictable, or is male projection at work, displacing attention from the plight of males onto the very personage of females? Furthermore, what is an unknown quantity but a woman? A man never knows what a woman wants. *She* never knows what she wants. A woman is a mystery. How is it possible to control something that is a mystery and that does not even know itself? Here the question of lack of knowledge is raised to the heights of the mysterious. But is woman really an unknown quantity, or is it the mystery of phallos and a male's lack of knowledge of phallos that is put onto her? Again, it is a question of projection.

Several of Jung's remarks on psychoanalytic theory and projection are of great interest in this context. He said many times over in one way and another that psychologists must recognize and acknowledge that "their theories are an expression of their own subjectivity."[61] Of his own work he wrote, "I consider my contribution to psychology to be my subjective confession. It is my personal psychology, my prejudice that I see psychological facts as I do." He nonetheless believed also that *"So far as we admit our personal prejudice,* we are really contributing towards an objective psychology"[62] (italics added). The remarks are in keeping with his delineation of an archetype, namely, that it is a general image or form that is filled in with respect to content. Moreover he affirmed that the way in which we see and understand the world is a function of our unconscious, in other words, that "the primary method by which the unconscious manifests itself is by projection."[63] Certainly in addition to his clinical writings, his studies of alchemy and of myth lend weight to his theses. His remarks should thus give pause for thought. Reflection in fact shows that introjection as well as projection is at work in the way attributions and counterattributions are made to females. In particular, positive female attributions are taken over by phallos and in turn become the basis for negative counterattributions.[64] What is in excess but female orgasms? Once males appropriate a natural excess of their own—in the manner described above—female excess is then denigrated, as are its female "results." A female, in other words, has more than what she needs to keep her happy. Yet proverbially, she is a constant nag; she complains incessantly. She is a harpy, an incorrigible bitch. Clearly, attributes are appropriated, and then denounced in proverbial ways in the name of excess.

We can incidentally ask in the same context whether the mystery of phallos that is negatively projected onto females is not linked to the male tendency to read out the ambiguity of penile erection. This possible linkage is suggested by the very character of the Lacanian phallus. It is a known quantity. It is the anchor of all other signifiers, the latter being thoroughly dependent on it. It is thus present in every word we speak. It is the very power of meaning. There is indeed absolutely nothing equivocal about Lacan's phallus, neither in terms of its being all powerful nor in terms of its never being satisfied; it *always* signifies desire because it always signifies lack. Accordingly, you can trust it to say what it means and to mean what it says. Just so with typical male readings of penile erection; ambiguity is unrecognized. Furthermore, an all-knowledgeable, circulating agent goes along with an all-knowledgeable, circulating signifier; that is, a male too knows what he means and means what he says. Whether it be food that arouses him (it), excitement that arouses him (it), or a woman that arouses him (it)—his (its) meaning is always the same. There is but one sexual desire after the next, and correspondingly, one lack after the next, as the chain of signifiers ordains. Ambiguity is in consequence out of the question. While tumescences come and go in a variety of situations—now of fear, now of anger, now of keen alertness, now of general excitement—there is no ambiguity about meaning because there is no ambiguity about male sexuality. There is only ambiguity about *female* sexuality. Consistent with that ambiguity is the characterization of a female as an unknown sexual quantity, even to herself. Proverbially, she says "yes" but means "no" and says "no" but means "yes." She herself does not know what she means. In consequence, a man never knows where he stands with a woman. He certainly knows where he stands with himself however.

Now all of the above female attributions, even her "positive" ones, are a "repudiation of femininity." The phrase comes from Freud.[65] He declared that two themes that are "tied to the distinction between the sexes"—penis envy and a male's "struggle against his passive or feminine attitude toward another male"—are especially prominent and troublesome in psychoanalysis. "What is common to the two themes," Freud said, is "an attitude toward the castration complex." He elaborated on this attitude saying that "I think that, from the start, 'repudiation of femininity' would have been the correct description of this remarkable feature in the psychical life of human beings." He even went so far as to say that this feature, best described as "repudiation of femininity," constitutes the "bedrock" of psychoanalysis.[66]

Freud's judgment corroborates in a most striking way Zilboorg's thesis regarding the androcentricity of psychoanalytic theory, its evolutionary roots, and the need for a rethinking and re-evaluation. Monick comments on this passage in Freud specifically, saying that "to Freud, the possession of phallos is the same as repudiation of the feminine, and the loss of phallos . . . is equated with feminization." In brief, "Without phallos, all is feminine."[67] Lacan's blockage of the psychoanalytic door by phallic excess makes the reverse point emphatically: without the feminine, all is phallos. He carries Freud's bedrock psychoanalytic principle to its psychoanalytic ultimate. Femininity is repudiated not only in "the psychical life of human beings" but in psychoanalysis.

In sum, unpredictability, castration, and excess are all deeply interrelated. Lacan's substantive removal of the living body from the psychoanalytic scene, most specifically his categorical placement of female *jouissance beyond the phallus,* is mandated by excess, an excess that is psychologically tied to an unspoken male lack of control and to an unspoken fear of lack. Female excess, in other words, endangers the whole of Lacan's psychoanalytic. On the one hand, a hole, rather than being the site of lack, appears to be the site of *fulfilled* desire; on the other hand, the phallus, while signifying authoritative power, is in actuality at the mercy of a higher power: it is in truth subject to forces outside its control. Its totemic status and its priapism do not and cannot actually protect it. Put in this perspective, it is clear that Lacan's problem with female excess—*actual* orgasms—comes too close to exposing the problem of male autonomy in the form of phallic unpredictability—*actual* risings and *actual* fallings of the penis. It comes too close to the real male living body. When Lacan raises the question about female *jouissance* toward the end of his life (1972–73), and when the living body in fact threatens to undo his psychoanalytic, Lacan takes a derisive stance toward females and attempts (not necessarily wittingly by any means) to avoid acknowledging the unpredictability of phallos by speaking of impotence, as we have seen, and in so doing to come up to, or even to outdo, female excess by affirming repression of sexual desire to be stronger (hence "more important") in males and the centrifugal tendency of males to substitute one female for another. Of course to speak of female excess as Lacan does is basically to reduce a female to her genitals, or more specifically, to *a* genital, since, as pointed out earlier, her "charming sheath" is not up to any signifying status. (Lacan in fact specifically designates female *jouissance* as *"vaginal."*)[68] Even Bernini's statue of Saint Theresa is described as "coming," Lacan thereby inferring (in snickering schoolboy fashion) that no matter who the female, she is first and foremost a sexual being, thus her excess is always present even if inappropriately so, and by extension, that female bodies in their excesses behave in ways that are out of control.[69] But we must remember that, with the silencing of living bodies, males fare no better within the Lacanian psychoanalytic. In order to establish and to maintain power, in order to achieve excess on par with female *jouissance,* a male must be equally reduced to his genitals and shown to be superior. Indeed, Lacan states that the phallus is chosen as a Signifier of signifiers not only "because it is the most tangible element in the real of sexual copulation" but because of its "turgidity." That bodily character "is the image of the vital flow . . . transmitted in generation."[70] However veiled it is in its signifying mode, the phallus is here recognized as a palpable, flesh and blood reality. There is of course a decisive difference between Lacan's female and male genitalic reductions. Lacan derides females—whether in terms of a hole that is an utter lack and nothing more, or in terms of what a female experiences but does not know. These are, for Lacan, precisely reasons why a female with her excesses is placed outside the psychoanalytic. *The* phallus, in contrast—and as its definite article already indicates—is elevated to a place of distinction, and even the penis through a kind of veneration by association enjoys a place of prominence. In fact, a mere hole is never spoken of as an object of desire. That privilege is reserved to the phallus. One turns what one cannot control into one's dearest object. In turn, one turns one's dearest object into the object of one's desire: in Lacan's psychoanalytic, a female *is* the phallus.

II. Being the Phallus and Having the Phallus: A Question of Imposture and of Lacan's Phallic Farce

Now although the Lacanian phallus is purely symbolic—or so Lacan would have us believe—it is not on that account to be written off simply as the cornerstone of a phallogocentric psychoanalytic theory. The Lacanian phallus can teach us something about linguistic totemism and linguistic priapism (or a priapic linguistics), for the Lacanian phallus, as has been shown, is a highly reworked and intricate corporeal archetype. A particularly striking instance of this intricate reworking is apparent in Lacan's personal concerns about "imposturing." When Lacan asks us if he is not an imposter,[71] he is asking us to judge whether he has blown himself up out of all proportion, and whether he is not dissembling before us, purporting to be something he is not. Were we to answer his question positively, that is, were we to affirm that he *is* an imposter, we would be affirming that he is suffering from both linguistic priapism and totemism, for both in blowing himself up out of all proportion and in dissembling before us, he is precisely playing the role of a Lacanian phallus; he both *has* and *is* the object of desire. Having it, he inflates as only a male does, and, in an archetypal sense, makes a grand spectacle of himself; being it, he parades as the totemic figure he would like to be but is not since *he* is not his dearest object nor is he the object of his desire: females are the phallus. If in this Lacanian context of imposture the question were raised as to how males elevate themselves to a position of power, the answer would be "Through making a spectacle of themselves in the imposture of phallic display." We should note that in this respect, there are strong affinities in imposturing between Lacan and the Trickster in trickster mythologies, mythologies that are found "in clearly recognizable form among the simplest aboriginal tribes and among the complex." Anthropologist Paul Radin, for example, in introducing his study of American Indian trickster mythologies, points out that "we encounter [the myth] among the ancient Greeks, the Chinese, the Japanese and in the Semitic world." The penis is a stock figure in such myths. In one myth, for example, the Trickster's penis is a detachable organ that on one occasion is "coiled up . . . and put in a box," and on another, circulates in the water where young women are swimming, pursues the chief's daughter, and catching her, "lodge[s] squarely in her."[72]

Imposturing aside, it is clearly actual experiences of male living bodies that move Lacan to elevate the phallus to a place of privilege in his psychoanalytic. To expose the lowly corporeal origins from which Lacan's indomitable phoenix arises, we can consult Lacanian texts themselves, for however cryptic and erudite, they leave no doubt but that linguistic totemism and priapism have their roots in the autonomous—and troubling —vagaries of male sexuality. An examination in this section of Lacan's descriptive drama of being and having the phallus in the act of copulation— what we may call Lacan's "Phallic Farce"—will show this unequivocally in an experiential sense. An examination in the following section of Lacan's retroactive psychoanalytic in terms of a well-known tenet of Freudian psychoanalysis should foreclose any vestige of theoretical doubt.

Lacan's introduction of the phallic qualifiers *being* and *having* is casual in the extreme. It occurs in the context of his explanation of "the division immanent in desire," that moment of experience in which one learns that one's mother does not have the phallus. "This is the moment of the experience without which no symptomatic consequence (phobia) or structural consequence (*Penisneid*) [penis envy] relating to the castration complex can take effect." In brief, it is at this moment of experience that one realizes the conjunction of *desire* and *lack* in the phallic signifier. Lacan goes on to say that what results in the future on the basis of the experience depends on the "law introduced by the father," but that *"simply by reference to the function of the phallus,* [one may] indicate the structures that will govern the relations between the sexes"[73] (italics added). It is at this point that he quite casually states "Let us say that these relations will turn around a 'to be' and a 'to have', which, by referring to a signifier, the phallus, have the opposed effect, on the one hand, of giving reality to the subject in this signifier, and, on the other, of derealizing the relations to be signified."[74] He cryptically adds that the opposed effect "is brought about by the intervention of a 'to seem' that replaces the 'to have'." The intervention of 'seeming', he says, has "the effect of projecting in their entirety the ideal or typical manifestations of the behaviour of each sex, including the act of copulation itself, into the comedy."[75] As we shall see, Lacan controls what he cannot actually control by enshrining it in eternal tumescence. Anything less than eternal tumescence is laughed off.

Precisely in view of the fact that a "seeming" enters into "the relation between the sexes," it is apparent that the act of copulation is on its way to being transformed into the phallic paradigm Lacan wants to spell out and establish. The act will become exemplary of the fundamental reduction to being and having the phallus that typifies females and males respectively, but with the added qualifier that "seeming" will replace "having." *When this Lacanian qualifier is exactingly translated into corporeal terms,* the puzzling sexual psychodrama is resolved. In the act of copulation, a male no longer *has* the phallus since, the phallus being inside a female, he *is* it; conversely, a female *is* no longer the phallus since, the phallus being inside her, she *has* it. Hence the seeming. "The relations between the sexes" and "the division immanent in desire," being brought down to the level of an act in which one thing is simply put inside another, a comedy of sorts indeed becomes apparent. Lacan's version of love-making is a game of "Who has/is the phallus?" There is no *body* home, so to speak. It is a game for a pure signifier. The act of copulation reduces to a linguistic amusement of "hide and seek the totem," an amusement that goes on and on in a comedy of misses, a rollicking diversion that is always foiled and has no end: if you have it, you want to be it; if you are it, you want to have it. Presumably, this is what makes the act of copulation so laughable, a phallic farce.

Once the words Lacan writes are properly fleshed out and copulation is properly reduced to the whereabouts of the phallus, there is no doubt but that genitalia are alone on the scene, literally caught in the act. They reduce *love-making* to "an act of copulation." There are in fact no loving living bodies present—no enfolding arms or entwined legs, no caresses, no felt bodily heat. There is only a view of isolated body parts acting up as if under a microscope. The "act" is indeed nothing more than the mechanical one of inserting and emptying a "turgid" penis into a vagina. Interestingly enough, Lacan speaks specifically of "the genital act," as if it were indeed simply a question of an autonomous genital and its turgid powers of insertion and ejaculation. He does so in conjunction with "the importance of preserving the place of desire in the direction of treatment"; specifically, he states, "that the genital act should, in effect, have found its place in the unconscious articulation of desire is the discovery of analysis."[76] Now Freud not only openly deplored "emphasizing exclusively the somatic factor in sexuality"—in which, he says, "nothing is meant but the need for coitus or analogous acts producing orgasm and emission of sexual secretions"—but he labelled it " 'wild' psycho-analysis." " 'Wild' psycho-analysis," he affirmed, misses the point of *"psycho*-sexuality."[77] In the Lacanian psychoanalytic, "the genital act"—the mechanical coupling of parts and the phenomenon of hydraulic relief—defuses the erstwhile power of penile display, turning it into a comedy on the order of "much ado about nothing." The literal disappearance of one thing inside another, and its transformation inside the other, creates the seeming by which an erstwhile phallus is reduced to playing the part of comic relief in an otherwise purely mechanistic drama. However much it wants to say what it means and mean what it says, it is incapable of doing so. Penile display in the form of a signifying phallus has in fact disappeared, its archetypal presence being no more than a memory of power past and a hope of power to come. It is only fitting, then, that in Lacan's psychoanalytic, death comes with sex. They are joined, we might say, in the "holey" matrimony of copulation, but only on the male side. When a male is spent, "deprived of what he gives," he "falls under the sway of individual death," says Lacan.[78] Insofar as he may fail to re-become his past, i.e., to rise again, a male lives perpetually under "the sway of individual death"; the waxing and waning of the phallus is a constant reminder of castration. On the female side, this joining of sex and death is experienced at a distance, witnessed rather than personally felt— though Lacan does not say as much. On the contrary, we can presume that because the phallus holds dominion over all signification, death equally defines *her* being in "the genital act." Placed within the Lacanian psychoanalytic, females conform no less to life under phallic law than to the law of phallic life.

What we come to realize is how much the autonomy of the Lacanian phallus actually derives from the very *real* corporeal autonomy of a penis, or as Monick would say, of phallos. We come to realize, in other words, that what Lacan says of the phallus has its origin in what he knows—but does not tell—of the transformative powers of phallos. When he says that the unconscious is the discourse of the Other, and that the Other is the subject of the unconscious, there is no doubt but that that Other is ultimately and crucially phallos: the Lacanian phallus is the linguistified psychoanalytic stand-in of phallos, complete with the latter's "inherent need to disappear." When Lacan says that "the motives of the unconscious are limited . . . to sexual desire,"[79] there is no doubt but that archetypal phallos grounds the Lacanian phallus as it plays its absolutely central role in articulating desire. It presents itself—*displays* itself—across all bodily holes and slits: "The Other, the capital Other, is already there in every opening, however fleeting it may be, of the unconscious."[80] It instantiates the "true subject of the unconscious"; it "speaks the subject."[81] The subject is thus secondary to the signifier in exactly the same way a male is secondary to phallos. He is consistently and to begin with at its mercy. In a special sense, but in exactly this way too, a female is at the mercy of phallos in terms of the possibility of phallos becoming inopportunely impotent. Phallos, as Monick tells us, "has a mind of its own." Just so with Lacan's signifier that "has a life of its own."[82] The corporeal archetype that consistently structures Lacanian psychoanalysis is in fact "The Freudian thing" that is in pursuit of Diana.[83] Indeed, the Freudian thing is that "thing" on which the whole of Lacan's psychoanalytic is modeled—but of course without "the thing's" tantalizing uncontrollability. By bringing the unconscious into the light and formulating its rules, the ups and downs of a penis are mathematically calculated and conquered by language. Phallos thereby becomes *the* Phallus, Signifier of signifiers. In turn, Lacan is in complete command. His psychoanalytic has achieved consummate prediction and control of the unconscious. It predicts when, why, and how the unconscious will appear; it specifies how the subject can be "liberated," that is, cured. Control and predictability of phallos are in fact twin themes of Lacan's psychoanalytic. Insofar as control and predictability are precisely the hallmarks of Western science, the twin thematic lends considerable weight to Roustang's claim that the central project underlying the whole of Lacan's work was a desire "to turn psychoanalysis into a science."[84]

In sum, the relationship between Lacan's linguistic phallus and a corporeal phallos is transparent and unmistakable. Wittingly or not, the Lacanian phallus is modeled archetypally on phallos both in terms of its self-autonomy—it obeys its own intricate formulaic laws and in so doing speaks the subject—and in terms of its one-dimensional significance and invincible power: it is invariably and uncompromisingly read as sexual desire. Archetypal aspects of phallos are thus psychoanalytically reworked into laws of the phallus. This fundamental insight into the archetypal structure of Lacanian psychoanalysis provides a significant perspective on the retroactivity that consistently structures—and, to some commentators' minds, plagues—Lacan's psychoanalytic.[85] Retroactive structuring is apparent not only in the stages of infant development where, as we have seen, a later stage determines the nature of the earlier one; it is apparent in the whole of his psychoanalytic: "The legibility of sex in the interpretation of the unconscious mechanisms is always retroactive."[86] This Lacanian rule is most prominently exemplified in Lacan's own reinterpretation of a well-known Freudian dictum. His reinterpretation, based in good part on a dream of one of Freud's patients, is a metaphorical refiguring of the fear of castration. But it is also, by that very refiguring, a radical restructuring of the archetype of the unconscious.

III.   Dreams of Death/Dreams of Glory

Freud's words—*Wo es war, soll Ich werden*—which Lacan painstakingly analyzes and reinterprets, are traditionally translated as "Where the id was, there the ego shall be." For Lacan, Freud's words become an invocation to what Monick would describe as the spirit of phallos. Lacan's ultimate translation reads, "There where it was, . . . it is my duty that I should come to being." Understood in one sense from the viewpoint of phallos, the self-proclaimed duty appears a virtual self-entreaty to rise again. Understood in another sense from the viewpoint of phallos, the self-proclaimed duty appears a self-entreaty to "come to being" *as subject,* that is, to be brought up to the light of day, to be in control, to rise above spent phallos, to hail the Signifier of signifiers, to escape death, in a word, to be liberated through Lacanian psychoanalysis. In explicating "there where it was," Lacan writes, "There where it was just now, there where it was for a while, between an extinction that is still glowing and a birth that is retarded, 'I' can come into being and disappear from what I say."[87] The two levels of meaning—the self-entreaty to rise again and the self-entreaty to come to being—are clearly embedded, the one allegorically, the other psycho-symbolically, in the explication. At the heart of Lacan's psychoanalytic "being," there is both a phallos that has risen and fallen, and a subject that is subject to death. Both are in need of, and can only be cured by, phallic healing. But of course phallic healing is a ready-made in Lacanian psychoanalysis. Risings, fallings, and death notwithstanding, the Lacanian subject is conceived from the beginning to be through and through a phallic subject. In effect, it will always live.

When Lacan, in order to explicate Freud's words further, turns to recounting a dream about a dead father, a dream Freud had commented upon, he writes, "If the figure of the dead father survives only by virtue of the fact that one does not tell him the truth of which he is unaware, what, then, is to be said of the I, on which this survival depends? . . . He [the dead father] did not know. . . . A little more and he'd have known. Oh! let's hope that never happens! Rather than have him know, *I*'d die. Yes, that's how *I* get there, there where it was. . . . Being of non-being, that is how *I* as subject comes on the scene."[88] The allegory is transparent: *rather than let phallos know its death, "I" will assume it.* In Lacan's psychoanalytic, the phallus is father to the son; and the son assumes his father's death. The urgent, anguished—one might say even near hysterical tone—of the passage is affecting. The "I" that is the survivor, Lacan tells us, "is abolished by knowledge of itself, and by a discourse in which it is death that sustains existence." Death—existential and psychoanalytic—is of a piece with the death of phallos. It looms ahead as behind, in the knowledge of one's mortality as in the knowledge of detumescence, in Lacan's formulaic discourse as in corporeal fact. The retroactive "there where it was just now" thus describes both the phallus and the human subject of Lacanian psychoanalysis. When Lacan affirms that "The legibility of sex in the interpretation of the unconscious mechanisms is always retroactive,"[89] he is dramatically reaffirming his reinterpretation of *Wo es war, soll Ich werden:* "there where it was" is now nothing; there where it was is a denouement— or castration—of phallos. Lacanian psychoanalysis demands that we always read backward to that moment.

The backward reading might seem to affirm that the power of the phallus is not invincible. At the same moment that Lacan acknowledges its vincibility, however, he rises above it. The self-entreaty that looks back to that place where in the past there was something—and that might be glossed, "May it (I) be there again!"—follows from both an acknowledgment of the death of phallos and the keen, even passionate, desire for its renewed life. This entreaty—*this "want-to-be"*—is given all the more force when Lacan speaks of the subject that *was* as "a locus of being."[90] Where the phallus *was* is where being *is*. Thus, Lacan's briefer translation of Freud's dictum, "where the subject was, there must I become," is doubly meaningful. Lacan in effect articulates the aim of his psychoanalytic at the same time that he metaphorically turns the unconscious into "the Freudian thing." The Freudian *thing* is and always has been the phallus which, as Freud himself tells us, is primary from the very beginning. The phallus as Signifier of signifiers is the voice of the unconscious, the *I* that speaks the subject and that is wholly distinct from the ego. Of the passage in which Lacan is reinterpreting Freud's words, Muller and Richardson write (in part quoting Lacan) that what Lacan is identifying is "a process somehow on the level of being, rather than ego consciousness, whereby the 'I' as subject comes to being in the unconscious, comes to 'emerge . . . from this very locus in so far as it is a locus of being'."[91] In Lacan's psychoanalytic, the unconscious is indeed a locus of being, but it is articulated by the phallus; the unconscious is a *phallic* unconscious. It irremediably splits all subjects at the same time that it legitimates them as subjects, subjects, that is, of a signifier that precisely, being "a locus of being," "has a life of its own."[92] Clearly, Lacan's psychoanalytic is patterned on the autonomous ups and downs of male bodies, and the ups and downs, and all the fears that go with them, are projected in one way and another on all psyches indiscriminately.

It is no surprise then that in Lacan's psychoanalytic, the subject is secondary to the privileged signifier. S/he is both beholden to it and bends to its autonomous rules, which is why Lacan can say that the signifier speaks the subject. What is utterly radical about this psychoanalytic is its utterly individual way of refiguring the archetype of the unconscious. When Lacan speaks of himself becoming where the subject was, he is recognizing precisely that *Other* subject that is the unconscious in its mode *not as classically female but as phallus.*[93] It is *that* Other, not the dark continent but the thoroughly exposed phallic one, that can be and is brought up into the light of day in Lacan's psychoanalytic. Indeed, as we have seen, Lacan banishes the feminine from his psychoanalytic when he dismisses female sexuality—female *jouissance.* That he has exiled female sexuality does no damage to his psychoanalytic, however, for the veiled signifier has been in full command all along. The whole of Lacan's psychoanalytic is in fact through and through a metaphor for the phallus, and that metaphor, dressed up in its rhetorical formulae, rules, algorithms, and vectors, is nothing more than a corporeal archetype in linguistic disguise. Whatever its façade of the moment, it is still recognizable for what it is. Penile display is the *real* bedrock of Lacanian psychoanalysis. Archetypally, and by turns, that display engages our attention, intimidates, taunts, beguiles, threatens, invites, fascinates, magically transforms—just like that other archetype Sartre characterized as being "in the form of a hole."[94] As "the hole" is radicalized within Sartre's psychoanalytic ontology, so penile display is radicalized within Lacan's psychoanalytic. Therein lie both the source and ongoing power of patriarchy. Totemically hoisted on high and bulging out of all proportion, Lacan's archetype is no mere and simple personification of a penis. Quite the contrary. Within a patriarchal culture the phallus does not stand for a male individual; *males stand for the phallus*. The terms of the metonymy are reversed because the archetype has been so radically intensified. A neutral phallus—a phallus females *are* and males *have*—is a myth; males alone, not males *and* females, are emblematic of Lacan's phallic order. Indeed, this is why Lacan's psychoanalytic is patriarchal and why classical psychoanalytic theory is androcentric: *without the feminine, all is phallos*. If we consider the phallus's symbolic autonomy to be an allegory of the autonomy of phallos, a linguistic totem to be an allegorical apotropaic against castration, and a priapic condition to be an allegorical figuration of excess, then Lacan's psychoanalytic is an allegorical psychoanalytic. In this guise, it is an archetypal fable of our time. The need for the feminine to disappear "seems to be inherent in it."[95]

Epilogue

The more we ignore corporeal matters of fact, the less we understand ourselves. The less we understand ourselves, the more prone we are to perpetuating the very oppressions that constrict our movement, blinker our vistas, and whittle away at our possibilities in the world. If these statements are true, then surely our point of departure must be to fathom what it is to be the bodies we are, for the roots of power wend their way through our bodies and direct us to the study of animate form. The task is in truth quasi-Heideggerian; it is to understand what thinking is, but in the radical sense of letting Being speak to us in the authentic voice of animate form. Thinking is consummately wedded to bodily life; indeed it begins in, and with, bodily experience. In the most fundamental sense, it is modeled along the living lines of our bodies.

The foregoing analyses of corporeal archetypes of power implicitly exemplify this analogical patterning of thought. At both cultural and individual levels, they show that analogical thinking structures the elaboration (exaggeration, suppression, neglect, distortion) of corporeal archetypes. The elaboration is apparent all the way from the cultural attire of Dani males—who by their near-permanent covering of their penes with a *holim* from a very early age put them out of sight and touch and thereby obviate (suppress) any possible cultural incorporations of the archetype of the phallus—to Lacan's psychoanalytic—which glorifies (exaggerates) a particular face of the archetype of the phallus in a prodigally individual way. Insofar as thinking in the most fundamental sense begins in bodily experience, it is not surprising that corporeal archetypes are reworked analogically, and that, in effect, corporeal meanings and values are conceptually transferred—or in psychological terms, projected—to other domains, including the entire domain that is a culture itself. Our bodies are in fact the *Ur*-archetype, the primal form on which we model our thinking and the artifacts to which that thinking gives rise: the cultural worlds we create are ones that conform first of all to our own bodies and then to the possibilities of bodies in general, especially their movement and their spatiality. They conform to animate form.

To say that we need to educate ourselves in the ways of animate form and its history is to say that we need to take our samenesses as well as our differences seriously. Elliott Liebow, noted author of *Tally's Corner,* a book on black street-corner life that soon became a classic after it was published in 1967, was interviewed recently in conjunction with his new book on homeless women. "I don't want to overlook the differences among us," he said, "but I don't think they're as important as the samenesses in us. And I think it's the samenesses that really need pointing out."[1] Given the pervasive contemporary focus on difference in our culture, a plea for urgency can easily be appended to the need's being met. The urgency of the plea is clearly all the more pressing given extremist forms of cultural relativism that bolster the notion that we are not natural bodies but cultural objects—texts that are differentially inscribed—and thus incommensurably and irremediably unlike one another. This inordinate bewitchment by culture results in a reductionism that is as pernicious and costly as its biological corollary. Cultural reductionism keeps us from taking evolution seriously. It in fact quickens the passing of natural history. It precludes our recognizing that, our individual and great historico-cultural diversities notwithstanding, we humans are basically the same. Though we speak in different tongues, speaking tongues are part of our evolutionary heritage; though we explain the world in different ways, explaining the world is part of our evolutionary heritage; though we dance, sing, tell stories, and paint differently, such creations are part of our evolutionary heritage. Similarly, though we eat different foods, we all eat—we all bite, we all chew; though we laugh at different things, we all laugh—we all grin, we all smile. Similarly too, though we all have different hands, different beds for sleeping, different walking gaits, and the like, we all use our hands to make things, we all lie down to sleep, we all walk the earth, and in the same binary patterning of our feet. When we ignore these ties that bind us in a common humanity and that articulate a very human repertoire of 'I can's,' we put ourselves out of reach of our own history, insulating ourselves from corporeal matters of fact and the archetypal forms latent within them. We proportionately distance ourselves from our own human nature; we proportionately distance ourselves from the task of thinking through, and ultimately understanding, the roots of power.

Our differences must be equally understood in bodily terms. Male and female bodies *are* different. Their corporeal differences give rise both to different experiences, including different experiences of power, and to different possibilities, including different possibilities of power. If we are to arrive at bona fide *human*—and not merely personal—self-understandings, then we must fathom not only what it is to be the bodies we are; we must attempt to fathom what it is to be the bodies we are not. We must, in effect, ask ourselves, what is it like to be female? what is it like to be male? not reflecting exclusively upon the kind of human animate form we happen to be and leaving the other kind for others to puzzle out, but reflecting as well upon *what it is like to be that other body*. This reflective challenge is not an intellectual exercise or a parlor game. Neither is it a mere cultural listing of what one "can" and "cannot" do. Neither yet is it a mere biological listing of what one can and cannot do. This reflective challenge is precisely reflective; it requires patience and time. This is because the challenge to think our way into our own animate form and into an animate form that is the same yet different from our own is a challenge to *think* and, in thinking, awaken bodily sources of meaning and value, of fear and delight, of images, symbols, and possibilities. The challenge in this sense is a challenge first of all to psychological honesty, perhaps especially for males in cultures that promote and treasure nothing less than always bravado male bodies. Moreover, to fathom the nature of the bodies we are is not to get a quick fix on them such that we merely reiterate the corporeal precepts of our individual and cultural groomings. On the other hand, and in broader perspective, the challenge is precisely to gain insight into our own and other cultural practices and dispositions in the process of educating ourselves in the deepest possible sense as to what it is to be a human body. Put in this broader perspective, the challenge is clearly not narcissistic but global as well as historical. The peacefulness of the world does not depend upon knowing the truth about who has sired whom, upon "guarantees of paternity" as some sociobiologists believe.[2] Events such as those recent ones in Bosnia decisively belie this doctrinaire assertion. Moreover epistemologically, it makes little sense to think that the peacefulness of the world depends on knowing what another knows rather than on knowing oneself. It makes still less sense ethically. The peacefulness of the world does not depend on shunting personal responsibility onto another person; it depends on individuals taking responsibility for themselves, which means, in part if not basically, understanding the roots of power and the kind of power one's actions instantiate and support. In broad terms, it depends on understanding the archetypal possibilities of a human body to the limit of one's capacities by reflecting upon what it is like to be the particular body one is and the particular body one is not; more broadly still, the particular *bodies* one is not.

By consistently attending to corporeal matters of fact that are the source of our samenesses and our differences, we begin to re-think power. We begin consistently to think in terms of animate form and its behavioral and experiential possibilities, thus in terms of that repertoire of 'I can's' that defines us as a species. By the same token, we begin consistently to think in terms of animate form and the corporeal archetypes it engenders, and in turn of the multiple ways in which these archetypes are and can be culturally reworked. The examination of optics of power and of the power of optics that opened onto an evolutionary genealogy and in turn onto corporeal and intercorporeal archetypes showed clearly that the personal and political are joined in the body, that however shaped by our diverse cultural heritages, power and power relations are built into the bodies we are. To have seen this is to have seen that "theory" in the most fundamental sense has been there all along, lying latent in the data, in animate form. Surely theories addressing inequities of power should begin with those original corporeal and intercorporeal archetypes that not only tie us to our primate history but bind us from the beginning in a common humanity even as in their differential reworkings they separate us in ways archetypally peculiar to a particular culture.

A second challenge attaches to this theoretical realization and builds on the insights of the first. This further challenge is to ask ourselves a two-stage question, the second half of which is similar to the one we earlier wanted Sartre to ask himself, namely, "What am I that I experience females as a hole, an obscene hole?" The first half of the question is: "what *are* the archetypal images I envision with respect to males and females?"; the second half is: "what am *I* that I experience males and females in this way?" When we reflect upon these questions and our answers to them, we reflect upon the ways in which we have individually fashioned and supported a particular archetypal content. We in fact discover firsthand the ways in which as individuals and as acculturated beings, individuals and acculturated beings no less than Lacan, we too elaborate corporeal archetypes, though far less flamboyantly and undoubtedly to far less baroque extremes than he. We begin in this way to take responsibility for our own understandings and attitudes, and to implement changes in our own lives commensurate with our insights. Surely the patriarchal conception of power as control can give way to more life-enhancing conceptions of power only if we examine and verify for ourselves the corporeal roots of power, reflect upon the ways in which we ourselves perpetuate the archetype that grounds the conception of power as control, and in turn cultivate a more consentient archetypal form. Because we cannot do away with archetypal forms, any more than we can do away with our bodies, we need to understand them, which means discovering and understanding the animate form and bodily experiences that are their source.

To meet this second challenge is to open ourselves to broader as well as deeper understandings of power as control, to an appreciation, for example, of how the conception is both fundamental to, and strongly supported by, the ingrained mind/body dichotomy of Western culture[3]—albeit in what might be called psychologically Machiavellian ways. What is at first denigrated within the dichotomy—namely, the body—is itself afterward dualistically conceived and then differentially valued. This is because, to begin with, there are male bodies and female bodies and because, in turn, one of these bodies becomes the standard upon which positive values are placed and an overarching symbol is created. Male bodies are conceived as powerful, strong, as having the right stuff; they stand firm and solid. Female bodies, in contrast, are conceived as soft, weak, unreliable; push them and they yield. Minds in Western societies are modeled along these bodily lines. Male minds are powerful, rigorous; they see the point; they are unflappable, orderly, controlled. Hence they are logical, rational, and the like. Female minds on the other hand are wifty, vague, jumbled, erratic—hence illogical and irrational. The striking conceptual similarity between male bodies and male minds and female bodies and female minds is clearly evident in Western patriarchal conceptions of "the stronger and weaker sex"—the body conceptually leading the way.[4]

Given elevated minds and low-life bodies as portrayed in the classic Western mind/body split, one would barely think that a scuffling body could play a central role in structuring human thought in this way. But we not only think analogically along corporeal lines using our body as the standard, we create meanings by way of our bodies. Corporeal representations of power as control in the form of a commanding, intimidating presence are in fact readily apparent in the artifactual staples of American culture. Elongate, solid objects—missiles, guns, rockets of all kinds— attest straightaway to their phallic origins. (It is interesting in this regard to note that interplanetary aliens are pictured as arriving on earth in friendly rather than menacing shapes. They do not fly to us in missiles, rockets, and the like, but in saucers—rounded, circular forms that can't harm or penetrate. Of course, outward appearances might deceive; and one could easily connect the deception to *vagina dentata* possibilities, but still . . .) Corporeal representations of power as control are similarly apparent in

primary Western behavioral symbols. A particularly striking example is the
Nazi salute: a dangling arm is briskly and promptly transformed into a solid
mass as it comes up to a diagonal ending position. Not only this, but there is
no hesitation in its upward movement; however flaccid the arm to begin
with, it is ever-ready to rise to the occasion. The dynamics and readiness of
the salute mirror the dynamics and readiness of "the virile member." Both
are symbolic of a *real* man. It is just this corporeal representation that made
the movie *Dr. Strangelove* the comical satire it was. Dr. Strangelove—a Nazi
political "doctor" who had lost his arm in the war—could not control his
prosthetic arm. Using all the force of his intact arm, he would try in vain to
bring his wayward, rigid member down to his side, for it would snap up
quite autonomously—and at most inappropriate times—its unpredictability unmistakably reflecting the sometimes embarrassing unpredictability of
penile erection.

Especially with respect to the representation of power as control,
national gestures can have considerable impact on the basis of their
archetypally gendered corporeal meanings. The decisive change in our own
pledge of allegiance in the early 1950s appears unquestionably linked to a
concern for what the pledge corporeally represented as it was practiced until
then. Originally, there was an opening gesture of the right arm outward
toward the flag from its initial placement close to the body, the hand being
held in the beginning over the heart. In this original version, almost the
entire pledge was recited with one's right arm extended outward. Such a
gesture is open—and was undoubtedly seen as *too* open, too vulnerable,
particularly as a war, even a cold one, was taking firm hold of the national
psyche. Indeed, from a patriarchal political perspective, what the gesture
represented was a kind of "female" body—a relating body, a body that
exposes itself to others, a body that gives of itself, and so on. Such a
"female" gesture left the body unprotected. In effect, what the gesture
represented was not a body preparing to fight or a body ready to fight: there
was no display of *power* in such a gesture. It was all open and flimsy, hardly a
symbol of national virility.

Clearly our twentieth-century Western concept of power is tethered to a
male body. But that concept articulates only one dimension of our natural
history: an optics of power in which precisely *control* reigns supreme. Power
is in consequence conceived in terms of sovereignty and, in turn, of the
ability to command, to dominate, to intimidate, to threaten, to assault.[5] The
exaggeration of this archetype within our culture tends to exaggerate a
concern with control to the point of obsession, leading to a denigration of
whatever is involuntary. In effect, wherever and whenever nature reigns with
respect to our own bodies, those preeminently and actively concerned with

control conceive themselves on the receiving end; they are akin to the penetrated; they experience themselves as powerless. Nature is conceived as having the upper hand; nature is in the driver's seat. Even those who may not be so preeminently and actively concerned but who have been similarly groomed to equate power with control—and control with power—easily if unwittingly conceive themselves as being on the receiving end and dutifully fall under the sway of negative appraisals of nature. Women in the throes of childbirth, for example, whose "involuntary uterine contractions" make them powerless, who hear themselves described as powerless and are duly treated as powerless, readily experience their bodies as being *out of control*.[6] What "involuntary uterine contractions" actually preclude, however, is *not* an experience of power but an experience of the miraculous. In particular, uterine contractions can be experienced as involuntary happenings—as per the standard Western medical description—*or* they can be experienced as the thoroughly *natural phenomena* they are, a testimonial to the wonders of nature. If a woman knew nothing of uterine contractions but only of a birthing process in which one living body brings forth another of its own kind, the sheer bodily dynamics of the process, however painful or easy, would be the all-encompassing reality. What the label "involuntary uterine contractions" does is set up an adversarial relationship. Rather than being wonders of her flesh, those rhythmic abdominal contractions are intrusions into her self-directed world, violators of her power. Experienced in the former rather than latter way, a woman giving birth—even a difficult and painful birth—might well respond to "involuntary uterine contractions" with the thought: "How wondrous that, at a particular time approximately nine months after conception, the uterus 'knows' what to do and does it!"

Cultural groomings obviously carry a powerful impress. When we fall under their sway and denigrate the involuntary, we miss the miraculousness of nature. We place nature oppositionally somewhere "out there" when in fact we *are* nature. Being part of nature, we are both in and out of control. Surely this fact is evident for all to appreciate: it is fundamentally there in our experience, male and female alike. A fanatical concern with control is a reflection of patriarchal thinking. Such concern is precisely a denial of the miraculous, a denial of what is involuntary in living bodies, a denial of the bodies we are.[7] Catching glimpses of the miraculous moves us—males and females alike—in the opposite direction, to a yea-saying of our bodies. This is because, in catching glimpses of the miraculous, we bear witness to the transformative wonders of being a body. When we awaken to the miraculous, we are no longer in patriarchal bondage. We no longer cling to a need for a thoroughly predictable and thoroughly controlled world. We in fact no longer equate power with control—or control with power. Instead, we sense the power of our bodies to transform us and at the same time sense the power of our own changing possibilities—our 'I can's'—with respect to those transformations.

When we begin to think in terms of 'I can's' rather than in terms of control, we recognize our powers as quintessentially human ones. Such powers are what we all—male and female alike—initially discovered: I can close and open my eyes; I can make sounds; I can hug; I can kiss; I can clench my fist; I can turn my head; I can stand; I can walk. It is in recalling and reclaiming these 'I can's' that we awaken the most fundamental understandings of power. They are grounded in the body. In this very sense, the miraculous is not basically sexual but human. Though surely there are sexual differences, the miraculous itself is corporeal. Once we realize this pan-human fact, we are already on the way to re-thinking power; we realize another dimension of our 'I can's' that has nothing to do with threat and vulnerability, with control and docile bodies. In that dimension of ourselves lie our creative powers, our powers of regeneration, of self-transformation, of imagination, of fecundity. We neglect valuing these powers. This is because an intensified power of optics has riveted our eyes and consumed our attention. As we have seen, this intensified power of optics is modeled on the corporeal archetypes of spectacle-making, of size, of certain intercorporeal bodily positionings, and most prominently, of penile display not as invitation but as threat. With respect to the latter, it is modeled not on the natural tumescent possibilities of a penis—on miraculous transformative powers—but on that unnatural and unquenchable organ, the phallus. Where power is modeled on an indomitable and priapic phallus, glimpses of the miraculous fade away. Not only this but as we have also seen, there may be a coveting of the possibilities of the Other. Rather than vying with females, however, resenting their miraculous possibilities, males have the possibility of turning away from phallically-ordered gendered bodies. Instead of funneling their energies into condensations of power and control, males no less than females have the possibility of discovering and affirming those life-enhancing archetypes that are latent in animate form. They have the possibility of discovering in themselves, too, miraculous dimensions of being that are "beyond the phallus."

Animate forms are not machines. They are not something we have at our disposal, something to use, manipulate, and master. Animate forms are living creatures; they are experiencing beings. They are wondrous creations, ourselves included. We need to turn our attention to these forms and articulate the corporeal archetypes they engender, archetypes that are part of the evolutionary semantics that postmodernists, sociobiologists, and others leave behind. These archetypes come first of all with each creature's being the species-specific body it is. In effect, the idea that we are a unity of opposites does not mandate a dichotomization of male and female powers so much as aver potentialities that stretch across a gamut of human possibilities that one can develop as one chooses or not. When an Other-subjugating conception of power is culturally exaggerated, when it is given inordinate and *exclusive* attention, when it is catapulted to a privileged and *gendered* position on high, the gamut of human possibilities shrinks; divergent powers are proportionately devalued and neglected. In consequence, an extraordinary effort is needed to rethink power and to recover other possible conceptions. Failing to make the effort, we perpetuate by omission the very conception of power that oppresses us. The result is that we *all* become lop-sided; we are all of us—male and female alike—in thrall to a concept of power that constricts our movement, blinkers our vistas, and whittles away at our possibilities in the world.

A corporeal turn can awaken us to the full and complex challenge of animate form just as it has awakened us here to corporeal archetypes of power, in particular, to those that instantiate the conception of power as control. It has led us to see how those archetypes are anchored in our primate history and to trace the ways in which cultures and individuals work and rework them in life-enhancing and life-destroying ways, which is to say, into intricate socio-political tapestries. Insofar as these tapestries engender a *corpus moralis,* they should lead ultimately to the roots of morality. A further challenge of animate form is thus linked to an evolutionary ethics. Given species identity and attachment and the procreational basis of life, primary among the corporeal analyses that would elucidate this evolutionary ethics are ones that would describe the foundations and rationality of caring. Caring is a miraculous intercorporeal power, a form of sense-making whose roots are as embedded in our phylogeny and ontogeny as the roots of power elucidated here.

In sum, touches of the miraculous are present in all bodies—bodies that become tumescent, bodies that become pregnant, bodies that menstruate, bodies that grow, bodies that heal, bodies that cry, smile, quiver, wince, blush, walk, speak, play—bodies that quicken to the pulse of life in all its changing forms. Like uterine contractions, the miraculous eludes our control. That is indeed what makes it miraculous. It is what male and female bodies are; it is what *all* bodies are. We can see this the moment we remove the phallic prisms that, like cataracts, cloud our sight and shroud our vision.

NOTES

PROLOGUE

1. Carol Hanisch's essay, "The Personal is Political," appeared in *Notes from the Second Year: Women's Liberation, Major Writings of the Radical Feminists,* ed. Shulamith Firestone and Anne Koedt (New York: Radical Feminism, 1970), pp. 76–78. Earlier, in 1969, Charlotte Bunch wrote, "There is no private domain of a person's life that is not political and there is no political issue that is not ultimately personal." See "A Broom of One's Own" in *Passionate Politics* (New York: St. Martin's Press, 1987), pp. 27–45, p. 29.

2. See Maxine Sheets-Johnstone, *The Roots of Thinking* (Philadelphia: Temple University Press, 1990), especially chapter 11.

3. See Sheets-Johnstone, *The Roots of Thinking.*

4. Catherine A. MacKinnon, "Feminism, Marxism, Method and State: An Agenda for Theory," *Signs* 7 (Spring 1982): 515–544, pp. 534–35.

5. Boston Women's Health Book Collective, *Our Bodies, Ourselves* (New York: Simon and Schuster, 1971).

6. The phrase "the common denominator of cultures" comes from George Peter Murdock, "The Common Denominator of Cultures," in *Readings in Introductory Anthropology,* 2 vols., ed. Richard G. Emerick (Berkeley: McCutcheon Publishing, 1969), vol. 1: 323–26.

7. Donald Symons, *The Evolution of Human Sexuality* (New York: Oxford University Press, 1979). See this book, chapter 3, for a critical analysis and discussion of the characterization of female sexuality as a "service" to males.

8. Symons, *The Evolution of Human Sexuality,* p. 239.

9. Ibid., p. 106.

10. Chris Weedon, *Feminist Practice and Poststructuralist Theory* (New York: Basil Blackwell Ltd., 1987), p. 12.

11. Joan W. Scott, "Deconstructing Equality-Versus-Difference: Or, the Uses of Poststructuralist Theory for Feminism," *Feminist Studies* 14/1 (Spring 1988): 33–50, p. 33.

12. Barbara Christian, "The Race for Theory," in *Gender and Theory,* ed. Linda Kauffman (New York: Basil Blackwell, 1989), pp. 225–237. This essay is a revised version of an essay that first appeared in *Cultural Critique* 6 (Spring 1987): 51–63.

13. Ibid., p. 227.

14. Mary Poovey, "Feminism and Deconstruction," *Feminist Studies* 14/1 (Spring 1988): 51–65, p. 63.

15. Ibid.

CHAPTER 1: OPTICS OF POWER AND THE POWER OF OPTICS

1. Jean-Paul Sartre, *Being and Nothingness,* trans. Hazel E. Barnes (New York: Philosophical Library, 1956), p. 266.

2. Michel Foucault, *Discipline and Punish,* trans. Alan Sheridan (New York: Vintage Books, 1979), p. 200.

3. Hubert L. Dreyfus and Paul Rabinow, *Michel Foucault: Beyond Structuralism and Hermeneutics,* 2nd ed. (Chicago: University of Chicago Press, 1983), p. 112. Dreyfus and Rabinow actually point out specifically that "Reading Merleau-Ponty one would never know that the body has a front and a back and can only cope with what is in front of it, . . . and so on."

4. Cf. Nancy Fraser, "Foucault's Body-Language," *Salmagundi* 61 (Fall 1983): 55–70. If there is no "identifiable common referent" underlying all of Foucault's socio-politically inscribed bodies, Fraser writes, then "one may ask with what right Foucault continues to speak of the body *simpliciter* at all?" (p. 64).

5. Dreyfus and Rabinow, *Michel Foucault,* p. 125.

6. Foucault, *Discipline and Punish,* p. 25. Dreyfus and Rabinow, *Michel Foucault,* p. 112.

7. Dreyfus and Rabinow, *Michel Foucault,* p. 112.

8. Ibid.

9. Foucault, *Discipline and Punish,* p. 48.

10. Michel Foucault, *The History of Sexuality,* vol. 1, trans. Robert Hurley (New York: Vintage Books, 1980), Part Three, "Scientia Sexualis."

11. Dreyfus and Rabinow, *Michel Foucault,* p. 233.

12. Michel Foucault, *The Use of Pleasure* (Vol. 2 of *The History of Sexuality*), trans. Robert Hurley (New York: Vintage Books, 1986), p. 146.

13. Sartre, *Being and Nothingness*, p. 266.

14. Ibid., p. 265.

15. Ibid.: See respectively part 3, chapter 1, section 4; and part 3, chapter 2, all sections.

16. Ibid., p. 258.

17. Ibid., p. 254.

18. Ibid., p. 255.

19. Ibid.

20. Ibid., p. 256.

21. Ibid., p. 257.

22. Ibid., p. 261.

23. Ibid., p. 262.

24. Ibid., p. 261.

25. Ibid., p. 267.

26. Ibid., pp. 267–68.

27. Ibid., pp. 262–63.

28. Ibid., p. 268.

29. Compare the end statement of Sartre's play *Huis Clos* (*No Exit*): "L'enfer, c'est les Autres."

30. Sartre, *Being and Nothingness*, p. 263.

31. Foucault, *Discipline and Punish*, p. 25.

32. Ibid., pp. 293–95.

33. Ibid., p. 173.

34. Ibid., pp. 195–96.

35. Ibid., p. 199.

36. Ibid.

37. Ibid., p. 200.

38. Ibid.

39. Ibid., p. 201.

40. Ibid.

41. Sartre, *Being and Nothingness*, p. 268.

42. See Foucault, *Discipline and Punish*, p. 207.

43. Ibid., pp. 201–2.

44. Sartre, *Being and Nothingness*, p. 257.

45. Ibid., p. 295.

46. Foucault, *The History of Sexuality*, vol. 1, pp. 17–35.

47. Ibid., p. 83.

48. Ibid.

49. Ibid., p. 63.

50. The fact that Foucault later recants "what I said about this *ars erotica*" is a reflection of his concern with a more livable ethics, specifically, a more livable

sexual ethics, and with a more livable sexual ethics rooted in our own Western history. Whatever he recants of what he wrote, it is not a reflection of an error in specifying an *ars erotica* itself nor a reflection of errors in the differences he briefly and in general terms spells out between an *ars erotica* and a *scientia sexualis*. See Dreyfus and Rabinow, *Michel Foucault*, pp. 234–35. See also note 51 and this continuing text.

51. Foucault, *The History of Sexuality*, vol. 1, p. 57; but see Dreyfus and Rabinow, *Michel Foucault*, pp. 234–35, where, in an interview, Foucault states that what he said about an *ars erotica* in *The History of Sexuality*, vol. 1, was "wrong." He states that "I should have opposed our science of sex to a contrasting practice in our own culture," and goes on to say that "the Greeks and Romans did not have any *ars erotica* to be compared with the Chinese *ars erotica*." This "cross-cultural correction" is not a retraction of what Foucault originally meant and substantively means by an *ars erotica*.

52. Foucault, *The History of Sexuality*, vol. 1, p. 157.

53. Ibid., p. 156.

54. The phrase is taken from Maurice Merleau-Ponty, *Phenomenology of Perception*, trans. Colin Smith (London: Routledge & Kegan Paul, 1962), p. 431.

55. Sartre, *Being and Nothingness*, p. 386.

56. Ibid., p. 385.

57. Ibid.

58. Ibid., p. 396.

59. Sartre inadvertently calls attention to a second and neglected meaning of that phrase when he writes that "To have sex means . . . to exist sexually for an Other who exists sexually for me." (*Being and Nothingness*, p. 384.) Sartre's point is that sexuality is apprehended in *desire*, not in the having of such and such anatomical parts. It is, as translator Hazel Barnes more pointedly puts it, "the deep-seated impulse of the For-itself to capture the Other's subjectivity" (p. xli).

60. Sartre, *Being and Nothingness*, p. 390.

61. Ibid.

62. Sartre (*Being and Nothingness*, p. 387–88) clearly distinguishes sexual desire from "other appetites" such as eating and drinking when he writes that while "the Other can note various physiological modifications"—"turgescence of the nipples of the breasts," for example—these modifications are not descriptive of sexual desire. His account of sexual desire as something other than a desire for certain body parts or for a certain state of certain parts of the body contrasts markedly with the bodily appearances that a well-known ethologist lists as sexual signals, i.e., sexual "releasers" or sexual "arousers"—

thus stimulators or objects of desire—and which include areolar tumescence as a "close visual signal" of women for men. See J. H. Crook, "Sexual Selection, Dimorphism, and Social Organization in Primates," *Sexual Selection and the Descent of Man, 1871–1971,* ed. B. Campbell (Chicago: Aldine, 1972), pp. 231–81; see particularly p. 251.

63. Sartre, *Being and Nothingness,* p. 389.

64. Ibid.

65. Ibid., p. 395.

66. Ibid., p. 393.

67. Foucault, *The History of Sexuality,* vol. 1, p. 71.

68. Ibid.

69. Ibid.

70. Ibid., p. 72.

71. Sartre's remarks on why "the true caress is the contact of two bodies in their mostly fleshy parts, the contact of stomachs and breasts[,]" (*Being and Nothingness,* p. 396) speak to this same point.

72. Bernard Berenson, "The Central Italian Painters of the Renaissance," in *The Bernard Berenson Treasury,* ed. H. Kiel (New York: Simon Schuster, 1962), pp. 97–101.

73. Aristotle. See *De Anima* 421ᵃ 16–19.

74. There is compelling evidence that in both our earliest experiences and prior to language, we perceive things in nonspecialized ways, that is, synaesthetically or intermodally. See Daniel N. Stern's *The Interpersonal World of the Infant* (New York: Basic Books, 1985) and his *Diary of a Baby* (New York: Basic Books, 1990). Stern's work and the work of others in psychology and psychiatry are discussed in detail in chapter 8 of this book. Cf. too Bishop Berkeley's *Essay Toward a New Theory of Vision* (1709) in *Berkeley Selections,* ed. Mary W. Calkins (New York: Charles Scribner's Sons, 1929).

75. Sartre, *Being and Nothingness,* p. 385.

76. Ibid.

77. Ibid., p. 398.

78. Sartre writes that where one's own pleasure becomes all-consuming, "the Other-as-object collapses, the Other-as-look appears, and my consciousness is a consciousness swooning in its flesh beneath the Other's look" (Ibid.).

79. Ibid., p. 386.

80. Ibid., p. 390.

81. Ibid.

82. Ibid.

83. Ibid., p. 391.

84. Ibid., p. 392. One can, of course, readily recognize this theme of the flesh in Merleau-Ponty's writings and appreciate its source.

85. Ibid., pp. 380–81.

CHAPTER 2: AN EVOLUTIONARY GENEALOGY

1. Charles Darwin, *Charles Darwin's Notebooks, 1836–1844,* ed. Paul H. Barrett et al. (Ithaca: Cornell University Press, 1987), p. 300.

2. Sarel Eimerl and Irven DeVore, *The Primates* (New York: Times, Inc., 1965), pp. 106, 108, 109.

3. Claud A. Bramblett, *Patterns of Primate Behavior* (Palo Alto, California: Mayfield, 1976), p. 47.

4. Michel Foucault, *Discipline and Punish,* trans. Alan Sheridan (New York: Vintage Books, 1979), p. 26–27.

5. Ibid.

6. Ibid., p. 27.

7. Thelma Rowell, *The Social Behaviour of Monkeys* (Middlesex: Penguin, 1974), p. 94.

8. Foucault, *Discipline and Punish,* p. 48.

9. Ibid., pp. 48–49.

10. See Maxine Sheets-Johnstone, "Animate Form and the Corporeal Turn," a paper presented at a panel session titled "The Corporeal Turn" at the thirty-first annual meeting of the Society for Phenomenology and Existential Philosophy, October 1992, Boston.

11. K. R. L. Hall and Irven DeVore, "Baboon Social Behavior," in *Primate Patterns,* ed. Phyllis Dolhinow (New York: Holt, Rinehart and Winston, 1972), pp. 125–180, p. 169.

12. Judith Shirek-Ellefson, "Social Communication in Some Old World Monkeys and Gibbons," in *Primate Patterns,* ed. Phyllis Dolhinow (New York: Holt, Rinehart and Winston, 1972), pp. 297–311, p. 307.

13. Jane van Lawick-Goodall, "A Preliminary Report on Expressive Movements and Communication in the Gombe Stream Chimpanzees," in *Primate Patterns,* ed. Phyllis Dolhinow (New York: Holt, Rinehart and Winston, 1972), pp. 25–84, p. 45.

14. John H. Kaufmann, "Social Relations of Adult Males in a Free-Ranging Band of Rhesus Monkeys," in *Social Communication Among Primates,* ed. Stuart A. Altmann (Chicago: University of Chicago Press, 1967), pp. 73–98, p. 79.

15. Donald Stone Sade, "Determinants of Dominance in a Group of

Free-Ranging Rhesus Monkeys," in *Social Communication Among Primates,* ed. Stuart A. Altmann (Chicago: University of Chicago Press, 1967), pp. 99–114, p. 100.

16. J. A. R. A. M. van Hooff, "The Facial Displays of the Catarrhine Monkeys and Apes," in *Primate Ethology,* ed. Desmond Morris (Garden City, NY: Anchor Books, 1969), pp. 9–88. Descriptions of these facial expressions are on pp. 22–38, quotations are on pp. 23 and 32 respectively.

17. Eugene Linden, "A Curious Kinship: Apes and Humans," *National Geographic* 181/3 (1992): 2–45, p. 36.

18. Stuart A. Altmann, "The Structure of Primate Social Communication," in *Social Communication Among Primates,* ed. Stuart A. Altmann (Chicago: University of Chicago Press, 1967), pp. 325–362, pp. 331–32.

19. Rowell, *The Social Behaviour of Monkeys,* pp. 85–86.

20. Phyllis Dolhinow, "The North Indian Langur," *Primate Patterns,* ed. Phyllis Dolhinow (New York: Holt, Rinehart and Winston, 1972), pp. 181–238, pp. 223–224. It should be noted that the two tables separate male and female gestures: male aggressive gestures begin with the stare; female aggressive gestures begin with a gesture simply termed "tense," the stare being the second gesture. It is furthermore of interest to note that the second submissive gesture for both male and female is "Turn head; look away," and that the third submissive gesture is "Turn back."

21. Sade, "Determinants of Dominance in Rhesus Monkeys," p. 102.

22. Ibid.

23. van Lawick-Goodall, "A Preliminary Report," p. 54.

24. Hall and DeVore, "Baboon Social Behavior," p. 174.

25. For an initial discussion of the significance of posture in intercorporeal relationships, see Maxine Sheets-Johnstone, *The Roots of Thinking* (Philadelphia: Temple University Press, 1990), p. 103.

26. It might be observed that were the look or the glance not thrown, one might be undaunted by the exposure and its abusive, even frightening, character, undaunted perhaps even to the point of trusting the Other, since indeed by not looking over one's shoulder or by not glancing at the tower, one would appear fully to entrust oneself bodily to the desires and intentions of the Other, whatever they might be. In either case, of course, one remains a docile body. Whether acknowledged by a look or a glance, or not, surveillance by another is not easily dismissed. Indeed, so long as the mere possibility of another's gaze exists, an Other is unremittingly present.

27. van Hooff, "Facial Displays," pp. 38–39.

28. Elinor Ochs and Bambi B. Schieffelin, "Language Acquisition and Socialization," in *Culture Theory: Essays on Mind, Self, and Emotion,* eds.

Richard A. Shweder and Robert A. LeVine (Cambridge: Cambridge University Press, 1984), pp. 276–320, p. 289.

29. Power relations are clearly tied here not just to vision, i.e., to witnessing the excretory bodily act of another, but to intercorporeal positionings, i.e., being in an elevated bodily position relative to a crouching other.

30. For an interdisciplinary and comprehensive accounting, see *The Evil Eye: A Folklore Casebook*, ed. Alan Dundes (New York: Garland Publishing, 1981); see also *The Evil Eye*, ed. Clarence Maloney (New York: Columbia University Press, 1976).

31. Sartre, *Being and Nothingness*, trans. Hazel E. Barnes (New York: Philosophical Library, 1956), p. 263 and *Huis Clos*, Scene V, respectively.

32. Foucault, *Discipline and Punish*, p. 136.

33. Frans de Waal, *Chimpanzee Politics* (New York: Harper Colophon Books, 1982), pp. 87–88.

34. Charles Darwin, *The Origin of Species* (1859), ed. J. W. Burrow (Middlesex: Penguin, 1968), and *The Descent of Man and Selection in Relation to Sex* (1871) (Princeton: Princeton University Press, 1981).

35. Regarding European toads, see David Attenborough, *Life on Earth* (Boston: Little, Brown and Company, 1979), p. 140; regarding male sea-elephants, see Charles Darwin, *The Descent of Man and Selection in Relation to Sex*, vol. II, p. 278.

36. Raymond Firth, " 'Postures and Gestures of Respect' " in *The Body Reader*, ed. Ted Polhemus (New York: Pantheon Books, 1978), pp. 88–108, p. 96.

37. Ibid., p. 104.

38. Ibid., p. 105.

39. Ibid., pp. 105–6.

40. Ibid., p. 102.

41. See Hans Hass, *The Human Animal* (New York: G. P. Putnam's Sons, 1970), pp. 146–47 for a brief discussion of the similarity between human and nonhuman appeasement gestures.

42. Firth, "Postures and Gestures of Respect," p. 94.

43. Yukimaru Sugiyama, "Social Behavior of Chimpanzees in the Budongo Forest, Uganda," *Primates* 10 (1969): 197–225, p. 199.

44. van Lawick-Goodall, "A Preliminary Report," p. 42.

45. Judith Butler, *Gender Trouble* (New York: Routledge, 1990), specifically, pp. 134–141.

46. Adolph Portmann, *Animal Forms and Patterns*, trans. Hella Czech (New York: Schocken Books, 1967).

47. Ibid., p. 31.

48. Ibid., p. 33.

49. Ibid.
50. Ibid., p. 70.
51. Ibid., p. 122.
52. Ibid., p. 35.
53. Ibid.
54. Ibid., pp. 183, 184, respectively.
55. Ibid., pp. 195–96.
56. Ibid., p. 184.
57. Ibid., p. 183.
58. Sartre, *Being and Nothingness,* p. 403.

59. Portmann's insights into the visible actually challenge us to re-think "outsides." They require us to pay attention to *form.* Because they specify no inherent distinction in kind between human and nonhuman, they put the question of interanimate relations in a decidedly ethical context. A human vision bereft of form values, a vision that distorts the natural ratio among the senses and is indifferent to natural semantic dimensions of intercorporeal life, is a vision that can sweep blindly over nonhuman as well as human forms of life. In each case, a similar kind of power relations is enforced: Others are mere instruments, if anything at all; they are accorded no status as individuals; their "inwardness" or felt life goes unacknowledged. Cast in Portmann's evolutionary perspective, the power of optics thus takes on a peculiar but incisive ethical significance. It is one thing to subjugate other humans through panoptical practices and institutions; it is one thing to enter into a visually anchored discourse with other humans in ways which sweep away their felt life. It is another thing to subjugate nonhumans and to appropriate them for use in a discourse of which they have no knowledge and from which they cannot escape. Moreover, just because there is no reciprocal inwardness—what grounds are there for thinking that a cat, for example, recognizes its human "owner" through a process of "reciprocal incarnation"?—is no justification for conscripting nonhuman animals into disciplinary technologies and denying them their own freedom and independence.

60. Lest there be any doubt or misunderstanding about the matter, perhaps it is necessary to note explicitly that not all form values are in the service of sweetness and light. Some form values are incitements to aggression, as when, for example, in interspecies terms, one creature hunts a particular other, or in intraspecies terms, when one creature threatens another. What is seen bodily of the other is in each case agonistically meaningful, whether a certain patterning, posture, gesture, or behavior.

61. A notable exception, of course, is Edmund Husserl's treatment of intersubjectivity—the constitution of an intersubjective world—in his fifth

Cartesian Meditation. See *Cartesian Meditations,* trans. Dorion Cairns (The Hague: Martinus Nijhoff, 1973), pp. 89–157. The term is anchored in decisively bodily ways.

62. Edward T. Hall, *The Hidden Dimension* (New York: Doubleday, 1966), p. ix.

63. Heini P. Hediger, "The Evolution of Territorial Behavior," in *Social Life of Early Man,* ed. Sherwood L. Washburn (New York: Viking Fund Publications in Anthropology, no. 31, 1961), pp. 34–57, p. 36.

64. Heini P. Hediger, *Studies of the Psychology and Behavior of Captive Animals in Zoos and Circuses* (London: Butterworths, 1955), p. 123.

65. J. B. Calhoun, "A Comparative Study of the Social Behavior of Two Inbred Strains of House Mice," *Ecological Monographs* 26 (1956): 81–103; "A Behavioral Sink," in *Roots of Behavior,* ed. Eugene L. Bliss (New York: Harper and Bros., 1962): 295–315; "Population Density and Social Pathology," *Scientific American* 206 (1962): 139–148.

66. O. Michael Watson, *Proxemic Behavior: A Cross-Cultural Study* (The Hague: Mouton & Co., 1970), p. 30.

67. W. John Smith, Julia Chase, Anna Katz Lieblich, "Tongue Showing: A Facil (sic) Display of Humans and Other Primate Species," in *Nonverbal Communication, Interaction, and Gesture,* ed. Adam Kendon, *Approaches to Semiotics,* no. 41 (The Hague: Mouton, 1981), pp. 503–548.

68. David Givens, "Greeting a Stranger: Some Commonly Used Nonverbal Signals of Aversiveness," in *Nonverbal Communication, Interaction, and Gesture,* ed. Adam Kendon, *Approaches to Semiotics,* no. 41 (The Hague: Mouton, 1981), pp. 219–250, p. 231.

69. Ibid., p. 230.

70. See, for example, Daniel C. Dennett, "Intentional Systems in Cognitive Ethology: The 'Panglossian Paradigm' Defended." *The Behavioral and Brain Sciences* 6 (1983): 343–390.

71. Sartre, *Being and Nothingness,* p. 258.

CHAPTER 3: CORPOREAL ARCHETYPES AND POWER: PRELIMINARY CLARIFICATIONS AND CONSIDERATIONS OF SEX

1. Ladelle McWhorter, "Culture or Nature? The Function of the Term 'Body' in the Work of Michel Foucault," *Journal of Philosophy* 86/11 (1989): 608–614, p. 612.

2. Carol C. Gould, "The Woman Question: Philosophy of Liberation

and the Liberation of Philosophy," in *Philosophy of Woman,* ed. Mary B. Mahowald (Indianapolis: Hackett Publishing, 1983), pp. 415–452, p. 418.

3. Alison M. Jaggar, "Love and Knowledge: Emotion in Feminist Epistemology," in *Gender/Body/Knowledge,* eds. Alison M. Jaggar and Susan R. Bordo (New Brunswick: Rutgers University Press, 1989), pp. 145–171, p. 150.

4. Donna Haraway, "A Manifesto for Cyborgs: Science, Technology, and Socialist Feminism in the 1980s," *Socialist Review* 15/2 (1985): 65–107, p. 72; and *Primate Visions* (New York: Routledge, 1989), p. 290.

5. Judith Butler, "Gendering the Body: Beauvoir's Philosophical Contribution," in *Women, Knowledge, and Reality: Explorations in Feminist Philosophy,* eds. Ann Garry and Marilyn Pearsall (Boston: Unwin Hyman, 1989), pp. 253–273, p. 261.

6. McWhorter, "Culture or Nature?" p. 612.

7. Ibid., p. 614.

8. Maxine Sheets-Johnstone, *The Roots of Thinking* (Philadelphia: Temple University Press, 1990).

9. Gould, "The Woman Question," pp. 418–19.

10. Charles Darwin, *Charles Darwin's Notebooks: 1836–1844,* eds. David Kohn and Paul H. Barrett (Ithaca: Cornell University Press, [1836–1844] 1987), p. 542.

11. Irenaus Eibl-Eibesfeldt, "Similarities and Differences Between Cultures in Expressive Movements," in *Nonverbal Communication,* ed. R. A. Hinde (New York: Cambridge University Press, 1972).

12. Paul Ekman, "Facial Expression," in *Nonverbal Behavior and Communication,* eds. Aron W. Siegman and Stanley Feldstein (Hillsdale, NJ: Lawrence Erlbaum, 1978), pp. 97–116; "The Argument and Evidence about Universals in Facial Expressions of Emotion," in *Handbook of Social Psychophysiology,* eds. H. Wagner and A. Manstead (New York: John Wiley and Sons, 1989), pp. 143–164; and Paul Ekman, W. V. Friesen, M. O'Sullivan, A. Chan, I. Diacoyanni-Tarlatzis, K. Heider, R. Krause, W. A. Le Compte, T. Pitcairn, P. E. Ricci-Bitti, K. R. Scherer, M. Tomita, and A. Tzavaras, "Universals and Cultural Differences in the Judgements of Facial Expressions of Emotion," *Journal of Personality and Social Psychology* 53 (1987): 712–17.

13. Charles Darwin, *The Expression of the Emotions in Man and Animals* (Chicago: University of Chicago Press, [1872] 1965).

14. Jane Flax, *Thinking Fragments* (Berkeley: University of California Press, 1990), pp. 218–19.

15. Susan Bordo, "Feminism Reconceives the Body," paper presented at Bates College, Lewiston, ME, April 1991.

16. Sheets-Johnstone, *Roots of Thinking,* pp. 19, 382.

17. Donna Haraway, "Situated Knowledges: The Science Question in Feminism and the Privilege of Partial Perspective," *Feminist Studies* 14/3 (1988): 575–99, p. 579.

18. Ibid., p. 580.

19. Ibid., p. 582.

20. See Sheets-Johnstone, *Roots of Thinking*, pp. 303–4.

21. Haraway, "Situated Knowledges," p. 583.

22. Ibid., p. 585.

23. Ibid., p. 588.

24. Ibid.

25. Helen Longino, *Science as Social Knowledge* (Princeton: Princeton University Press, 1990), p. 221.

26. Ibid., p. 222.

27. Ibid.

28. This phrase—"to the things themselves" (*zu den Sachen selbst*)—is a well-known Husserlian dictum. Philosopher Edmund Husserl used it to plead the need for, and the way to, a sound and consummate epistemology.

29. In her earlier discussion of human evolution, Longino unfortunately accepts without question standard fragmentary practice in evolutionary biology. She writes that "the main questions addressed in the search for human origins are standardly grouped into two categories: anatomical evolution and social evolution," and goes on to note that "there are some changes central to human development that are captured by neither of these categories . . . like locomotion" (*Science as Social Knowledge*, p. 104). That something as fundamental as locomotion fits into neither basic category should give pause enough for questioning them. Clearly the dichotomy as well as the categories themselves obliterate the possibility of examining animate form. Accordingly, a retreat to a common level of experience depends on prior clarification of what is being described. If one starts with a sundered specimen—or with synecdochic renditions of the body—one can hardly arrive at understandings of animate form.

30. Lorraine Code, *What Can She Know?* (Ithaca: Cornell University Press, 1991), p. 148.

31. Ibid., p. 149.

32. As noted in the prologue, an overemphasis by many feminists on theory, especially to the exclusion of practice, is viewed by other feminists as deleterious. Barbara Christian, for example, writes that "when Theory is not rooted in practice, it becomes prescriptive, exclusive, elitist," and she goes on to voice strong concerns about the way in which "Theory" can become *"authoritative discourse."* "The Race for Theory," in *Gender and Theory*, ed.

Linda Kauffman (New York: Basil Blackwell, 1989), pp. 225–237, pp. 231 and 233, respectively.

33. Code, *What Can She Know?*, p. 295.

34. Ibid., p. 148. Re 'proprioception': Code's quote is from Oliver Sacks's *The Man Who Mistook His Wife for a Hat*.

35. Judith Butler, *Gender Trouble* (New York: Routledge, 1990), pp. xi, xii.

36. This phrase comes from William Saroyan's play *The Time of Your Life*. The words are uttered several times over in the course of the play by an otherwise near-mute character. The phrase, of course, could be uttered with equal conviction by a foundationalist to a relativist or by a relativist to a foundationalist. It might be noted that the rhetoric of the relativist's anti-foundationalist claims can at times exceed the bounds of truth, as when Judith Butler (*Gender Trouble*, p. 12) includes Edmund Husserl in her list of philosophers who are part of the dualistic philosophical tradition "that supports relations of political and psychic subordination and hierarchy." The inclusion is clearly without foundation, as any familiarity with Husserl's *Ideas II* and his consistent concern with the fundamental significance of what he calls 'I can's' and 'kinestheses' attests. See *Ideas Pertaining to a Pure Phenomenology and to a Phenomenological Philosophy*, Second Book: *Studies in the Phenomenology of Constitution* (commonly known as *Ideas II*), trans. R. Rojcewicz and A. Schuwer (Boston: Kluwer Academic, 1989).

37. Butler, *Gender Trouble*, p. 139.

38. Ibid., p. 129. Arleen Dallery makes a not dissimilar claim on behalf of *écriture feminine* in "The Politics of Writing (the) Body: *Écriture Feminine*," in *Gender/Body/Knowledge*, eds. Alison M. Jaggar and Susan R. Bordo (New Brunswick: Rutgers University Press, 1989), pp. 52–67. However she contradicts her own onto-essentialist denials when she insists that "a 'real' body prior to discourse is meaningless" (p. 59) and that "there is no fixed, univocal, ahistorical woman's body as the referent of this [*écriture feminine*] discourse" (p. 63) at the same time that she celebrates "woman's erotic embodiment" (p. 56) complete with, e.g., labia (p. 59) and speaks of "feminine structures of erotic embodiment where self and other are continuous, in pregnancy, childbirth, and nursing" (p. 54). Unless the stuff of erotic "embodiment" and the designated feminine "structures" exist only on paper, or unless we grant philosophic license along with poetic, it is difficult to know how to reconcile an actual living body—a moving, resonating, species-specific form (and, in fact, in terms of pregnancy, childbirth, and nursing, a *sex*-specific animate form)—with something that has no mean-

ing outside of language and that, even as languaged, has no historically invariant identity.

39. I am referring to Charles Darwin's *The Expression of the Emotions in Man and Animals* as well as to *The Origin of Species* (Middlesex: Penguin, [1859] 1968) and *The Descent of Man and Selection in Relation to Sex* (Princeton: Princeton University Press, [1871] 1981).

40. Butler, *Gender Trouble*, p. 132.

41. Ibid., pp. 140–41.

42. This phrase is sometimes erroneously attributed to Maurice Merleau-Ponty. It in fact comes from the somatic studies of Edmund Husserl, in particular his insightful and seminal descriptive analyses of "the kinestheses." See especially Husserl's *Ideas II;* but see also, for example, his *The Crisis of European Sciences and Transcendental Phenomenology,* trans. David Carr (Evanston: Northwestern University Press, 1970), pp. 106–8, 161, 217, 331–32, his *Cartesian Meditations,* trans. Dorion Cairns (The Hague: Martinus Nijhoff, 1973), p. 97, and his *Ideas Pertaining to a Pure Phenomenology and to a Phenomenological Philosophy,* Third Book: *Phenomenology and the Foundations of the Sciences* (commonly known as *Ideas III*), trans. T. E. Klein and W. E. Pohl (Boston: Martinus Nijhoff, 1980), pp. 106–12.

43. See Darwin, *The Descent of Man and Selection in Relation to Sex.*

44. William G. Eberhard, *Sexual Selection and Animal Genitalia* (Cambridge: Harvard University Press, 1985).

45. For a discussion of these aspects of animate form, see Sheets-Johnstone, *Roots of Thinking,* particularly chap. 7, "Hominid Bipedality and Sexual Selection Theory."

46. See, for example, Sandra Harding, "The Instability of the Analytical Categories," in *Sex and Scientific Inquiry,* eds. Sandra Harding and Jean F. O'Barr (Chicago: University of Chicago Press, 1987), pp. 283–302; Ruth Hubbard, *The Politics of Women's Biology* (New Brunswick: Rutgers University Press, 1990); Anne Fausto-Sterling, *Myths of Gender* (New York: Basic Books, 1985); Evelyn Fox Keller, "The Gender/Science System; or, Is Sex to Gender as Nature Is to Science?" *Hypatia* 2/3 (1987): 37–49, reprinted in *Feminism and Science,* ed. Nancy Tuana (Bloomington: Indiana University Press, 1989); Marion Lowe, "Social Bodies: The Interaction of Culture and Women's Biology," in *Biological Woman—The Convenient Myth,* eds. Ruth Hubbard, Mary Sue Henifer, and Barbara Fried (Cambridge, MA: Schenkman, 1982), pp. 91–116.

47. See, for example, Butler, "Gendering the Body" and *Gender Trouble;* Mary Poovey, "Feminism and Deconstruction," *Feminist Studies* 14/1 (1988): 51–65; Dallery, "The Politics of Writing (the) Body."

48. Humans belong to the biological order called primates, to the biological family called hominidae (which includes the great apes), and to the genus *Homo.*

49. Carl G. Jung, *On the Nature of the Psyche,* trans. R. F. C. Hull, (Princeton: Princeton University Press, 1960).

50. See, for example, Suzanne Chevalier-Skolnikoff, "Heterosexual Copulatory Patterns in Stumptail Macaques (*Macaca arctoides*) and in Other Macaque Species," *Archives of Sexual Behavior* 4/2 (1975): 199–220; Jane van Lawick-Goodall, "The Behaviour of Free-Living Chimpanzees in the Gombe Stream Reserve," *Animal Behaviour Monographs,* vol. 1, part 3 (1968); Yukimaru Sugiyama, "Social Behavior of Chimpanzees in the Budongo Forest, Uganda," *Primates* 10 (1969): 197–225; K. R. L. Hall and Irven DeVore, "Baboon Social Behavior," in *Primate Patterns,* ed. Phyllis Dolhinow (New York: Holt, Rinehart and Winston, 1972), pp. 125–180 Tomas Enomoto, "The Sexual Behavior of Japanese Monkeys," *Journal of Human Evolution* 3 (1974): 351–72.

51. "Presenting, which is usually a gesture of submission, is often accompanied by nervous, even fearful, behavior on the part of the presenting animal, whereas mounting, conversely, in baboons is usually an indicator of relative dominance" (Hall and DeVore, "Baboon Social Behavior," p. 174).

52. It is notable that a woman primatologist, Mireille Bertrand, who did her studies and fieldwork in the 1960s in India, France, and Thailand, is an exception. In her book-length monograph on stumptail macaques (*The Behavioral Repertoire of the Stumptail Macaque* [Basel, Switzerland: Karger, 1969]; published simultaneously as *Bibliotheca Primatologica* 11), she writes: "Social presenting has generally been assumed to originate from the sexual one: a monkey may avoid the aggression of a superior by taking the posture of the receptive female, which is supposed to be both submissive and sexually arousing" (pp. 185–86). She goes on to say that "This interpretation involves certain difficulties," and, after itemizing them, concludes, "Thus it is possible that social presenting does not derive from sexual presenting" but is rather connected with perineal grooming: "Perineal presenting may have acquired a submissive connotation because the performer stands still and waits, does not take the initiative of the next move, and shows its rump and back, that is, what it would show if it were fleeing" (p. 186). It is furthermore of interest to note Bertrand's initial remarks on sex and genital displays: "The importance of sex among monkeys and apes is in general grossly overestimated by the layman, as it was by some of the first primatologists." She writes that the misconception may be ex-

plained by two reasons: first, "infra-human primates use genitalia in a variety of social signals outside a sexual context, and these are interpreted as sexual by human observers"; and second, "in captivity, sex and genital displays may be exaggerated when animals are confined, kept in artificial, sexually imbalanced groupings" (p. 178). Bertrand's analysis says as much about human primates, and obviously male human primates in terms of the history of primatology and zoology, as it does about nonhuman primates.

53. Konrad Lorenz, *On Aggression,* trans. Marjorie K, Wilson (New York: Bantam, 1967), pp. 130–31.

54. J. R. Napier and P. H. Napier, *The Natural History of the Primates* (London: British Museum [Natural History], 1985), p. 75.

55. Hans Kummer, *Primate Societies* (Chicago: Aldine, 1971), p. 109.

56. Gordon W. Hewes, "World Distribution of Certain Postural Habits," *American Anthropologist* (n.s.) 57/2 (1955): 231–44.

57. Ibid., p. 238.

58. Ibid.

59. Ibid.

60. Ibid., p. 239.

61. Gordon W. Hewes, "The Anthropology of Posture," *Scientific American* 196 (1957): 123–32.

62. Clellan S. Ford and Frank A. Beach, *Patterns of Sexual Behavior* (New York: Harper Torchbooks, 1951), pp. 1, 94.

63. Ibid., p. 95.

64. van Lawick-Goodall, "Behaviour of Free-Living Chimpanzees," p. 217.

65. Sheets-Johnstone, *Roots of Thinking,* p. 101.

66. C. M. Rogers, "Implications of Primate Early Rearing Experiment for the Concept of Culture," in *Precultural Primate Behavior,* ed. Emil Menzel (Basel [Switzerland]: Karger, 1973), pp. 185–191, p. 188.

67. Ibid.

68. Frans de Waal, *Chimpanzee Politics: Power and Sex Among the Apes* (New York: Harper Colophon Books, 1982), p. 159.

69. G. Mitchell, *Human Sex Differences: A Primatologist's Perspective* (New York: Van Nostrand Reinhold, 1981), p. 44.

70. See, for example, K. R. L. Hall, "Social Vigilance Behaviour of the Chacma Baboon (*Papio ursinus*)," *Behavior* 16 (1960): 261–94; Wolfgang Wickler, "Socio-sexual Signals and Their Intra-specific Imitation Among Primates," *Primate Ethology,* ed. Desmond Morris (Garden City, NY: Anchor Books, 1969), pp. 89–189; D. W. Ploog, J. Blitz, and F. Ploog, "Studies on Social and Sexual Behavior of the Squirrel Monkey (*Saimiri sciureus*)," *Folia*

Primatologica 1 (1963): 29–66. For detailed discussions of this behavior and its import, see Sheets-Johnstone, *Roots of Thinking,* chapters 4, 7.

71. van Lawick-Goodall, "Behaviour of Free-Living Chimpanzees," p. 217.

72. For a justification of this methodological practice, see Sherwood Washburn, "The Analysis of Primate Evolution with Particular Reference to the Origin of Man," *Cold Spring Harbor Symposia on Quantitative Biology* 15 (1950): 57–78 and Jane Lancaster, *Primate Behavior and the Emergence of Human Culture* (New York: Holt, Rinehart and Winston, 1975).

73. J. Hanby, "Sociosexual Development in Primates," in *Perspectives in Ethology,* vol. 2, eds. P. P. G. Bateson and Peter H. Klopfer (New York: Plenum Press, 1976), pp. 1–67, p. 37.

74. de Waal, *Chimpanzee Politics.*

75. The shift may, of course, be related in deep and poignant ways to many feminists' felt lack of, and present-day search for, "a desire of their own." See in particular Jessica Benjamin, "A Desire of One's Own: Psychoanalytic Feminism and Intersubjective Space," in *Feminist Studies/Critical Studies,* ed. Teresa de Lauretis (Bloomington: Indiana University Press, 1986), pp. 78–101.

76. Insofar as male genitalia are regularly utilized as a taxonomic standard, the assumption of similarity is in each case a reasonable assumption.

77. For a full discussion of the radical shift and its evolutionary significance, see Sheets-Johnstone, *Roots of Thinking,* chapter 7.

78. As noted in chapter 2, the difference between signal and display is succinctly defined by primatologist Claud Bramblett: "Some signals have been affected by natural selection, i.e., *ritualized,* so that the signal is exaggerated and incorporates several elements that make it more complex. We call such a signal a *display"* (Bramblett, *Patterns of Primate Behavior,* p. 47). Female primate "sexual swellings," as vulval changes are called, and presenting aptly exemplify the basic difference: the signal is exaggerated by the act.

79. Adolf Portmann, *Animal Forms and Patterns,* trans. Hella Czech (New York: Schocken Books, 1967).

80. Linda Lopez McAlister (pers. com., 1991) reminded me that "in some centuries [males] have taken to stuffing their cod-pieces," and she drew my attention to the fact that "cosmetic surgeons in Miami now do penile-enhancement surgery."

81. Certainly a prime issue revolves about the notion of biology as an objective science untainted by cultural biases. Within this issue lies also the question of the way in which, and the degree to which, biologists play into an already established cultural *eidos* and the degree to which they help create it. (Discussion of the issue addressed in this chapter emphasizes the latter dimension but is not meant to suggest that the former dimension is nonexist-

ent.) Another prime issue involves a synecdochic rendition of *female*. For a detailed discussion of this rendition and a critical analysis of Sartre's characterization of females as being "in the form of a hole," see chapter 6: "Corporeal Archetypes: Penetration and Being 'in the Form of a Hole'."

82. Donald Symons, *The Evolution of Human Sexuality* (New York: Oxford University Press, 1979), p. 284.

83. Ibid., p. 177.

84. It is of considerable interest to note that in Symons's book, there is no entry in the index for *genitals*, for *penis*, or for *vagina*. For a book on the evolution of human sexuality, this would seem to be an extremely serious omission.

85. See Stephen Jay Gould and Richard Lewontin, "The Spandrels of San Marco and the Panglossian Paradigm: A Critique of the Adaptationist Programme," *Proceedings of the Royal Society of London,* Series B, Biological Science 205 (1979): 581–98. See also Symons's optimistic view that "if humans exhibit relatively uniform dispositions under a wide range of environmental conditions, these dispositions probably were uniformly adaptive among our Pleistocene ancestors and hence develop in a relatively stereotyped manner" (*Evolution of Human Sexuality,* p. 71). Symons casually links present-day sexual behavior with sexual behavior originating three million and more years ago, and this without any reference to the body itself.

86. In this context, perhaps the foremost positive value of the doctrine of cultural relativism is to caution us about where we look for human universals. Cultural practices are no match for corporeal archetypes.

87. Carl G. Jung, *Psychological Types,* trans. R. F. C. Hull (Princeton: Princeton University Press, 1976), p. 236.

88. Symons, *Evolution of Human Sexuality,* chapter 8.

89. Darwin, *The Descent of Man,* p. 262.

90. Eberhard, *Sexual Selection and Animal Genitalia.*

91. Ibid., p. 71.

92. Symons, *Evolution of Human Sexuality,* p. 94.

93. The phrase, which Symons uses (*Evolution of Human Sexuality,* p. 106), comes originally from Frank A. Beach ("Human Sexuality and Evolution," in *Reproductive Behavior,* eds. William Montagna and William A. Sadler [New York: Plenum Press, 1974], pp. 333–65, p. 357).

94. Symons, *Evolution of Human Sexuality,* p. 22.

95. Darwin, *The Descent of Man,* p. 278.

96. Ibid., p. 279.

97. Symons, *Evolution of Human Sexuality,* p. 203.

98. In addition to Eberhard's work (*Sexual Selection and Animal Genita-*

lia), see Ronald A. Fisher's original formulation of runaway selection in his *The Genetical Theory of Natural Selection,* 2nd ed. (New York: Dover, 1958), pp. 150–53.

99. Darwin, *The Descent of Man,* p. 262.

100. Lest it be thought that such discrimination is farfetched, it might be helpful to point out that field and experimental evidence show that female bowerbirds, for instance, "show a strong preference for particular males," and this on the basis of the bowers the males build. The females discriminate with fine attention the number of decorations on the bower platform, the color of the decorations, the "well-built[ness]" of the bower, and the "overall quality" of the bower. Gerald Borgia, "Sexual Selection in Bowerbirds," *Scientific American* 254 (1986): 92–100, pp. 92, 100, 99, respectively.

101. Klaus Theweleit, *Male Fantasies,* 2 vols., trans. Stephen Conway (Minneapolis: University of Minnesota Press, 1987).

102. Symons, *Evolution of Human Sexuality,* p. 284.

103. Ibid., p. 284. Symons actually goes on to gloss this appraisal with the statement that "these impulses [of males to rape females] are part of human nature because they proved adaptive over millions of years." Not only does he claim that rape is a *hominid male genetic endowment,* but he goes on to say that any change in the arrangement "might well entail a cure worse than the disease": "Where males can win females' hearts through tears rather than spears, through a show of vulnerability rather than strut and swagger, many will do so, but the desires persist and the game remains essentially the same. Given sufficient control over rearing conditions, no doubt males could be produced who would want only the kinds of sexual interactions that women want; but such rearing conditions might well entail a cure worse than the disease" (p. 285).

104. The conception of female sexuality as an anatomical passageway or opening recalls Sartre's famous statement "sex is a hole" (*Being and Nothingness,* trans. Hazel Barnes [New York: Philosophical Library, 1956], p. 614). For critical assessments of Sartre's psychological ontology of sex, see Margery Collins and Christine Pierce's "Holes and Slime: Sexism in Sartre's Psychoanalysis," *Philosophical Forum* 5/1–2 (1973–74): 112–27; Christine Pierce's "Philosophy" (a review essay), *Signs* 1/2 (1975): 487–503; and this book, chapter 6. It is of interest to note that Sartre's psychological ontology of sex appears actually to contradict his earlier phenomenological ontology of sex as desire. In his phenomenological analysis of desire as *the* mode of being by which we realize sex, i.e., come to know ourselves and others as sexed beings, Sartre is at pains, one might say, to emphasize, on the one hand, the distinctions between

desire and pleasure and between desire and sexual act, and, on the other hand, the goal of desire as *reciprocal* incarnation. Desire has clearly nothing to do with *holes*. See Sartre, *Being and Nothingness*, pp. 382–98.

105. The literature on exhibitionism is of interest to review in this context. See, for example, Robert J. Stoller, "Sexual Deviations," in *Human Sexuality in Four Perspectives*, ed. Frank A. Beach (Baltimore: Johns Hopkins University Press, 1976), pp. 190–214, specifically, pp. 202–3 and Ivor H. Jones and Dorothy Frei, "Exhibitionism—A Biological Hypothesis," *British Journal of Medical Psychology* 52 (1979): 63–70. The literature on penile display as threat display is of equal interest to review. See, for example, Wickler, "Socio-sexual Signals," pp. 164–67, Irenaus Eibl-Eibesfeldt, *Ethology*, 2nd ed. (New York: Holt, Rinehart and Winston, 1975), pp. 428–31, and Sheets-Johnstone, *Roots of Thinking*, pp. 189–91.

CHAPTER 4: CORPOREAL ARCHETYPES AND POSTMODERN THEORY

1. Jill Johnston, *Marmalade Me* (New York: E. P. Dutton & Co., 1971), pp. 64, 112–13, 113–14. (Jill Johnston was the dance critic's dance critic in the age of New Dance in New York City, 1960–1970.)

2. Jacques Derrida, *Of Grammatology* (Baltimore: The Johns Hopkins University Press, 1976), Translator's Preface, xxxv.

3. Ibid., xlix.

4. Ibid.

5. Ibid., liii–liv.

6. Derrida, *Of Grammatology*, p. 85.

7. André Leroi-Gourhan, *Le geste et la parole*, vol. 1 (Paris: Albin Michel, 1964), p. 272.

8. See Maxine Sheets-Johnstone, *The Roots of Thinking* (Philadelphia: Temple University Press, 1990).

9. Derrida, *Of Grammatology*, p. 84.

10. Ibid., p. 85.

11. Ibid., pp. 84, 85.

12. Ibid., p. 85.

13. Ibid.

14. One might initially wonder: why a *precarious* balance, and if precarious how tenable could it be? But wonder ceases upon recognizing the necessity of a *precarious* balance to the history of writing: only a precarious balance can lead the way to a transfiguration of human animate form.

15. When he writes that "What always threatens this balance is confused with the very thing that broaches the *linearity* of the symbol" (*Of Grammatology*,

p. 85), Derrida means that in actuality, in order to write, one cannot be upright, i.e., one must be bent over—an interpretation we will presently consider when we examine Derrida's reference to Nietzsche vis-à-vis upright posture and writing.

16. My translations from *Le geste,* pp. 182–83.

17. "Force et signification" and "La parole soufflée," in *Writing and Difference,* trans. Alan Bass (Chicago: University of Chicago Press, 1978), pp. 3–30 and 169–195 respectively.

18. For a discussion of the significance of penile display as sexual signalling behavior and as intraspecific aggressive display behavior in the evolution of primates—including human primates—see Sheets-Johnstone, *Roots of Thinking.*

19. Derrida, "Parole," p. 183.

20. Ibid., 184.

21. Ibid.

22. Ibid.

23. Ibid., 185.

24. Ibid., p. 188.

25. Derrida, "Force," p. 29. The passage comes from Nietzsche's *Twilight of the Idols.*

26. Ibid. Note the queer non-parallel usage of a verbal noun with a common noun. Why is not the choice between writing and danc*ing*? Clearly a lively body is immediately associated with the latter. *Dance,* in contrast, specifies the choreographed piece, and only secondarily the dynamic moving bodies that bring the piece to life.

27. Ibid., pp. 29–30.

28. Perhaps also the difference between a creative act—even including the creative act of *making* love—and a deconstructive one.

29. Sigmund Freud, "A Note upon the 'Mystic Writing Pad'." *Standard Edition of Complete Works* XIX, trans. James Strachey (London: The Hogarth Press, 1955), pp. 227–232. It might be worth pointing out that the use of Freud in such fundamental, uncritical ways is culturally myopic. For example, Kinji Imanishi, dean of Japanese ethology remarked, ". . . we very much doubt whether it is possible to judge the human race by the limited data obtained from persons in need of psychoanalysis." The statement is quoted in anthropologist W. Arens's *The Original Sin* (New York: Oxford University Press, 1986), p. 89. Arens goes on to remark that, "Failing to give an inch to Western intellectual fashions," Imanishi finds "that far too much credence has been given in general to Freudian ideas."

30. Sigmund Freud, "The Ego and the Id," *Standard Edition of Complete Works* XIX, trans. James Strachey (London: Hogarth Press, 1955), pp. 12–59, p. 26. The phrase is repeated on p. 27.

31. Derrida, "Freud and the Scene of Writing," in *Writing and Difference,* p. 196.

32. Ibid., p. 231.

33. No wonder either then that Derrida speaks of the coincidence of critical and clinical discourse. See "Parole," p. 189.

34. See also Sheets-Johnstone, *The Roots of Thinking.*

35. Or, to paraphrase brain neurologist Jason W. Brown, "one representation after the next." See his provocative critique of efforts to understand the mind as a purely cognitive device in *The Life of the Mind* (Hillsdale, NJ: Lawrence Erlbaum, 1988), p. 3.

36. See, for example, Donna Haraway, "Situated Knowledges: The Science Question in Feminism and the Privilege of Partial Perspective," *Feminist Studies* 14/3 (Fall, 1988): 575–599.

37. For examples and a full discussion of nonhuman primate deception, including commentary by philosophers (e.g., Daniel Dennett), see A. Whiten and R. W. Byrne, "Tactical Deception in Primates," *The Behavioral and Brain Sciences* 11 (1988): 233–273.

38. See Sheets-Johnstone, *The Roots of Thinking,* for corporeal analyses of fundamental human practices and beliefs. Were we to take Derrida's history seriously, we would have to ask how, if meaning were always deferred, a verbal language ever got started since, in the real life world of living creatures, there is no repressed underside replete with marginalia, accidents, supplements, and the like. Moreover, it would be difficult to explain how always deferrable meanings could possibly be anything other than risky if not fatal. Precise intercorporeal understandings are of critical significance in the everyday living world. One can easily see this by watching female lionesses who hunt together.

39. Maxine Sheets-Johnstone, "Taking Evolution Seriously," *American Philosophical Quarterly* 29/4 (1992): 343–52. An earlier version of this paper was given at the American Philosophical Association Pacific Division meeting in March 1991, in San Francisco.

40. Edmund Husserl, *Ideas Pertaining to a Pure Phenomenology and to a Phenomenological Philosophy,* Second Book, trans. R. Rojcewicz and A. Schuwer (Dordrecht: Kluwer Academic, 1989), p. 61.

41. An unnamed reviewer of this chapter (when it was written as a paper in submission to a philosophy journal) wrote that "By neglecting these [*Geschlecht*] essays, the author is forced to speculate on Derrida's view of the body on the basis of passages that are not immediately related to the topic." This charge is puzzling if not strongly debatable, especially when put into the context of the tenets of deconstruction, which, insofar as they legitimate multiple readings

of the works of others, also legitimate multiple readings of Derrida. In any event, this brief afterword is written to address the significance of the *Geschlecht* essays.

42. Jacques Derrida, *"Geschlecht:* Sexual Difference, Ontological Difference," *Research in Phenomenology* 13 (1983): 65–83, p. 75.

43. Ibid., p. 78.

44. Ibid., p. 76.

45. Ibid.

46. Ibid., p. 77.

47. Ibid., p. 74.

48. Ibid., p. 78.

49. Ibid., p. 75.

50. Ibid., p. 77.

51. Bettyann Kevles, *Females of the Species* (Cambridge: Harvard University Press, 1986).

52. Derrida, *"Geschlecht,"* pp. 82, 83.

53. Jacques Derrida, *"Geschlecht II,"* in *Deconstruction and Philosophy,* ed. John Sallis (Chicago: University of Chicago Press, 1987), pp. 161–196, p. 165.

54. Ibid., p. 171.

55. Ibid., p. 172.

56. Ibid., p. 173.

57. Ibid., p. 174.

58. Ibid., pp. 173, 174.

59. Ibid., p. 174.

60. Ibid.

61. Ibid.

62. For paleoanthropological references and a discussion of the origin of language, see Sheets-Johnstone, *The Roots of Thinking,* chapter 6.

63. Derrida, *"Geschlecht II,"* pp. 174, 175.

64. Jane van Lawick-Goodall, "A Preliminary Report on Expressive Movements and Communication in the Gombe Stream Chimpanzees," in *Primate Patterns,* ed. Phyllis Dolhinow (New York: Holt, Rinehart and Winston, 1972), pp. 25–84, p. 74.

65. Derrida, *"Geschlecht II,"* p. 178.

66. For a thorough description and discussion of termite-fishing, see Geza Teleki, "Chimpanzee Subsistence Technology: Materials and Skills," *Journal of Human Evolution* 3 (1974): 575–594.

67. Derrida, *"Geschlecht II,"* p. 178.

68. Ibid., p. 179.

CHAPTER 5: CORPOREAL ARCHETYPES: SEX AND AGGRESSION

1. Daniel Rancour-Laferrière, *Signs of the Flesh: An Essay on the Evolution of Hominid Sexuality* (New York: Mouton de Gruyter, 1985), p. 270.

2. Cathy Winkler, "Rape as Social Murder," *Anthropology Today* 7/3 (June 1991):12–14, p. 12.

3. Ruth Hubbard, *The Politics of Women's Biology* (New Brunswick: Rutgers University Press, 1990), p. 115.

4. Beryl Lieff Benderly, *The Myth of Two Minds* (New York: Doubleday, 1987), p. 52.

5. Hubert L. Dreyfus and Paul Rabinow, *Michel Foucault*, 2nd ed. (Chicago: University of Chicago Press, 1983), p. 233.

6. Sigmund Freud, "The Infantile Genital Organization," *Standard Edition* XIX, trans. James Strachey (London: The Hogarth Press, 1961), pp. 141–45, p. 145. See also chapter 6, this book.

7. Jean-Paul Sartre, *Being and Nothingness*, trans. Hazel E. Barnes (New York: Philosophical Library, 1956), p. 614.

8. N. Thompson-Handler, R. K. Malenky, and N. Badrian, "Sexual Behavior of *Pan paniscus* under Natural Conditions in the Lomako Forest, Equateur, Aire," in *The Pygmy Chimpanzee*, ed. R. L. Susman (New York: Plenum Press, 1984), pp. 347–368, p. 362.

9. Thelma Rowell, *The Social Behaviour of Monkeys* (Middlesex: Penguin, 1972), p. 123.

10. Donald Symons, *The Evolution of Human Sexuality* (New York: Oxford University Press, 1979), p. 106.

11. Frank A. Beach, "Cross-Species Comparisons and Human Heritage," in *Human Sexuality in Four Perspectives*, ed. Frank A. Beach (Baltimore: Johns Hopkins University Press, 1976), pp. 296–316, p. 303.

12. See Margery L. Collins and Christine Pierce's "Holes and Slime: Sexism in Sartre's Psychoanalysis," *Philosophical Forum* 5/1–2 (Fall 1973–Winter 1974): 112–127; and Christine Pierce's Review Essay, "Philosophy," *Signs* 1/2 (1975): 487–503.

13. Sartre, *Being and Nothingness*, p. 522.

14. Ibid.

15. Ibid.

16. See ibid.: "What would Descartes have been if he had known of contemporary physics? This is to suppose that Descartes possesses an *a priori* nature more or less limited and altered by the state of science in his time and that we could transport this *brute nature* to the contemporary period in which it would react to more extensive and more exact knowledge" (italics added).

17. See, for example, John Money and Patricia Tucker, *Sexual Signatures* (Boston: Little, Brown & Co., 1975), pp. 31–35.

18. Dreyfus and Rabinow, *Michel Foucault,* pp. 242–43.

19. For a discussion of the significance of this point, see Maxine Sheets-Johnstone, "Evolutionary Residues and Uniquenesses in Human Movement," *Evolutionary Theory* 6 (April 1983): 205–209.

20. Peggy Reeves Sanday, "Rape and the Silencing of the Feminine," in *Rape,* eds. Sylvana Tomaselli and Roy Porter (New York: Basil Blackwell, 1986), pp. 84–101, p. 89. For a detailed description of the young boys' initiation, see the original study of the Sambia by Gilbert Herdt: *The Sambia: Ritual and Gender in New Guinea* (New York: Holt, Rinehart and Winston, 1987).

21. Thomas Gregor, *Anxious Pleasures: The Sexual Lives of an Amazonian People* (Chicago: University of Chicago Press, 1985), p. 8.

22. "After instruction on how to magically replace [their semen] by consuming tree sap, they accept and enjoy this new form of homosexual relations." It might be of interest to note that the older boys who become men by assuming the masculinizing, penetrative role are in turn betrothed to preadolescent girls and are bisexual until the girl matures. They then become exclusively heterosexual. Ibid.

23. A recent book bears this very title. See Margaret T. Gordon and Stephanie Riger, *The Female Fear* (New York: The Free Press, 1989).

24. Dolf Zillmann, *Connections Between Sex and Aggression* (Hillsdale, NJ: Lawrence Erlbaum Associates, 1984), p. 44.

25. See Gordon and Riger, *The Female Fear.*

26. Ibid., p. 60.

27. Ibid., p. 93.

28. Ibid., p. 97.

29. On its evolutionary ties, see, for example, R. Thornhill and N. W. Thornhill, "Human Rape: An Evolutionary Analysis," *Ethology and Sociobiology* 4 (1983): 137–173; Delbert Thiessen, "Rape as a Reproductive Strategy: Our Evolutionary Legacy," American Psychological Association Fellow's Address, 1983 (see *Violence Against Women: A Critique of the Sociobiology of Rape,* eds. Suzanne R. Sunday and Ethel Tobach [New York: Gordian Press, 1985] for a full discussion of the controversy surrounding Thiessen's address); W. M. Shields and L. M. Shields, "Forcible Rape: An Evolutionary Perspective," *Ethology and Sociobiology* 4 (1983): 115–136. On its incidence in large American cities, see, for example, Menachem Amir, *Patterns in Forcible Rape* (Chicago: University of Chicago Press, 1971); *Defining Rape,* ed. Linda Brookover Bourque (Durham, NC: Duke University Press, 1989), particularly parts III and IV; *Forcible Rape: The Crime, the Victim, and the Offender,* eds. Duncan Chappell, Robley Geis, and

Gilbert Geis (New York: Columbia University Press, 1977); Lorenne M. G. Clark and Debra J. Lewis, *Rape: The Price of Coercive Sexuality* (Toronto: The Women's Press, 1977). On its central relationship to an understanding of the link between sex and aggression, see Zillmann, *Connections Between Sex and Aggression.* On its being a form of psychological perversion, see, for example, *The Rape Victim,* ed. Deanna R. Nass, (Dubuque, IA: Kendall/Hunt Publishing, 1977).

30. See, for example, Ann W. Burgess and Lynda L. Holmstrom, "Rape Trauma Syndrome," in *Forcible Rape,* eds. Chappell, Geis, and Geis, pp. 315–328, specifically pp. 325–327. See also Robert Kugelmann, "Life Under Stress: From Management to Mourning," in *Giving the Body Its Due,* ed. Maxine Sheets-Johnstone (Albany: State University of New York Press, 1992), pp. 109–131, for an interesting conceptual parallel with stress-*management* programs.

31. Gene G. Abel, Edward B. Blanchard, and Judith V. Becker, "Psychological Treatment of Rapists," in *Sexual Assault,* eds. Marcia J. Walker and Stanley L. Brodsky (Lexington, MA: Lexington Books, 1976), pp. 99–116, p. 106.

32. For a discussion of how castration was formerly a punishment for rape, see Susan Brownmiller, *Against Our Will* (New York: Simon and Schuster, 1975).

33. See Mary Koss and Mary Harvey, *The Rape Victim* (Lexington, MA: The Stephen Greene Press, 1987). For the original formulation of crisis theory, see E. Lindeman "Symptomatology and Management of Acute Grief," *American Journal of Psychiatry* 101 (1944): 141–148. The latter article was based on work with the survivors of the Coconut Grove fire in Boston in 1943.

34. Koss and Harvey, *The Rape Victim,* p. 26; Koss and Harvey are actually drawing on the writings of Morton Bard and Dawn Sangrey, *The Crime Victim's Book* (New York: Basic Books, 1979).

35. Koss and Harvey, *The Rape Victim,* p. 29.

36. Ibid., p. 33.

37. Carroll M. Brodsky, "Rape at Work," *Sexual Assault,* eds. Marcia J. Walker and Stanley L. Brodsky (Lexington, MA: Lexington Books, 1976), pp. 35–51, p. 45.

38. Koss and Harvey, *The Rape Victim,* p. 29.

39. Ibid.

40. Deena Metzger, "It Is Always the Woman Who Is Raped," in *The Rape Victim,* ed. Deanna R. Nass, pp. 9–10.

41. But see Rancour-Laferrière, *Signs of the Flesh.* Rancour-Laferrière gives a semiotic-sociobiological explanation of castration anxiety, one anchored both in the notion of males as naturally dominant over females and in the notion of

insemination as an altruistic gesture. "If [the woman] cannot accept a certain degree of power asymmetry in his favor, then he is in the position of feeling obliged to render altruism *without the normal compensatory return in the form of power and control*" (his italics), p. 326.

42. Surgical and chemical castration has been utilized as a deterrent to rapists, but *only* by their own choice. See Abel, Blanchard, and Becker, "Psychological Treatment of Rapists," p. 108.

43. What was originally a treatise on *sex* and its relation to reproduction has turned into treatises on the *sexual* as synonymous with reproduction. Darwin's original studies of selection in relation to sex focused primarily on morphological characters and mating behaviors of nonhuman males and females. This original work has been glossed by sociobiologists in terms of cost-benefit sexual, i.e., *reproductive*, strategies. Thus, in the same way that science has opened up the sexual lives of humans such that sexuality is part of our everyday discourse, it has opened up the sexual lives of nonhumans and made their sexuality topical to that of humans. The question is, is sexuality something that can be predicated of nonhuman animals? Some might want to answer, "No," and defend their answer on the grounds of clearly distinguishing sexual from reproductive behavior. Others might want to answer, "Yes—at least for some animals," and defend their answer on the grounds that sexuality individualizes. In other words, where there is a distinctive way of relating sensuously to particular other bodies and to one's own body such that one's choice of mating partner is determined by affinity and pleasure, then in some nonhuman animals at least, there is sexuality.

44. Randy Thornhill, Nancy Wilmsen Thornhill, and Gerard Dizinno, "The Biology of Rape," in *Rape,* eds. Sylvana Tomaselli and Roy Porter (New York: Basil Blackwell, 1986), pp. 102–121, p. 121.

45. Cheryl F. Harding, "Sociobiological Hypotheses About Rape: A Critical Look at the Data Behind the Hypotheses," in *Violence Against Women,* eds. Suzanne R. Sunday and Ethel Tobach (New York: Gordian Press, 1985), pp. 23–58, p. 33.

46. Martin Daly and Margo Wilson, *Sex, Evolution, and Behavior,* 2nd ed. (Belmont, CA: Wadsworth Publishing, 1983), p. 84.

47. Ethel Tobach and Suzanne R. Sunday, "Epilogue," in *Violence Against Women: A Critique of the Sociobiology of Rape,* eds. Suzanne R. Sunday and Ethel Tobach (New York: Gordian Press, 1985), pp. 129–156, p. 132.

48. Harding, "Sociobiological Hypotheses," p. 33. Harding is citing F. McKinney, S. R. Derrickson, and P. Mineau, "Forced Copulation in Waterfowl," *Behaviour* 86/3–4(1983): 250–294.

49. Daly and Wilson, *Sex, Evolution, and Behavior,* p. 117.

50. Ibid.

51. Ibid.

52. Ibid.

53. Ibid., p. 110: "[In the animal kingdom] we almost always find some more or less extreme form of the two basic sexual strategies—the nuturant female and the prodigal male."

54. Birute M. F. Galdikas, "Orangutan Adaptation at Tanjung Puting Reserve: Mating and Ecology," in *The Great Apes,* eds. David A. Hamburg and Elizabeth R. McCown (Menlo Park, CA: Benjamin/Cummings Publishing, 1979), pp. 195–233, p. 203.

55. Peter S. Rodman, "Individual Activity Patterns and the Solitary Nature of Orangutans," in *The Great Apes,* eds. David A. Hamburg and Elizabeth R. McCown (Menlo Park, CA: Benjamin/Cummings Publishing, 1979), pp. 235–255, p. 255.

56. Ibid., p. 195.

57. John MacKinnon, "Reproductive Behavior in Wild Orangutan Populations," in *The Great Apes,* eds. David A. Hamburg and Elizabeth R. McCown (Menlo Park, CA: Benjamin/Cummings Publishing, 1979), pp. 257–273. p. 258.

58. J. R. Napier and P. H. Napier, *The Natural History of the Primates* (London: British Museum [Natural History], 1985), p. 166.

59. Rodman, "Individual Activity Patterns, p. 250.

60. Ibid., p. 204.

61. J. R. Napier and P. H. Napier, *A Handbook of Living Primates* (New York: Academic Press, 1967), p. 272.

62. Ibid., p. 270.

63. Napier and Napier, *The Natural History,* p. 172.

64. Galdikas, "Orangutan Adaptation," p. 206.

65. MacKinnon, "Reproductive Behavior," p. 261.

66. Ibid., p. 262

67. Galdikas, "Orangutan Adaptation," p. 208.

68. Ibid.

69. Ibid., p. 207.

70. Ibid., p. 225. See also, MacKinnon, "Reproductive Behavior," p. 267.

71. Galdikas, "Orangutan Adaptation," p. 225.

72. MacKinnon, "Reproductive Behavior," p. 262.

73. Konrad Lorenz, *On Aggression,* trans. Marjorie Kerr Wilson (New York: Bantam Books, 1963).

74. Cathy Winkler, epigraphically quoted at the beginning of this chapter, presented a paper titled "Comparison of Legal and Physical Rape: The

PS Game" at the 93rd American Anthropological Association Meetings in San Francisco on 2 December 1992. In that paper, she analyzed "the cultural mindset and the strategies" of a District Attorney in Atlanta, Georgia, arguing that "his actions parallel those of rapists and that his motives and decisions center on *political status* issues or the PS Game." In fact, at the time of her presentation, she stated that "For six years I've been a victim of a protracted struggle to get the DA . . . to try the defendant." (Quotes from p. 1.)

75. Harding, "Sociobiological Hypotheses," see particularly p. 26.

76. Daly and Wilson, *Sex, Evolution, and Behavior,* p. 83.

77. Ibid., p. 88.

78. Ibid.

79. Ibid., p. 285.

80. Ibid., pp. 113 and 110 respectively.

81. David B. Miller, letter dated 4 August 1983, in Sunday and Tobach, *Violence Against Women,* p. 163.

82. Zillmann, *Connections,* p. 57.

83. The phrase comes from Delbert Thiessen's American Psychological Association Invited Fellows Address, "Rape as a Reproductive Strategy: Our Evolutionary Legacy," delivered in 1983. See Sunday and Tobach, *Violence Against Women* for a discussion of the controversy surrounding Thiessen's presentation.

84. Paul D. Maclean, "The Triune Brain, Emotion, and Scientific Bias," in *The Neurosciences: Second Study Program,* ed. F. O. Schmitt (New York: Rockefeller University Press, 1970), pp. 336–349.

85. Paul D. Maclean and Detlev W. Ploog, "Cerebral Representations of Penile Erection," *Journal of Neurophysiology* 25/1 (1962): 29–55, p. 40.

86. Zillmann, *Connections,* p. 64.

87. Ibid. In Maclean's original article, it is noted that the progression from chirping to cackling to piercing cries is in proportion to the lowering of electrodes into brain matter. See Maclean and Ploog, "Cerebral Representation," p. 47.

88. Zillmann, *Connections,* p. 64.

89. Ibid.

90. It is apposite to raise the question of nonhuman animal experimental studies in this context. Do they make, or have they ever made, a difference in the way humans comport themselves? Have they had any consequences whatsoever with respect to human cultural institutions and practices? If they have not, then the studies, with one crucial difference, are equivalent to studies in theoretical physics in that whatever their human interest, and however considerable that interest might be, nonhuman animal experimental studies have no real bearing on humans or human affairs. The crucial difference, of course, is that chimpan-

zees and rhesus monkeys, for example, are living creatures whereas electrons and photons are not. To experiment with living creatures and work for positive change in human affairs is a vocation that might at least be defended, whether one actually espouses and supports the experimentation or not. To experiment with living creatures for sheer intellectual sport is indefensible.

91. Zillmann, *Connections,* p. 67; italics added.

92. Ibid., 69.

93. Ibid., 70.

94. See studies cited in Zillmann, *Connections,* and in Lee Ellis, *Theories of Rape* (New York: Hemisphere Publishing, 1989).

95. See Benderly's discussion of this issue in *The Myth of Two Minds,* particularly pp. 200–1.

96. Chapter 6 of this book addresses this question.

97. Zillmann, *Connections,* p. 70.

98. Ibid.

99. Ibid.

100. Ibid.

101. Ibid., p. 150.

102. Ibid.

103. The phrase, "[sexual] access organs" would lead one to think that male and female alike have access organs. A moment's reflective thought, however, shows that this is not the case. As discussed earlier at length, females do not have a sexual *organ.* To speak of *access organs* as if they existed in both males and females is thus conceptually—and perceptually—wayward. With respect to females, access organs signify a misapprehension of animate form; they are an *animate* signification signifying nothing.

104. Zillmann, *Connections,* p. 150.

105. Maxine Sheets-Johnstone, *The Roots of Thinking* (Philadelphia: Temple University Press, 1990), p. 100. The specific references are: K. R. L. Hall, "Social Vigilance Behaviour of the Chacma Baboon *(Papio ursinus),*" *Behavior* 16 (1960): 261–94; Wolfgang Wickler, "Socio-sexual Signals and Their Intra-specific Imitation among Primates," in *Primate Ethology,* ed. Desmond Morris (Garden City, NY: Anchor, 1969), 89–189; Detlev Ploog, J. Blitz, and F. Ploog, "Studies on Social and Sexual Behavior of the Squirrel Monkey *(Saimiri sciureus),*" *Folia primatologica* 1 (1963): 29–66; Jane van Lawick-Goodall, "The Behaviour of Free-Living Chimpanzees in the Gombe Stream Reserve," in *Animal Behaviour Monographs,* vol. 1, pt. 3 (1968): 165–311; Tomas. Enomoto, "The Sexual Behavior of Japanese Monkeys," *Journal of Human Evolution* 3 (1974): 351–72; Frank E. Poirier, "Colobine Aggression: A Review," in *Primate Aggression, Territoriality, and Xenophobia,* ed. Ralph L. Holloway (New York: Academic Press, 1974), pp. 123–57; C. E. G. Tutin and

P. R. McGinnis, "Chimpanzee Reproduction in the Wild," in *Reproductive Biology of the Great Apes,* ed. C. E. Graham (New York: Academic Press, 1981), pp. 239–64; and Irenaus Eibl-Eibesfeldt, *Ethology,* 2nd ed. (New York: Holt, Rinehart, and Winston, 1975).

106. Zillmann, *Connections,* p. 181.

107. The linguistic act of course *assumes* that an accurate specification of the situation can be made. The question of accuracy of specification would figure focally in debates about whether rape is a violent criminal act or simply a sexual caper.

108. A recent article in *The Chronicle of Higher Education* aptly captures the seriousness of the question. The article is titled "A Pas de Deux of Sports and Ballet." It describes a course at Towson State in which male athletes partner ballerinas, with both prospering from the experience. In particular, male/female relationships are forged that are nonsexual, and this in spite of the intimate body contact. One of the male athletes is quoted: "One of the benefits of sports is having the companionship of teammates. . . . But I'd never been on a coed sports team, if you want to call this a coed sports team. You and your partner were teammates, but it wasn't a sex thing. It was like male-female partners—neutral almost. That was an interesting and new experience" (Vol. 40/12 (November 10, 1993): B4–5; p. B5).

109. Zillmann, *Connections,* pp. 178–79.

110. Ibid., p. 160.

111. Ibid.

112. Ibid., pp. 160–61.

113. A. Nicolas Groth (with H. Jean Birnbaum), *Men Who Rape* (New York: Plenum Press, 1979), pp. 106–7.

114. Ibid., p. 107.

115. See Linda Brookover Bourque's chapter, "Men Who Rape: Psychodynamic and Sociocultural Evidence," in Bourque, *Defining Rape,* pp. 59–76, p. 68.

116. Clark and Lewis, *Rape: The Price of Coercive Sexuality,* p. 134.

117. Sanday, "Rape and Silencing."

118. Clark and Lewis, *Rape,* p. 135.

119. *Daily Emerald,* University of Oregon, May 28, 1991, p. 8.

120. Zillmann, *Connections,* p. 191

121. Ibid., p. 192.

122. Symons, *Evolution of Human Sexuality,* p. 177.

123. Ibid.

124. Sanday, "Rape and Silencing," p. 86.

125. Ibid., p. 87.

126. Ibid., p. 88.

127. Ibid. Sanday quotes from Leslee Nadelson, "Pigs, Women, and the Men's House in Amozonia (sic): An Analysis of Six Mundurucu Myths," in *Sexual Meanings: The Cultural Construction of Gender and Sexuality,* eds. Sherry B. Ortner and Harriet Whitehead (New York: Cambridge, 1981), pp. 240–272, p. 270.

128. Ellis, *Theories,* p. 81.

129. Ibid., 84.

130. Ibid., 90.

131. Ibid., 86.

132. Ibid., 91.

133. Ibid.

134. Ibid.

135. Ibid., pp. 70–1.

136. H. H. Feder, "Specificity of Steroid Hormone Activation of Sexual Behaviour in Rodents," in *Biological Determinants of Sexual Behaviour,* ed. J. B. Hutchison (New York: John Wiley & Sons, 1978), pp. 395–424, p. 414.

137. John Bancroft, "The Relationship Between Hormones and Sexual Behaviour in Humans," in *Biological Determinants of Sexual Behaviour,* ed. J. B. Hutchison (New York: John Wiley & Sons, 1978), pp. 493–519, p. 514.

138. Linda Marie Fedigan, *Primate Paradigms* (Montreal: Eden Press, 1982), p. 282.

139. Tobach and Sunday, "Epilogue," p. 132.

140. Zillmann, *Connections,* p. 96.

141. Bancroft, "Hormones and Human Sexual Behaviour," p. 506.

142. Ellis, *Theories,* p. 71.

CHAPTER 6: CORPOREAL ARCHETYPES: PENETRATION AND BEING "IN THE FORM OF A HOLE"

1. Jean-Paul Sartre, *Being and Nothingness,* trans. Hazel E. Barnes (New York: Philosophical Library, 1956), p. 614.

2. Klaus Theweleit, *Male Fantasies,* 2 vols., trans. Stephen Conway (Minneapolis: University of Minnesota Press, 1987), vol. 1, pp. 200–201.

3. *Philosophical Forum* 5/1–2 (Fall 1973–Winter 1974): 112–127, p. 119; the original in *Being and Nothingness,* p. 613.

4. "Philosophy," *Signs* 1/2 (1975): 487–503, quote, p. 496.

5. Sartre, *Being and Nothingness,* p. 614.

6. Ibid., p. 600.

7. Ibid., p. 608.

8. Ibid., p. 609.

9. Ibid., p. 575.

10. Ibid., pp. 578 and 579–80 respectively.

11. Ibid., p. 580.

12. Reference is, of course, to Simone de Beauvoir's well-known *The Second Sex.*

13. See especially the writings of Evelyn Fox Keller for both a critical analysis of standard androcentric scientific practice and a liberating reconceptualization thereof: e.g., "Making Gender Visible in the Pursuit of Nature's Secrets," *Feminist Studies/Critical Studies,* ed. Teresa de Lauretis (Bloomington: Indiana University Press, 1986), pp. 67–77; "Feminism as an Analytical Tool for the Study of Science," *Academe* 69/5 (Sept.–Oct. 1983): 15–21; "Gender and Science," in *Discovering Reality,* eds. Sandra Harding and Merrill B. Hintikka (Boston: D. Reidel, 1983), pp. 187–205; "Feminism and Science," in *Women, Knowledge, and Reality,* eds. Ann Garry and Marilyn Pearsall (Boston: Unwin Hyman, 1989), pp. 175–188.

14. Sartre, *Being and Nothingness,* p. 578.

15. Ibid., p. 579.

16. Ibid.

17. Ibid.

18. Ibid., p. 614.

19. Ibid., pp. 614, 615 respectively.

20. Ibid., p. 615.

21. Ibid.

22. Leonardo da Vinci, *Philosophical Diary,* trans. Wade Baskin (New York: Wisdom Library, 1959), p. 19.

23. Sartre, *Being and Nothingness,* p. 613.

24. See, for example, T. G. R. Bower, *Development in Infancy* (San Francisco: W. H. Freeman, 1974), particularly p. 238.

25. Sartre, *Being and Nothingness,* p. 605.

26. Ibid., p. 606.

27. Ibid., p. 612.

28. Ibid., p. 613.

29. Maxine Sheets-Johnstone, *The Roots of Thinking* (Philadelphia, Temple University Press, 1990); see particularly chapter 9, "On the Origin and Significance of Paleolithic Cave Art."

30. David McKnight, "Men, Women, and Other Animals: Taboo and Purification among the Wik-mungkan," in *The Interpretation of Symbolism,* ed. Roy Willis (London: Malaby Press, 1975), pp. 77–97, p. 88.

31. Sartre, *Being and Nothingness,* p. 614.

32. See Jean-Paul Sartre, *La Nausée* (Paris: Gallimard, 1938).

33. Sartre, *Being and Nothingness*, p. 614.

34. The eye too is just such an archetypal form. See Maxine Sheets-Johnstone, "The Body as Cultural Object/The Body as Pan-Cultural Universal," in *Phenomenology of the Cultural Disciplines*, eds. Mano Daniel and Lester Embree (Boston: Kluwer Academic, 1994), pp. 85–118.

35. Sartre, *Being and Nothingness*, p. 615.

36. Ibid., p. 613.

37. Daniel Rancour-Laferrière, *Signs of the Flesh: An Essay on the Evolution of Hominid Sexuality* (New York: Mouton de Gruyter, 1985), p. 305.

38. Ibid., 317–330.

39. Sigmund Freud, "The Taboo of Virginity," *Standard Edition of Complete Works* XI, trans. James Strachey (London: The Hogarth Press, 1957), pp. 193–208, p. 199.

40. Ibid., pp. 198–99.

41. Carleton Gajdusek, "Physiological and Psychological Characteristics of Stone Age Man," in *Engineering and Science:* Symposium on Biological Bases of Human Behavior (Pasadena: California Institute of Technology, Alumni Association, March 1970), pp. 26–33, 56–62; p. 58.

42. See Sheets-Johnstone, *The Roots of Thinking*, chapters 4 and 7.

43. Ibid., pp. 190–191. See also examples given by Rancour-Laferrière in *Signs of the Flesh*, pp. 298–300. For a graphic illustration of Priapus, see Eugene Monick, *Phallos: Sacred Image of the Masculine* (Toronto: Inner City Books, 1987), p. 105.

44. For a recent report on priapism, see Rick Weiss, "Historic Priapism Pegged to Frog Legs," *Science News* 139/1 (January 5, 1991), p. 6.

45. Quoted in Rancour-Laferrière, *Signs of the Flesh*, p. 299. For original, see Robert J. Stoller, *Sexual Excitement: Dynamics of Erotic Life* (New York: Pantheon Books, 1979), p. xiii.

46. Robert J. Stoller, "Sexual Deviations," in *Human Sexuality in Four Perspectives*, ed. Frank A. Beach (Baltimore: Johns Hopkins University Press, 1976), pp. 202–3.

47. Ivor H. Jones and Dorothy Frei, "Exhibitionism—A Biological Hypothesis," *British Journal of Medical Psychology* 52 (1979): 63–70.

48. Ibid., p. 67.

49. Quoted by Roy Schafer, "Problems in Freud's Psychology of Women," in *Female Psychology: Contemporary Psychoanalytic Views*, ed. Harold P. Blum (New York: International Universities Press, 1977), pp. 331–360, p. 354. For original, see Sigmund Freud, "Infantile Genital Organization," *Standard Edition of Complete Works* XIX, trans. James Strachey (London: The Hogarth Press, 1961) pp.141–45, p. 145.

50. Freud goes on to remark following the above quoted statement that

"The vagina is . . . valued as the place of shelter for the penis; it enters into the heritage of the womb." He speaks of it as "the sanctuary." Freud, "Infantile Genital Organization," p.145. See also Schafer, "Problems in Freud's Psychology," p. 357.

51. Freud, "Infantile Genital Organization," p. 145.

52. Ibid.

53. Galen, *On the Usefulness of the Parts of the Body,* 2 vols., ed. and trans. Margaret T. May (Ithaca: Cornell University Press, 1968), vol. 2, p. 628.

54. Jacques Lacan, "La direction de la cure et les principes de son pouvoir," *Écrits* (Paris: Éditions du Seuil, 1966), p. 642.

55. David Barash, *Sociobiology and Behavior* (New York: Elsevier, 1982), p. 147.

56. William G. Eberhard, *Sexual Selection and Animal Genitalia* (Cambridge: Harvard University Press, 1985). In my own analysis and discussion of pleasure in the joining of flesh on flesh, I have cited precisely such biological studies. See *The Roots of Thinking,* chapter 7.

57. Sartre, *Being and Nothingness,* p. 392.

58. Ibid.

59. Ibid.

60. Ibid., p. 385.

61. Ibid.

62. Ibid., p. 388.

63. Ibid., p. 389.

64. See chapter 10 for an analysis and discussion of male autonomy.

65. On the body as a semantic template, see Sheets-Johnstone, *The Roots of Thinking.*

66. Karl G. Heider, *Grand Valley Dani* (New York: Holt, Rinehart and Winston, 1979), p. 56.

67. Ibid.

68. Ibid.

69. Ibid., p. 57.

70. Ibid., p. 56.

71. Ibid., p. 81.

72. Ibid., p. 79.

73. Gajdusek, "Physiological and Psychological Characteristics of Stone Age Man," pp. 58–59.

74. Ibid., p. 59.

75. Ibid., p. 61.

76. Ibid., p. 60.

77. See in particular *This Sex Which Is Not One,* trans. Catherine Porter (Ithaca: Cornell University Press, 1985).

78. Harriet E. Lerner, "Parental Mislabeling of Female Genitals as a Determinant of Penis Envy and Learning Inhibitions in Women," *Female Psychology: Contemporary Psychoanalytic Views,* ed. Harold P. Blum (New York: International Universities Press, 1977), pp. 269–283, p. 269.

79. Ibid., p. 70.

80. Ibid.

81. For interesting analyses and discussions of binary opposition, see *The Attraction of Opposites: Thought and Society in the Dualistic Mode,* eds. David Maybury-Lewis and Uri Almagor (Ann Arbor: University of Michigan Press, 1989); Rodney Needham, *Counterpoints* (Berkeley: University of California Press, 1987).

CHAPTER 7: CORPOREAL ARCHETYPES AND THE PSYCHOANALYTIC VIEW: BEGINNING PERSPECTIVES ON THE BODY IN LACAN'S PSYCHOANALYTIC

1. Mary Douglas, *Natural Symbols,* 2nd ed. (London: Barrie & Jenkins, 1973), p. 185.

2. Seyla Benhabib, "The Generalized and the Concrete Other," in *Feminism as Critique: On the Politics of Gender,* eds. Seyla Benhabib and Drucilla Cornell (Minneapolis: University of Minnesota Press, 1987), pp. 77–95, p. 95.

3. As Bice Benvenuto and Roger Kennedy (among other commentators) note, Lacan takes up Freud's earlier work and theory in which Freud distinguishes the unconscious from consciousness, and in which he founds psychoanalysis as a science of the unconscious—and of sexuality, i.e., unconscious and sexuality are the "basic terms of psychoanalytic experience." *The Works of Jacques Lacan* (London: Free Association Books, 1986), p. 11.

4. The uncanny confluence of ontologies—and the uncommon parallel with respect to the claim that all desire is basically sexual—is likely the result of the lectures of Alexandre Kojève at the École des Hautes Études in the 1930s. These lectures, which focused on Hegel's *Phenomenology of Mind* and which drew attention to Hegel's central thesis regarding desire and the subject of desire, were attended by Parisian intellectuals of the day. See John P. Muller and William J. Richardson's reader's guide to *Écrits* for a brief indication of the material focused upon in the lectures: *Lacan and Language* (New York: International Universities Press, 1985).

5. Ibid.

6. See, for example, Francois Roustang's fine analysis of the influence of Lévi-Strauss's work on Lacan in *The Lacanian Delusion,* trans. Greg Sims (New York: Oxford University Press, 1990).

7. Jacques Lacan, "On a Question Preliminary to any Possible Treatment of Psychosis," *Écrits: A Selection,* trans. Alan Sheridan (New York: W. W. Norton, 1977), p. 192; "D'une question préliminaire à tout traitement possible de la psychose," *Écrits* (Paris: Éditions du Seuil, 1966), p. 547.

8. One might say that in a similar way the body is precipitated out of an ontology in which desire is conceived as a project of the for-itself (a Sartrean ontology), and out of an anthropology in which cultural practices are conceived as sets of linguistic-like rules (a Lévi-Straussian anthropology). In contrast, however much Jung's analytic psychology stresses *psychic* archetypes, his psychology nonetheless acknowledges the body in fundamental if not fully clarified or elaborated ways.

9. Luce Irigaray, *This Sex Which Is Not One,* trans. Catherine Porter (Ithaca: Cornell University Press, 1985). For critical commentaries—both positive and negative—see, for example, Diana J. Fuss, " 'Essentially Speaking': Luce Irigaray's Language of Essence," *Hypatia* 3/3 (Winter 1989): 62–80; Arleen Dallery, "The Politics of Writing (the) Body: *Écriture feminine,*" in *Gender/Body/Knowledge,* eds. Alison M. Jaggar and Susan R. Bordo (New Brunswick: Rutgers University Press, 1989), pp. 52–67; Jane Gallop, *The Daughter's Seduction: Feminism and Psychoanalysis* (Ithaca: Cornell University Press, 1982).

10. As suggested in chapter 3, a mania for theory can obfuscate and paralyze rather than clarify and energize. As suggested in chapter 3, so also can an over-zealous preference for a particular theoretical stand or for a certain way of doing academic business. It is relevant to note in this context that the title of an early book of the women's movement in the United States, *Our Bodies Ourselves,* was changed to reflect more accurately the fusion of "facts and feelings." The original title was *Women and Their Bodies.* The new title staunchly if implicitly affirms the communally recognized truth of the watchwords of the feminist movement. (Boston Women's Health Book Collective, 2nd ed. [New York: Simon and Schuster, 1976], p. 11).

11. Robert S. Steele, *Freud and Jung: Conflicts of Interpretation* (Boston: Routledge & Kegan Paul, 1982), p. 71. In his thorough hermeneutical evaluation of Freud and Jung, Steele refers many times to Freud's and Jung's introspective techniques and the importance of their self-analyses, saying for example, that "[P]arthenogenetically, a great mind, through introspection, fertilized itself. In finding himself, Freud founded psychoanalysis" (p. 69); and that "What emerged from this experiment [Jung's experiment in active fantasizing] was a science which, like Freud's, was based upon introspection and self-analysis, a science in which a modicum of objectivity is gained only through an exhaustive exploration of one's own subjectivity" (p. 281).

12. Carl G. Jung, *Memories, Dreams, Reflections,* rev. ed., ed. A. Jaffe,

trans. R. and C. Winston (New York: Pantheon Books, 1973), p. 178; quoted in Steele, *Freud and Jung*, p. 281. Steele writes that "[Jung's] break with Freud was one of the incidents which precipitated Jung's intensive self-analysis. Jung's five-year period of introspection formed the experiential basis for his later psychological work" (*Freud and Jung*, p. 257).

13. I am well aware of revisionist formulations of Freud's psychoanalysis and of the charges against him. The most recent review of these charges appeared in the 18 November 1993 issue of *The New York Review of Books* (vol. XL, No. 19). See Frederick Crews's "The Unknown Freud," pp. 55–66.

14. William James, *Principles of Psychology*, 2 vols. (New York: Dover Publications, 1950), vol. 1, p. 239.

15. Jacques Lacan, "Variantes de la cure-type," *Écrits*, p. 352. (This paper has not been translated into English.)

16. Jacques Lacan, "Aggressivity in Psychoanalysis," *Écrits: A Selection*, p. 15; *Écrits*, p. 109.

17. "The Function and Field of Speech and Language in Psychoanalysis," in *Écrits: A Selection*, pp. 68–69; *Écrits*, p. 279–80. See Muller and Richardson, *Lacan and Language*, p. 79 for a brief commentary.

18. Commentators Muller and Richardson note at the conclusion of their book the progressive change in Lacan's psychoanalytic, from his speaking of the unconscious of the subject to his speaking of the subject of the unconscious. In terms of his writings, the changeover in theoretical formulation occurs between 1953 ("The Function and Field of Speech and Language in Psychoanalysis") and 1960 ("The Subversion of the Subject and the Dialectic of Desire in the Freudian Unconscious"). See *Lacan and Language*, pp. 416–17.

19. Transference in Lacanian psychoanalytic terms means a transference onto the analyst of all one's needs, demands, and desires, *all* in a psychohistorical sense up to the very present.

20. Jacques Lacan, "The Mirror Stage as Formative of the Function of the I as Revealed in Psychoanalytic Experience," *Écrits: A Selection*, p. 5; *Écrits*, p. 98.

21. Jane Gallop, *Reading Lacan* (Ithaca: Cornell University Press, 1985). See in particular chapter 3, "Where to Begin?"

22. For example, in "The Mirror Stage," he writes that the meaning of the infant leaning forward toward its reflection in the mirror "discloses a libidinal dynamism which has hitherto remained problematic, as well as an ontological structure of the human world that accords with my reflections on paranoiac knowledge." *Écrits: A Selection*, p. 2; *Écrits*, p. 94.

23. Ibid.

24. Lacan, "Function and Field," *Écrits: A Selection*, p. 86; *Écrits*, p. 300.

25. Gallop, *Reading Lacan;* see particularly, pp. 81–82.

26. Jacques Lacan, "The Agency of the Letter in the Unconscious or Reason Since Freud," *Écrits: A Selection*, p. 154; *Écrits*, p. 502.

27. Lacan, "The Mirror Stage," *Écrits: A Selection*, p. 2; *Écrits*, p. 94. Jane Gallop translates *souche* as "rootstock," which seems a more apt equivalent of the French in this instance (*Reading Lacan*, p. 81).

28. Jacques Lacan, "The Direction of the Treatment and the Principles of Its Power," *Écrits: A Selection*, p. 231; *Écrits*, p. 590; "Function and Field," *Écrits: A Selection*, p. 41; *Écrits*, p. 248.

29. Roustang, *The Lacanian Delusion*, p. 19.

30. Ibid. pp. 19–20.

31. Lacan, "Function and Field," *Écrits: A Selection*, p. 73; *Écrits*, pp. 284–85.

32. Roustang, *The Lacanian Delusion*, p. 20.

33. Jacques Lacan, "Au-delà du 'Principe de réalité'," *Écrits* (Paris: Éditions du Seuil, 1966), pp. 73–92. The paper has not been translated into English.

34. Roustang, *The Lacanian Delusion*, pp. 22, 23.

35. Ibid., p. 23.

36. Ibid., p. 31.

37. Claude Lévi-Strauss, *Introduction à l'oeuvre de Marcel Mauss*, p. xix, quoted in Roustang, *The Lacanian Delusion*, p. 30.

38. Roustang, *The Lacanian Delusion*, p. 30.

39. Ibid., pp. 33–34. Lacan, "Function and Field," *Écrits: A Selection*, p. 68; *Écrits*, p. 279.

40. Roustang, *The Lacanian Delusion*, p. 38.

41. Ibid.

42. David Macey, *Lacan in Contexts* (New York: Verso, 1988), p. 121.

43. Roustang, *The Lacanian Delusion*, p. 34.

44. Ibid., p. 30. In discussing Lévi-Strauss's privileging of the signifier, Roustang says that "He thereby posits the autonomy of what Lacan will single out and substantify as the Symbolic."

45. "The Science of the Real" is in fact the name of the lengthy chapter in which Roustang discusses and analyzes Lacan's project of turning psychoanalysis into a science. Ibid. pp. 18–106.

46. Lacan, "Du sujet enfin en question," *Écrits*, p. 233. The paper has not been translated into English. I have used Roustang's translation (*The Lacanian Delusion*, p. 36).

47. Quoted in Roustang, *The Lacanian Delusion,* p. 38. From *Bulletin de l'association freudienne,* no. 1, pp. 4–5.

48. For a discussion of Lacan's views regarding session length, see "Function and Field," *Écrits: A Selection,* pp. 97–99; *Écrits,* pp. 312–314.

49. Roustang, *The Lacanian Delusion,* pp. 38–39. From *Bulletin de l'association freudienne.*

50. Ibid., p. 39.

51. Ibid., p. 104.

52. Jacques Lacan, *Le Séminaire de Jacques Lacan, Livre Trois, Les Psycoses* (Paris: Éditions du Seuil, 1981), p. 187. I have used Roustang's translation *(The Lacanian Delusion,* p. 41).

53. Jacques Lacan, *The Four Fundamental Concepts of Psycho-Analysis,* ed. Jacques-Alain Miller, trans. Alan Sheridan (New York: W. W. Norton, 1978), p. 157.

54. One wonders whether Lacan is confusing *symptom* with *symbol.* It is common to define a symbol as a kind of sign, in particular, a sign "intended for someone." The term "sign," in this instance, is the generic term for symbol and signal (symptom). See Susanne Langer, *Feeling and Form* (New York: Charles Scribner's Sons, 1953), p. 26.

55. Lacan, *Four Fundamental Concepts,* p. 158.

56. Lacan, "Function and Field," *Écrits, A Selection,* p. 59; *Écrits,* p. 269.

57. Lacan, *Four Fundamental Concepts,* p. 158.

58. Ibid., p. 184. See also Roustang, *The Lacanian Delusion,* p. 83.

59. Jacques Lacan, "The Subversion of the Subject and the Dialectic of Desire in the Freudian Unconscious," *Écrits: A Selection,* pp. 314–15; *Écrits,* p. 817.

60. Ibid., p. 315; p. 817.

61. Ibid.

62. Ibid., p. 315; p. 818.

63. Ibid.

64. Ibid., p. 316; p. 818.

65. Ibid. p. 314; p. 816.

66. Ibid. pp. 302–16; pp. 805–18.

67. Edmund Husserl, "The Origin of Geometry," trans. David Carr, in *Husserl: Shorter Works,* eds. Peter McCormick and Frederick A. Elliston (Notre Dame, IN: Notre Dame Press, 1981), pp. 255–270, p. 261.

68. Lacan, "Function and Field," *Écrits: A Selection,* p. 57; *Écrits.* p. 267. (See also Roustang, *The Lacanian Delusion,* p. 38.)

69. Ibid., p. 87; p. 301.

70. Ibid., pp. 90–91; pp. 304–5.

71. Ibid., p. 92; p. 306. Lacan immediately adds, "Before long we'll be preaching the Gospel according to Lévy-Bruhl to him," making allusion, of course, to Lévy-Bruhl's thoroughly scorified (by Lévi-Strauss) philosophical "anthropology."

72. Lacan, *Les Psychoses*, p. 202. I have used Roustang's translation *(The Lacanian Delusion*, p. 42).

73. Roustang, *The Lacanian Delusion*, p. 59. Macey, *Lacan in Contexts*, p. 132.

74. *Les Psychoses*, p. 202. I have used Roustang's translation *(The Lacanian Delusion*, p. 42).

75. Roustang, *The Lacanian Delusion*, p. 27.

76. Lacan "Aggressivity," *Écrits: A Selection*, p. 10; *Écrits*, p. 103.

77. We will consider Lacan's flimsy answer to the question in chapter 10.

78. Jacques Lacan, "The Signification of the Phallus," *Écrits: A Selection*, p. 287; *Écrits*, p. 692.

79. See William Lyons, *The Disappearance of Introspection* (Cambridge: MIT Press, 1986).

80. Benvenuto and Kennedy, *The Works of Jacques Lacan*, p. 19.

81. Lacan, "The Mirror Stage," *Écrits: A Selection*, p. 5; *Écrits*, p. 98.

82. Lacan, "Aggressivity," *Écrits: A Selection*, p. 17; *Écrits*, p. 111.

83. Ibid., p. 4; p. 97.

84. Benvenuto and Kennedy, *The Works of Jacques Lacan*, p. 12.

85. Carl G. Jung, "The Transcendent Function," in *The Portable Jung*, ed. Joseph Campbell, trans. R. F. C. Hull (New York: Viking Press, 1971), pp. 273–300, p. 274.

86. Roger Kennedy in Benvenuto and Kennedy, *The Works of Jacques Lacan*, p. 81.

87. Juliet Mitchell, "Introduction—I," in *Feminine Sexuality*, eds. Juliet Mitchell and Jacqueline Rose (New York: Pantheon Books, 1982), p. 4.

88. Catherine Clément, *The Lives and Legends of Jacques Lacan*, trans. Arthur Goldhammer (New York: Columbia University Press, 1983), p. 123.

89. Ibid.

90. Other commentators view the return as highly idiosyncratic. Benvenuto and Kennedy, for example, point out that unlike other psychoanalytic schools of thought, Lacanian psychoanalysis concentrates on Freud's earliest formulations of psychoanalysis in which "[the] basic terms of psychoanalytic experience—the unconscious and sexuality—were first evolved and elaborated." Freud's later theoretical revisions of the unconscious and sexuality and his later formulations of psychoanalysis in terms of ego, id, and super-ego are not the generative source of Lacan's central aim.

91. Ibid., p. 12.

92. Lacan, *Four Fundamental Concepts*, p. 149.

93. Immanuel Kant, *Metaphysical Foundations of Natural Science* [1786], trans. J. Ellington (Indianapolis: Bobbs-Merrill, 1970), Preface.

94. In preface to that examination, attention should be parenthetically called to Lacan's own positive estimation of Kant's contribution to our understanding of the imaginary. In one of his discussions, Lacan describes the imaginary as the "source and storehouse of the preconscious," and states "[that] it has already fortunately been approached in the philosophical tradition, and . . . that the idea-schema of Kant situate themselves at the border of this field—at least this is where the idea-schema find their most brilliant formulations" (*leurs plus brillantes lettres de créance*). Prefatory attention should furthermore be called to the fact that even with these praising words, Lacan actually connects the imaginary with the *pre*verbal, hence with the *pre*conscious. What is imaginary is thus totally outside the Symbolic. It is defined in terms of the inner world of the subject, in particular, the inner world where "the subject is the infantile doll that he has been, he is the excremental object, he is a sewer, he is a windpipe." The definition, of course, is far afield of Kant's description of the imagination, as we shall see. Lacan's estimation of the imaginary is in fact coincident with that of Lévi-Strauss. Though more graphically denigrating than that of Lévi-Strauss, Lacan's conception of the imaginary basically follows Lévi-Strauss, who emphatically distinguishes all that is symbolic from all that is not—including anything on the order of personal images. It is of interest to point out that in making the categorical distinction, Lévi-Strauss, and Lacan following him, carve a path which theoretically links them to a Jungian perspective in that, once *laws* of the unconscious are claimed—once "the unconscious . . . is reducible to a function—the symbolic function . . . [that ordains] the same laws among all men, and actually corresponds to the aggregate of these laws," then a collective unconscious is catapulted into being. For Lacan, it should be added, this symbol-producing "collective unconscious" has absolutely nothing to do with "affective signification" (Lacan, *Psychoses*, p. 185).

95. Lacan, "Aggressivity" and "The Mirror Stage," *Écrits: A Selection*, p. 19, p. 2; *Écrits*, p. 113, p. 95.

96. "The Mirror Stage," *Écrits: A Selection*, p. 4; *Écrits*, p. 96.

97. Ibid., p. 2; p. 94.

98. Ibid.

99. Ibid.

100. Ibid., p. 4; p. 97.

101. See Muller and Richardson, *Lacan and Language*, p. 39 for other possible meanings of "quadrature." Quadrature might also have to do with the

paranoiac alienations which are double with respect to the subject and double with respect to the external world: a misidentification of the subject with his own reflection and the misidentification of this reflected image with the image of the other in the process of transitivism, which leads to interpersonal confusion. There is similarly a fundamental paranoiac miscognition of external things.

102. For a thorough treatment of this topic, see Albert A. Johnstone, *Rationalized Epistemology: Taking Solipsism Seriously* (Albany: State University of New York Press, 1991).

103. For a provocative account of dreams, their imagic nature and their evolutionarily rooted meaning, see Jonathan Winson, "The Meaning of Dreams," *Scientific American* 263/5 (November 1990): 86–96.

104. *Les Psychoses*, p. 202. I have used Roustang's translation (*The Lacanian Delusion*, p. 42).

105. For a full analysis and discussion of the origin of language, see Maxine Sheets-Johnstone, "On the Origin of Language," chapter 6, *The Roots of Thinking* (Philadelphia: Temple University Press, 1990).

106. Paul Schilder, *The Image and Appearance of the Human Body* (New York: International Universities Press, 1950).

107. I owe this term to Walter J. Ong. See his essay, " 'I See What You Say': Sense Analogues for Intellect," in his book *Interfaces of the Word* (Ithaca: Cornell University Press, 1977), pp. 121–144.

108. "The Mirror Stage," *Écrits: A Selection*, pp. 2–3; *Écrits*, p. 95.

109. Immanuel Kant, *Critique of Pure Reason*, trans. Norman K. Smith (London: Macmillan, 1953), p. 183.

110. Cited in Roustang, *The Lacanian Delusion*, p. 38; from *Bulletin de l'association freudienne*.

111. "Aggressivity," *Écrits: A Selection*, p. 9; *Écrits*, p. 102. By closing off the body, Lacan effaces any grounds on which to secure the *real* (or *Real*) a place in the analytic. That introspection is said to provide "abstractly isolated moments" within the analytic situation is akin to affirming an original *corps morcelé*. "Abstractly isolated moments," like a body in bits and pieces, are in absolute antithesis to a living body. An inept specimen which needs a mirror to jar it into wholeness (however illusory), and which until such jarring remains statically inept over a period of months, could hardly be looked to to provide anchorage for a developing psychoanalytic, whether practical or theoretical. Language, in contrast, provides ready anchorage.

112. "The Mirror Stage," *Écrits: A Selection*, pp. 3, 5; *Écrits*, pp. 95–96, 98; "Aggressivity," *Écrits: A Selection*, pp. 17–18; *Écrits*, pp. 111–12.

113. One could characterize this continuum at its poles as being Jungian at

the creative (e.g., artistic) end and Lacanian at the delusional (i.e., psychotic) end.

114. Sartre's *Being and Nothingness* was published in France in 1943. Lacan's activities during the wartime years are not wholly filled in, but certainly they were not productive writing and speaking years. It would be difficult to doubt, however, that they were reading years, specifically years during which Lacan at some point read Sartre's *magnum opus*. The "transparent allusions" to *Being and Nothingness* in Lacan's essay, "The Mirror Stage," delivered at the International Psychoanalytic Congress in Zurich in 1949, are undeniable.

115. Jean-Paul Sartre, *Being and Nothingness*, trans. Hazel E. Barnes (New York: Philosophical Library, 1956), p. 565.

116. Ibid., p. 21.

117. Lacan, "Function and Field," *Écrits: A Selection*, p. 103; *Écrits*, p. 319.

118. Lacan, *Four Fundamental Concepts*, p. 154.

119. Lacan, "Function and Field," *Écrits: A Selection*, p. 104; *Écrits*, p. 319.

120. Lacan, *Four Fundamental Concepts*, p. 131.

121. Ibid., p. 130.

122. Muller and Richardson, *Lacan and Language*, p. 21–22.

123. Ibid., p. 22.

124. Benvenuto and Kennedy, *The Works of Jacques Lacan*, p. 170.

125. See, for example, Macey, *Lacan in Contexts*.

126. Lacan, "Direction of the Treatment," *Écrits: A Selection*, p. 264; *Écrits*, p. 628.

127. Lacan, "Agency of the Letter," *Écrits: A Selection*, p. 151; *Écrits*, p. 500.

128. Sartre, *Being and Nothingness*, p. 598.

129. Ibid., pp. 590–91.

130. Ibid., p. 591.

131. Ibid., p. 592.

132. Ibid.

133. Ibid.

134. Ibid.

135. Ibid.

136. Ibid., p. 593.

137. Lacan, "Aggressivity," *Écrits: A Selection*, p. 19; *Écrits*, pp. 113–14.

138. Muller and Richardson, *Lacan and Language*, p. 338. Lacan original in "Signification of the Phallus," *Écrits: A Selection*, pp. 289–90; *Écrits*, p. 694.

139. Jacques Lacan, *Écrits* (Paris: Éditions du Seuil, 1966), pp. 889–892; the two paragraphs are on p. 892. There is no English translation of this essay.

140. Ibid.

141. Lacan, "Agency of the Letter," *Écrits: A Selection*, p. 156; *Écrits*, p. 506. We should note Lacan's informative and at the same time guileful preamble to his quotation and discussion of the "sheaf" metaphor. He explains his choice of the metaphor as follows: "Let us immediately find an illustration; Quillet's dictionary seemed an appropriate place to find a sample that would not seem to be chosen for my own purposes, and I didn't have to go any further than the well-known line of Victor Hugo . . ." (Ibid.).

142. Lacan, "La Métaphore du Sujet," p. 892.

143. In another sense, of course—an experiential sense—it is the only object that can be the source of its own nothingness.

144. Lacan, "Agency of the Letter," *Écrits: A Selection*, p. 157; *Écrits*, p. 507.

145. Ibid.

146. Ibid., p. 157; pp. 507–8.

147. Irigaray, *This Sex Which Is Not One*.

148. A *point de caption* is an upholsterer's pin that effectively ties all together. Lacan uses the phrase in "Agency of the Letter," *Écrits: A Selection*, p. 154; *Écrits*, p. 503. Lacan also uses the phrase *point de caption* to describe "les noeuds"—the anchor points in any network of associations. He refers to "les noeuds" in "Function and Field," *Écrits: A Selection*, p. 59; *Écrits*, p. 269. Jane Gallop reminds us at the end of her chapter "Reading the Phallus" that "'noeud', the French word for knot, is a well-known crude term for 'penis'" (*Reading Lacan*, p. 156).

149. See Steele, *Freud and Jung*, especially pages 132–152, 259–269.

150. Sartre, *Being and Nothingness*, p. 574.

151. Ibid.

152. Ibid.

153. Ibid.

154. Ibid.

155. Ibid., p. 575.

156. Benvenuto and Kennedy, *The Works of Jacques Lacan*, pp. 160–161.

CHAPTER 8: CORPOREAL ARCHETYPES AND POWER: A CRITICAL EXAMINATION OF LACAN'S PSYCHOANALYTIC OF THE INFANT-CHILD

1. George Butterworth, "Structure of the Mind in Human Infancy," *Advances in Infancy Research,* vol. 2, ed. Lewis P. Lipsitt and Carolyn K. Rovee-Collier (Norwood, NJ: Ablex Publishing Corporation, 1983), pp. 1–29, p. 25.

2. James A. Kleeman, "A Boy Discovers His Penis," *The Psychoanalytic Study of the Child* 20 (1965): 239–266, p. 241.

3. See David Macey, *Lacan in Contexts* (New York: Verso, 1988), chapter 5 titled "Linguistics or 'Linguisterie'." In brief, Lacan himself admits (in Seminar XX, *Encore*) that his work, encapsulated by the dictum, "the unconscious is structured like a language," does not belong within the field of linguistics. Rather, it belongs in the field of "linguisterie," a term he himself coins.

4. Jacques Lacan, "The Mirror Stage as Formative of the Function of the I as Revealed in Psychoanalytic Experience," *Écrits: A Selection,* trans. Alan Sheridan (New York: W. W. Norton, 1977), p. 4; *Écrits* (Paris: Éditions du Seuil, 1966), p. 96.

5. Ibid.

6. Ibid., p. 4; p. 97.

7. Ibid., p. 4; p. 96.

8. Maurice Merleau-Ponty, "The Child's Relations with Others," trans. William Cobb, in *The Primacy of Perception,* ed. James M. Edie (Evanston: Northwestern University Press, 1964), pp. 96–155, p. 136.

9. Lacan, "The Mirror Stage."

10. Merleau-Ponty, "The Child's Relations With Others, p. 136.

11. Jacques Lacan, "Aggressivity in Psychoanalysis," *Écrits: A Selection,* p. 19; *Écrits,* p. 113.

12. Merleau-Ponty, "The Child's Relations with Others," p. 133.

13. Wallon's text was published in 1949.

14. Daniel N. Stern, *The Interpersonal World of the Infant* (New York: Basic Books, 1985), pp. 78–79.

15. Lacan, "The Mirror Stage," *Écrits: A Selection,* p. 2; *Écrits,* p. 94.

16. Ibid.

17. Insofar as the visual comes to dominate, it is indeed because we come to cue our movement by what is visually about us. Our visual environs cue the direction of our movement, the site of placement of our feet, the degree of extension of our arm, the range of our trunk movement, and so on. We move in

accord with what is visually about us, yielding to the visual over the tactile-kinesthetic to avoid hurting ourselves not only in a culturally cluttered world, but in such activities as hiking and climbing. What we forget, however, is that what we know visually in all such situations is what we originally learned from touch and movement, from our tactile-kinesthetic body.

18. Stern, *The Interpersonal World of the Infant*, p. 14.

19. Ibid., pp. 39–40.

20. Ibid., pp. 41–42.

21. Ibid., pp. 50–51.

22. Andrew N. Meltzoff, "Imitation, Intermodal Co-ordination and Representation in Early Infancy," in *Infancy and Epistemology: An Evaluation of Piaget's Theory*, ed. George Butterworth (Brighton, [England]: Harvester Press Ltd., 1981), pp. 85–114, pp. 96–103.

23. Ibid., p. 103.

24. Stern, *The Interpersonal World of the Infant*, p. 51.

25. Meltzoff, "Imitation, Intermodal Co-ordination and Representation in Early Infancy," p. 109.

26. Maxine Sheets-Johnstone, "The Materialization of the Body: A History of Western Medicine, A History in Process," in *Giving the Body Its Due*, ed. Maxine Sheets-Johnstone (Albany: State University of New York Press, 1992), pp. 132–158.

27. Essays such as those in *Advances in Infancy Research* that focus on intermodal perception validate what some might otherwise castigate as Stern's "poetic" language. In fact, lest Stern's descriptions be disregarded as so much linguistic flowering over infant capabilities, attention should be drawn specifically to the sizable literature on infant research that shows through myriad and varied experimental studies how interlaced actual perception is from the very beginning. As noted, our adult tendency is to divide the senses: academic texts are testimonials to this sensory separation. In short, there is every reason to believe that we indoctrinate ourselves into a belief system not only counter to the way in which we first came into the world and the way the world first appeared to us, but in fact counter to the way in which we lead our everyday lives—amidst a sensory medley and with no experiential distinction of mind and body.

28. Stern, *The Interpersonal World of the Infant*, p. 63.

29. Ibid.

30. Ibid., p. 69.

31. Ibid.

32. Ibid., p. 70.

33. Ibid.

34. Ibid., p. 71.

35. Ibid.

36. Jane Flax, *Thinking Fragments* (Berkeley: University of California Press, 1990), p. 113.

37. Daniel N. Stern, *Diary of a Baby* (New York: Basic Books, 1990), p. 77.

38. Daniel Dennett, "Intentional Systems in Cognitive Ethology: The 'Panglossian Paradigm' Defended," *The Behavioral and Brain Sciences* 6 (1983): 343–390, p. 384.

39. Stern, *Diary of a Baby*, p. 21.

40. Ibid., p. 18.

41. Ibid., p. 25.

42. Ibid., p. 29.

43. Ibid.

44. Ibid.

45. Ibid., p. 39.

46. Ibid., pp. 31–43.

47. Butterworth, "Structure of the Mind in Human Infancy," p. 2.

48. Ibid., p. 3.

49. See J. J. Gibson, *The Senses Considered as Perceptual Systems* (Boston: Houghton Mifflin, 1966); *The Ecological Approach to Visual Perception* (Boston: Houghton Mifflin, 1979). For a discussion of how Gibson's approach to development differs from that of Piaget, see Butterworth, "Structure of the Mind in Human Infancy," pp. 5–6.

50. Butterworth, "Structure of the Mind in Human Infancy," p. 12.

51. Ibid.

52. Ibid., p. 16.

53. Ibid., p. 20.

54. Ibid., p. 25.

55. Ibid., p. 80.

56. Ibid.

57. Stern, *Diary of a Baby*, p. 99.

58. See, in contrast, Stern's descriptive account of infant delight which includes highlighting of the following: the infant's desire to communicate his feelings, the idea that delight is an experience that can be shared, the notions of synaesthesia, "psychic intimacy" (p. 107), and "analogic matching" (p. 106). Ibid., pp. 101–7.

59. Jacques Lacan, "The Subversion of the Subject and the Dialectic of Desire in the Freudian Unconscious," *Écrits: A Selection*, p. 297; *Écrits*, p. 799.

60. Ibid., p. 318; p. 820. The context from which this phrase is drawn centers on the castration complex. It begins, "Certainly there is in all this

what is called a bone. Though it is precisely what is suggested here, namely, that it is structural of the subject . . ." In other words, the castration complex structures the subject, in both the imaginary and symbolic orders, but it structures the imaginary only in retrospect.

61. Ibid., p. 310; p. 812.

62. We might add that an infant's affectivity is not either the servant of its cognitions. As Stern points out and documents, while developmental psychologists "have tended to stress the cognitive capacities required for an infant to have an affective experience . . . the realization is now occurring that not all affective life is the handmaiden to cognition, either for infants or for adults. . . . [I]infants' *feelings,* especially in the beginning, can and must be considered irrespective of what they *know.*" Stern, *The Interpersonal World of the Infant,* p. 66n.

63. Macey, *Lacan in Contexts,* p. 99. Quote from Jacques Lacan, "Some Reflections on the Ego," *International Journal of Psycho-Analysis* 34 (1953), Part I, p. 14.

64. For a discussion of the origin of language in phylogenetic terms, see Maxine Sheets-Johnstone, *The Roots of Thinking* (Philadelphia: Temple University Press, 1990). See especially chapter 6, "On the Origin of Language."

65. Jacques Lacan, "The Direction of the Treatment and the Principles of Its Power," *Écrits: A Selection,* p. 253; *Écrits,* p. 616.

66. See Stern, *The Interpersonal World of the Infant,* particularly chapter 8, and *Diary of a Baby.*

67. For a further justification of this way of referring to Lacan's evidence, see chapter 10.

68. Jacques Lacan, "The Agency of the Letter in the Unconscious or Reason Since Freud," *Écrits: A Selection,* p. 171; *Écrits,* p. 523.

69. *Kern unseres Wesen:* Lacan says that Freudian psychoanalysis leads us to "the nucleus of our being." Ibid. p. 173; p. 526.

70. John P. Muller and William J. Richardson, *Lacan and Language* (New York: International Universities Press, 1982), p. 167.

71. René Descartes, *Meditations on First Philosophy,* trans. John Cottingham, Robert Stoothoff, Dugald Murdoch (New York: Cambridge University Press, 1984), p. 17.

72. I asked a blind student (Angela Schneidecker) in an introductory philosophy course I taught at Western Oregon State College ("Being and Knowing," Spring 1992) whether she had a face. She replied that because she was blind not from birth but from early childhood, she knew what a face was, but that friends of hers who are blind from birth know what a face is only by learning to correlate the word with a certain unknown object. In other words,

they have no idea what a face is. As an example, she said that when people speak to these friends of hers about putting on eye shadow to highlight their eyes, they have no idea what that means. It is nonsensical to them.

73. Lacan, "Aggressivity," *Écrits: A Selection*, p. 18; *Écrits*, p. 112.

74. Lacan takes Freud's explanation of his grandson's "Fort! Da!" game— it is the renunciation of instinctual satisfaction—and turns it into the child's moment of truth: his entrance into the Symbolic order. "The Function and Field of Speech and Language in Psychoanalysis," *Écrits: A Selection*, p. 103; *Écrits*, pp. 318–19.

75. Jacques Lacan, "The Freudian Thing, or the Meaning of the Return to Freud in Psychoanalysis," *Écrits: A Selection*, p. 141; *Écrits*, p. 431.

76. Lacan, "Direction of the Treatment," *Écrits: A Selection*, p. 236; *Écrits*, p. 597.

77. Ibid., p. 243; p. 605.

78. Ibid.

79. "Subversion of the Subject," *Écrits: A Selection*, p. 309; *Écrits*, p. 811.

80. Lacan, "Aggressivity," *Écrits: A Selection*, p. 19; *Écrits* p. 113.

81. Stern, *Diary of a Baby*, p. 53.

82. Stern, *The Interpersonal World of the Infant*, p. 80.

83. We need all the more to consider the clinical and developmental literature on infants and children since Freudian theory is based not on infant-child experience but on retrospective adult experience, specifically, retrospective neurotic adult experience. This is in part why the question as to whether one can rightfully map the normal on the basis of the abnormal surfaces in critical psychoanalytical studies.

84. Lacan, "Function and Field," *Écrits: A Selection*, p. 104; *Écrits*, p. 319.

85. Stern, *The Interpersonal World of the Infant*, p. 176.

86. Ibid.

87. Ibid., p. 181.

88. Ibid.

89. Ibid.

90. Ibid., p. 182.

91. Ibid.

92. Ibid., p. 226.

93. Carl G. Jung, *The Archetypes and the Collective Unconscious*, 2nd ed., trans. R. F. C. Hull (Princeton: Princeton University Press, 1990), p. 169.

94. Ibid.

95. Ibid., pp. 168, 164.

96. Ibid., p. 170.

97. Ibid.
98. Ibid., p. 175.

CHAPTER 9: CORPOREAL ARCHETYPES AND POWER: LACAN'S PSY-CHOANALYTIC OF THE UNCONSCIOUS

1. Jacques Lacan, *The Four Fundamental Concepts of Psycho-Analysis,* trans. Alan Sheridan, ed. Jacques-Alain Miller (New York: W. W. Norton, 1978), p. 150.

2. Sigmund Freud, "The Dissection of the Psychical Personality," (in "New Introductory Lectures on Psycho-Analysis"), *Standard Edition of Complete Works* XXII, trans. James Strachey (London: The Hogarth Press, 1964), pp. 57–80, p. 70.

3. Carl G. Jung, *The Archetypes and the Collective Unconscious,* 2nd ed., trans. R. F. C. Hull (Princeton: Princeton University Press, 1990), p. 3.

4. Ibid., pp. 172–73.

5. Sigmund Freud, "The Unconscious," *Standard Edition of Complete Works* XIV, trans. James Strachey (London: Hogarth Press, 1957), pp. 166–215, p. 187.

6. See Sigmund Freud, "Analysis Terminable and Interminable," *Collected Papers,* vol. V, trans. Joan Rivière (London: The Hogarth Press, 1957), pp. 316–357. It is worthwhile quoting the end passage of this essay (pp. 356–57) which deals with women patients who do not prosper—or who are not "cured" by analysis: "We can only agree with them when we discover that their strongest motive in coming for treatment was the hope that they might somehow still obtain a male organ, the lack of which is so painful to them. All this shows that the form of the resistance is immaterial: it does not matter whether it appears as a transference or not. The vital point is that it prevents any change from taking place—everything remains as it was. We often feel that, when we have reached the wish for a penis and the masculine protest, we have penetrated all the psychological strata and reached 'bedrock' and that our task is accomplished. And this is probably correct, for *in the psychological field the biological factor is really the rock-bottom.* The repudiation of feminity must surely be a biological fact, part of the great riddle of sex. Whether and when we have succeeded in mastering this factor in an analysis is hard to determine. We must console ourselves with the certainty that everything possible has been done to encourage the patient to examine and to change his attitude to the question" (italics added). What Freud calls "the masculine protest"—against passivity—"is in

fact nothing other than fear of castration." How male fear of castration can thus fail to be central—if not "bedrock"—in Freudian psychoanalysis, not simply in the sense of dismemberment but in the critical sense of failing to achieve erection, is inexplicable.

7. We can see this most readily in Lacan's equivocation of penis and phallus, an equivocation virtually all commentators of Lacan grapple with and which we will presently examine.

8. Lacan, *Four Fundamental Concepts*, pp. 23, 34.

9. Ibid., p. 24.

10. Ibid., pp. 25–26.

11. Ibid., p. 37.

12. Freud, "The Unconscious," p. 187.

13. François Roustang, *The Lacanian Delusion*, trans. Greg Sims (New York: Oxford University Press, 1990). The third chapter in Roustang's book (pp. 107–121) is titled "The Principle of Incoherence." Lacan's self-proclamations are quoted by Roustang on pp. 107 and 109.

14. Ibid., p. 109.

15. Jacques Lacan, "Aggressivity in Psychoanalysis," *Écrits: A Selection*, trans. Alan Sheridan (New York: W. W. Norton, 1977), p. 15; *Écrits*, (Paris: Éditions du Seuil, 1966), p. 109.

16. Ibid.

17. Jacques Lacan, "The Direction of the Treatment and the Principles of Its Power," *Écrits: A Selection*, p. 254; *Écrits*, p. 617.

18. See psychiatrist D. W. Winnicott on the usefulness of memories of birth and birth trauma in psychoanalysis in "Birth Memories, Birth Trauma, and Anxiety," in his *Collected Papers: Through Paediatrics to Psycho-Analysis* (New York: Basic Books, 1958), pp. 174–193.

19. Jacques Lacan, "Some Reflections on the Ego," *International Journal of Psycho-Analysis* 34 (1953), Part 1, p. 15.

20. Jacques Lacan, "The Mirror Stage as Formative of the Function of the I as Revealed in Psychoanalytic Experience," *Écrits: A Selection*, p. 3; *Écrits*, pp. 95–96.

21. Lacan, "Direction of the Treatment," *Écrits: A Selection*, p. 233; *Écrits*, p. 594.

22. Lacan, *Four Fundamental Concepts*, p. 209.

23. "What is the place of interpretation?" is the name of the subheading within the essay, "Direction of the Treatment," *Écrits: A Selection*, p. 232; *Écrits*, p. 592.

24. Ibid., p. 233; p. 593.

25. Ibid., p. 237; p. 598.

26. Lacan, *Four Fundamental Concepts*, p. 230.

27. Ibid., p. 274.

28. David Macey's translation of "Le sujet . . . commence l'analyse en parlant de lui sans vous parler à vous, ou en parlant à vous sans parler de lui. Quand il pourra vous parler de lui, l'analyse sera terminée" (*Lacan in Contexts*, p. 147). Original in "Introduction au commentaire de Jean Hyppolite sur la 'Verneinung' de Freud," *Écrits*, p. 373.

29. Lacan, "The Function and Field of Speech and Language in Psychoanalysis," *Écrits: A Selection*, pp. 40–56; *Écrits*, pp. 247–265.

30. Ibid., p. 50; p. 259.

31. Lacan, "Direction of the Treatment," *Écrits: A Selection*, p. 254; *Écrits*, p. 617.

32. Muller and Richardson, *Lacan and Language*, p. 269.

33. Lacan, *Four Fundamental Concepts*, p. 29.

34. Lacan, "Function and Field," *Écrits: A Selection*, p. 103; *Écrits*, pp. 318–19. The well-known passage reads: ". . . the moment in which desire becomes human is also that in which the child is born into language."

35. Ibid., p. 55; p. 265.

36. Ibid.

37. Muller and Richardson, *Lacan and Language*, p. 75.

38. Jacques Lacan, "Intervention on Transference," in *Feminine Sexuality*, eds. Juliet Mitchell and Jacqueline Rose (New York: Pantheon Books, 1982), p. 72.

39. Macey, *Lacan in Contexts*, p. 124.

40. Elizabeth Grosz, *Sexual Subversions* (Boston: Allen & Unwin, 1989), p. 19.

41. Elizabeth Grosz, *Jacques Lacan: A Feminist Introduction* (New York: Routledge, 1990), p. 170.

42. Sigmund Freud, "The Ego and the Id," *Standard Edition of Complete Works* XIX, trans. James Strachey (London: The Hogarth Press, 1961), pp.13–59, p. 13.

43. Cf. Jane Flax, *Thinking Fragments* (Berkeley: University of California Press, 1990), p. 242, note 31. In her text, relative to postmodernism's depriviledging of the mind, she states that "epistemologies that rely on the possibility of accurate self-observation and direct, reliable access to and control over the mind and its activities [become untenable]." Her related note reads: "For example, Husserl's transcendental phenomenology, especially the *epoché*. Cf. Edmund Husserl, *The Crisis of European Sciences and Transcendental Phenomenology*, . . . especially part 3B, #69." But see Edmund Husserl, *The Crisis of European Sciences and Transcendental Phenomenology*, trans. David Carr (Evanston: Northwestern University Press, 1970), p. 239.

44. Freud wrote of female sexuality—"the sexual life of adult women"—as "a 'dark continent' for psychology." The phrase is even written by Freud in English. See "The Question of Lay Analysis," *Standard Edition of Complete Works* XX, trans. James Strachey (London: The Hogarth Press, 1959), pp. 183–250, p. 212.

45. Lacan, *Four Fundamental Concepts*, p. 184.

46. Quoted in Macey, *Lacan in Contexts*, p. 195. Original in Jacques Lacan, *"Hamlet* III: Le Désir de la mère," *Ornicar?*, 25, Autumn 1982, p. 23.

47. Lacan, *Four Fundamental Concepts*, p. 144.

48. Ibid., p. 146.

49. Sigmund Freud, ("Femininity" in "New Introductory Lectures on Psycho-Analysis") *Standard Edition of Complete Works* XXII, trans. James Strachey (London: The Hogarth Press, 1964), pp. 112–135, p. 113.

50. David Macey makes this point in a related way when he shows how, in Lacan's psychoanalytic, psychoanalysis is joined to female genital determinism. See Macey, *Lacan in Contexts*, chapter 6, "The Dark Continent."

51. Ibid., p. 209.

52. Mary Ann Doane, "Veiling Over Desire: Close-ups of the Woman," in *Feminism and Psychoanalysis*, ed. Richard Feldstein and Judith Roof (Ithaca: Cornell University Press, 1989), pp. 105–141, pp. 127–28.

53. Jacques Lacan, "The Meaning of the Phallus," in *Feminine Sexuality*, eds. Jacqueline Rose and Juliet Mitchell (New York: Pantheon Books, 1982), p. 82. "The Meaning of the Phallus" is Mitchell and Rose's translation of "La Signification du Phallus." I use their title translation and text in the present context of examining the relationship between feminity and the unconscious rather than "The Signification of the Phallus" in *Écrits: A Selection* and its original in *Écrits*.

54. Jacques Lacan, "God and the *Jouissance* of The Woman," in *Feminine Sexuality*, eds. Juliet Mitchell and Jacqueline Rose, pp. 146–47.

55. Jacques Lacan, "Signification of the Phallus," *Écrits: A Selection*, pp. 289–290; *Écrits*, pp. 694–95.

56. Lacan, *Four Fundamental Concepts*, p. 193.

57. Ibid.

58. Jacqueline Rose, "Introduction—II," in *Feminine Sexuality*, p. 43.

59. Ibid., pp. 43, 44.

60. Ibid., p. 53.

61. Ibid.

62. Lacan, "Meaning of the Phallus," p. 84.

63. Ibid., pp. 84–85.

64. Ibid.

65. Quoted in Macey, *Lacan in Contexts*, p. 197.

66. Lacan, *Four Fundamental Concepts,* p. 43.

67. Luce Irigaray, *This Sex Which Is Not One,* trans. Catherine Porter (Ithaca: Cornell University Press, 1985).

68. Lacan devotes some time to "the subject who is supposed to know." See, for example, *Four Fundamental Concepts,* chapter 18. Interestingly enough, he brings up this very subject in the opening paragraphs of "God and the *Jouissance* of The Woman" (*Feminine Sexuality,* p. 139).

69. See Macey, *Lacan in Contexts,* chapter 6, for a detailed, documented account.

70. Lacan, "God and the *Jouissance* of The Woman," p. 145.

71. Lacan, *Four Fundamental Concepts,* p. 230.

72. Lacan, "God and the *Jouissance* of The Woman," p. 146.

73. Ibid., p. 145.

74. Ibid., p. 146.

75. Jung, *Archetypes and the Collective Unconscious,* p. 48.

76. Ibid., p. 66.

77. Jung's distinction between a personal and collective unconscious is fundamental to his analytic psychology. See his *Archetypes and the Collective Unconscious,* p. 42, for example, and his *Aion: Researches into the Phenomenology of the Self,* 2nd ed., trans. R. F. C. Hull (Princeton: Princeton University Press, 1968), p. 7.

78. See Lacan's own brief and passing reference to the copulatory aspect of the phallus in "Signification of the Phallus," *Écrits: A Selection,* p. 287; *Écrits,* p. 692. Lacan does not of course either recognize or draw out its relational significance.

79. Carl G. Jung, *Psychological Types,* trans. H. G. Baynes, revised by R. F. C. Hull (Princeton: Princeton University Press, 1976), p. 244.

80. Ibid., p. 457.

81. Macey, *Lacan in Contexts,* p. 124.

82. Jung, *Aion,* p. 15.

83. Lacan's "Encore" seminar papers were given in 1972–73. The seminar was his next to last seminar before his death.

84. Juliet Mitchell and Jacqueline Rose, introductory remarks to "God and the *Jouissance* of The Woman," and "A Love Letter," in *Feminine Sexuality,* p. 137.

85. Lacan, "God and the *Jouissance* of The Woman" and "A Love Letter," in *Feminine Sexuality,* chapter 6, beginning p. 140 and continuing.

86 Ibid., p. 145.

87. One is, of course, reminded of Descartes: thinking is a matter of consciousness and knowledge.

CHAPTER 10: CORPOREAL ARCHETYPES AND POWER: THE PHALLUS AS ARCHETYPE

1. Serge Leclaire, quoted in David Macey, *Lacan in Contexts* (New York: Verso, 1988), p. 189. Leclaire, Macey tells us, was at one time director of the Département de Psychanalyse at Vincennes. The quote is from a seminar he gave at Vincennes. Vindications of Leclaire's claim are evident even within the objectively hallowed halls of science. See, for example, recent *Science News* articles on "Mother Nature['s]. . . feminizing hormonal influence on the animal kingdom" (Janet Roloff, "The Gender Benders," 145/2:24–27, p. 24); "The phallus on males [male alligators] was one-half to one-third the normal size. . . . " (Ibid., p. 25); "Year-old male offspring [rats] exibited a range of reproductive abnormalities [such as]. . . a cleft phallus. . . and a partially unfused phallus. . . " ("Another Emasculating Pesticide Found," 146/1:15). Clearly, nowadays, not only do some biologists of rank apparently not think of the penis as anything less than the phallus, they do not think anything of endowing nonhuman creatures with the human cultural artifact.

2. Macey, *Lacan in Contexts,* p. 199.

3. Eugene Monick, *Phallos: Sacred Image of the Masculine* (Toronto: Inner City Books, 1987), p. 22.

4. Daniel Rancour-Laferrière, *Signs of the Flesh* (New York: Mouton de Gruyter, 1985), p. 305.

5. Monick, *Phallos,* p. 82.

6. Ibid., pp. 83–84.

7. Ibid., pp. 16–17.

8. Jacques Lacan, *The Four Fundamental Concepts of Psycho-Analysis,* ed. Jacques-Alain Miller, trans. Alan Sheridan (New York: W. W. Norton, 1978), p. 43.

9. Jacques Lacan, "The Signification of the Phallus," *Écrits: A Selection,* trans. Alan Sheridan (New York: W. W. Norton, 1977), pp. 287–88; *Écrits* (Paris: Éditions du Seuil, 1966), p. 692.

10. Jacques Lacan, "The Function and Field of Speech and Language in Psychoanalysis," *Écrits: A Selection,* pp. 103–5; *Écrits,* pp. 319–21.

11 François Roustang, *The Lacanian Delusion,* trans. Greg Sims (New York: Oxford University Press, 1990), p. 129.

12 Jacques Lacan, "Aggressivity in Psychoanalysis," *Écrits: A Selection,* p. 16; *Écrits,* p. 110.

13. Jacques Lacan, *Le Séminaire de Jacques Lacan: Les Psychoses, Livre III* (Paris: Éditions du Seuil, 1981), p. 111.

14. Lacan, "Signification of the Phallus," *Écrits: A Selection,* p. 288; *Écrits,* p. 692.

15. It might be noted that Lacan's phallic latency is easily conceived as pregnancy, that is, the phallus is latent with meanings—significations—as pregnant females are latent with new human life. In each case, something is not immediately visible but hidden; it is present, but without showing itself. Of course, females themselves are latent with the possibility of becoming pregnant in the first place.

16. Roustang, *The Lacanian Delusion*, p. 124.

17. Robert Stoller, "Facts and Fancies: An Examination of Freud's Concept of Bisexuality," in *Women and Analysis*, ed. Jean Strouse (Boston: G. K. Hall, 1985), pp. 343–364; quote, p. 353.

18. Monick, *Phallos*, p. 16.

19. Ibid.; see pp. 37, 28, 79, 105, respectively.

20. William I. Grossman and Walter A. Stewart, "Penis Envy: From Childhood Wish to Developmental Metaphor," in *Female Psychology: Contemporary Psychoanalytic Views*, ed. Harold P. Blum (New York: International Universities Press, 1977), pp. 193–212, pp. 193, 194.

21. Ibid., pp. 203, 211.

22. See, for example, Jane Gallop's chapter "Reading the Phallus," in *Reading Lacan* (Ithaca: Cornell University Press, 1985), pp. 133–156.

23. Sigmund Freud, "The Acquisition of Power Over Fire," in *Collected Papers*, Vol. V (*Miscellaneous Papers, 1888–1938*), ed. James Strachey (London: The Hogarth Press, 1957), pp. 288–294.

24. Ibid., p. 289.

25. Ibid., p. 291.

26. Ibid., pp. 291–92.

27. Ibid., p. 293.

28. Ibid., pp. 293–94.

29. Ibid., p. 294. There is no reason of course to believe that only "primitive" conceptual and comportmental practices are tied to animate form and to correlative tactile-kinesthetic experiences thereof, or in other words, that only "primitive people" think analogically. For a thorough examination of the origin of fundamental (i.e., primate-derived and pan-cultural) human concepts and a defense of the idea that thinking is modeled on the body, see Maxine Sheets-Johnstone, *The Roots of Thinking* (Philadelphia: Temple University Press, 1990).

30. Freud, "The Acquisition of Power," p. 288.

31. Lacan, *Four Fundamental Concepts*, p. 205. See also Roustang, *The Lacanian Delusion*, p. 85.

32. Roustang, *The Lacanian Delusion*, p. 85.

33. Lacan, *Four Fundamental Concepts*, p. 184.

34. To some extent at least, Lacan appears to avoid reference to the mouth,

that is, to the mouth as a *hole,* as if to protect the place of speech from any bodily associations. The omission is particularly noticeable not only when he discusses the *objet à* (in *Four Fundamental Concepts*), but when he itemizes the bodily slits and holes on his psychoanalytic body and refers to the "anatomical mark" of lips and to "the enclosure of the teeth" (in "The Subversion of the Subject and the Dialectic of Desire in the Freudian Unconscious"). The euphemisms appear to assure that language is untainted by any bodily associations.

35. Muller and Richardson, *Lacan and Language,* p. 22.

36. Monick, *Phallos,* p. 16.

37. Ibid., pp. 15–16.

38. Lacan, "Signification of the Phallus," *Écrits: A Selection,* p. 290; *Écrits,* p. 695.

39. Gregory Zilboorg, "Masculine and Feminine: Some Biological and Cultural Aspects," *Psychiatry* 7 (1944): 257–296, p. 262.

40. Ibid., p. 273.

41. Ibid., p. 268.

42. Ibid., p. 282.

43. Ibid.

44. Ibid., p. 287.

45. Ibid.

46. Ibid., pp. 287–88.

47. Ibid., p. 290.

48. Lacan, "Function and Field," *Écrits: A Selection,* p. 32; *Écrits,* p. 239.

49. For example, see *The City of God:* Book XIII (New York: Fathers of the Church, 1952), pp. 316–17; Book XIV (New York: Fathers of the Church, 1952), pp. 389–92.

50. Especially with respect to the medical study and popularization of female sexuality, males and medical science are also typically silent about wet dreams. They are also typically silent about penile erection on waking and during sleep.

51. Lacan, *Four Fundamental Concepts,* p. 290. The passage reads: "Yet it should not be thought that the sort of infidelity that would appear to be constitutive of the male function is proper to it. For if one looks more closely, the same redoubling is to be found in the woman, except that the Other of Love as such, that is to say, in so far as he is deprived of what he gives, finds it difficult to see himself in the retreat in which he is substituted for the being of the very man whose attributes she cherishes." "Looking more closely" clearly indicates looking at an excess of orgasms over detumescence.

52. Jacques Lacan, "The Meaning of the Phallus," in *Feminine Sexuality,* eds. Juliet Mitchell and Jacqueline Rose (New York: Pantheon Books, 1982), p. 84.

53. Ibid., p. 85.

54. William H. Masters, Virginia E. Johnson, Robert C. Kolodny, *Human Sexuality* (Boston: Little, Brown and Company, 1982), p. 368.

55. Ibid., p. 374. The authors state (pp. 373–74) that "Before the publication of *Human Sexual Inadequacy* in 1970, the term *frigidity* was generally used to describe a number of female sexual difficulties," specifically, those three specified in the text.

56. Jacques Lacan, "The Agency of the Letter in the Unconscious or Reason Since Freud," *Écrits: A Selection,* p. 157; *Écrits,* p. 507.

57. Macey, *Lacan in Contexts,* p. 155. (As far as I can tell, Macey's specific reference to Lacan, given as *Séminaire III,* p. 303, is incorrect; I have been unable to locate the passage in Lacan from which Macey quotes.)

58. Ibid.

59. See note 139 in chapter 7.

60. Sigmund Freud, "The Infantile Genital Organization," *Standard Edition of Complete Works* XIX, trans. James Strachey (London: The Hogarth Press, 1961), pp. 141–145, p. 142.

61. Robert S. Steele, *Freud and Jung: Conflicts of Interpretation* (Boston: Routledge & Kegan Paul, 1982), p. 312.

62. Carl G. Jung, "The Tavistock Lectures: On the Theory and Practice of Analytical Psychology," trans. R. F. C. Hull, *Collected Works,* vol. 18 (Princeton: Princeton University Press, 1980), pp. 5–182, p. 125. See also Steele, *Freud and Jung,* p. 312.

63. Steele, *Freud and Jung,* p. 307.

64. I use the term "attributions" rather than "attributes" not to avoid essentialism but because the very idea of excess is an attribution, that is, a valuation according to an unspoken but palpably present standard. As self-appropriated, however, excess and other attributions are spoken of as attributes, for the person appropriating them is taking them as essential features of him/herself.

65. Sigmund Freud, "Analysis Terminable and Interminable," *Standard Edition of Complete Works* XXIII, trans. James Strachey (London: The Hogarth Press, 1961), pp. 216–253.

66. Ibid., p. 252.

67. Monick, *Phallos,* p. 44.

68. Jacques Lacan, "God and the *Jouissance* of The Woman," in *Feminine Sexuality,* p. 146.

69. Ibid., p. 147.

70. Lacan, "Signification of the Phallus," *Écrits: A Selection,* p. 287; *Écrits,* p. 692.

71. Lacan, *Four Fundamental Concepts,* p. 263. Lacan actually asks whether

we, meaning all Lacanian psychoanalysts, are not impostors.

72. Paul Radin, *The Trickster: A Study in American Indian Mythology* (New York: Schocken Books, 1972), pp. xxiii, 19. (Commentaries by Karl Kerenyi and Carl Jung appear at the end of the book.)

73. Ibid., p. 289.

74. Lacan, "Signification of the Phallus," *Écrits: A Selection,* p. 289; *Écrits,* p. 694.

75. Ibid.

76. Jacques Lacan, "The Direction of the Treatment and the Principles of Its Power," *Écrits: A Selection,* p. 269; *Écrits,* p. 633.

77. Sigmund Freud, "Observations on 'Wild' Psycho-Analysis," trans. Joan Riviere, ed. Ernest Jones, *Collected Papers,* vol. 2 (London: The Hogarth Press, 1924), pp. 297–304, p. 299.

78. Lacan says further: "through being subject to sex, [the living being] has fallen under the blow of individual death." He also states that "The link between sex and death, sex and the death of the individual, is fundamental" (*Four Fundamental Concepts,* p. 150).

79. Jacques Lacan, "The Freudian Thing or the Meaning of the Return to Freud in Psychoanalysis," *Écrits: A Selection,* p. 142; *Écrits,* p. 432.

80. Lacan, *Four Fundamental Concepts,* p. 130.

81. Lacan, "The Freudian Thing," *Écrits: A Selection,* p. 128; *Écrits,* p. 417.

82. Bice Benvenuto and Roger Kennedy, *The Works of Jacques Lacan* (London: Free Association Books, 1986), p. 116.

83. Lacan, "The Freudian Thing," *Écrits: A Selection,* p. 145; *Écrits,* p. 436.

84. Roustang, *The Lacanian Delusion,* pp. 19–20.

85. See especially Gallop, *Reading Lacan,* chapter 3.

86. Lacan, *Four Fundamental Concepts,* p. 176.

87. Jacques Lacan, "The Subversion of the Subject and the Dialectic of Desire in the Freudian Unconscious," *Écrits: A Selection,* p. 300; *Écrits,* p. 801.

88. Ibid., p. 300; p. 802.

89. Lacan, *Four Fundamental Concepts,* p. 176.

90. Lacan, "The Freudian Thing," *Écrits: A Selection,* p. 128; *Écrits,* p. 417.

91. Muller and Richardson, *Lacan and Language,* p. 131.

92. Benvenuto and Kennedy, *The Works of Jacques Lacan,* p. 116.

93. Lacan at one point is in fact at the brink of recalling Sartre's psychoanalytic of things when he speaks of a desk as a thing, a thing that speaks, i.e., whose qualities are ontologically part of *it.* Lacan's use of a desk as an example is

interesting in that a desk is a place of writing and, given his ever close reading of Freud, it is unlikely that he would have missed Freud's writing-copulation analogy.

94. Jean-Paul Sartre, *Being and Nothingness,* trans. Hazel E. Barnes (New York: Philosophical Library, 1956), p. 614.

95. Lacan, *Four Fundamental Concepts,* p. 43. (See chapter 9, p. 286 for original context of quote.)

EPILOGUE

1. Elliott Liebow, quoted in *The Chronicle of Higher Education* 39/30 (31 March 1993), p. A1. (The title of Liebow's new book is *Tell Them Who I Am.*)

2. See chapter 5, p. 144.

3. This is not to say that only Western cultures are beset by the dichotomy. See, for example, J. S. La Fontaine, "Person and Individual: Some Anthropological Reflections," in *The Category of the Person: Anthropology, Philosophy, History,* eds. Michael Carrithers, Steven Collins, and Steven Lukes (New York: Cambridge University Press, 1985), pp. 123–40. La Fontaine writes that the Lugubra (Ugandans) view woman as behaving erratically because they lack "a spirit of responsibility" (p. 127); and that the Taita (Kenyans), who believe the head to be "the locus of consciousness, speech, memory and knowledge[,]" also believe '[w]omen's heads . . . to be weaker than men's" (p. 128).

4. Patriarchal corporeal standards can of course shift, and corporeal values in turn change, as witness more recently the positive image of "hard" female bodies.

5. That violence defines our society should come as no surprise given the intense preoccupation of our society with this kind of power. Violence establishes control. As pointed out in chapter 2, violent acts are spectacular: they rivet our attention; they make things explode before our eyes and roar in our ears. There is no doubt but that predominant symbols, metaphors, and images of Western life that are saturated in motifs of violence and control are corporeally tied to male bodies. Corporeal archetypes from size to penile display—not as invitation but as threat—come in many cultural forms. Susan Bordo recently noted that macho conceptions of power extend even to eating habits. Real males have huge appetites. Anything less than a he-man meal makes a man effeminate. As examples, Bordo points out not only Tommy La Sorda commercials but media portrayals of President Clinton's eating habits. "Clinton's love of food," she writes, "continually gets represented as embarrassing, out-of-control, feminine 'binge' behavior. He just can't resist those goodies,

like the rest of us girls." See Susan Bordo, "Reading the Male Body," *Michigan Quarterly Review* 32/4 (Fall 1993): 696–737, p. 723.

6. See Emily Martin, *The Woman In the Body* (Boston: Beacon Press, 1987).

7. The point is topical not only to "involuntary uterine contractions" and the like, but to anorexia. See Susan Bordo's excellent analysis of anorexia, especially her discussion of how "anorexia *begins in* . . . what is, in our time, conventional feminine practice." In the course of dieting, Bordo writes, "the young woman discovers what it feels like to crave and want and need and yet, through the exercise of her own will, to triumph over that need. In the process, a new realm of meanings is discovered, a range of values and possibilities that Western culture has traditionally coded as 'male' and rarely made available to women: an ethic and aesthetic of self-mastery and self-transcendence, expertise, and power over others through the example of superior will and control." *Unbearable Weight* (Berkeley: University of California Press, 1993), p. 178.

BIBLIOGRAPHY

Abel, Gene G., Edward B. Blanchard, and Judith V. Becker. "Psychological Treatment of Rapists." In *Sexual Assault,* eds. Marcia J. Walker and Stanley L. Brodsky, 99–116. Lexington, MA: Lexington Books, 1976.

Altmann, Stuart A. "The Structure of Primate Social Communication." In *Social Communication among Primates,* ed. Stuart A. Altmann, 325–362. Chicago: University of Chicago Press, 1967.

Amir, Menachem. *Patterns in Forcible Rape.* Chicago: University of Chicago Press, 1971.

Arens, W. *The Original Sin.* New York: Oxford University Press, 1986.

Aristotle. *De Anima.* Rev. trans. by J. A. Smith. *The Complete Works of Aristotle,* vol. 1. Edited by Jonathan Barnes. Princeton: Princeton University Press, 1984.

Attenborough, David. *Life on Earth.* Boston: Little, Brown and Company, 1979.

Bancroft, John. "The Relationship between Hormone and Sexual Behaviour in Rodents." In *Biological Determinants of Sexual Behaviour,* ed. J. B. Hutchison, 493–519. New York: John Wiley & Sons, 1978.

Barash, David. *Sociobiology and Behavior.* New York: Elsevier, 1982.

Bard, Morton, and Dawn Sangrey. *The Crime Victim's Book.* New York: Basic Books, 1979.

Beach, Frank A. "Human Sexuality and Evolution." In *Reproductive Behavior,* eds. William Montagna and William A. Sadler, 333–365. New York: Plenum Press, 1974.

———. "Cross-Species Comparisons and Human Heritage." In *Human Sexuality in Four Perspectives,* ed. Frank A. Beach, 296–316. Baltimore: Johns Hopkins Press, 1976.

Benderly, Beryl Lieff. *The Myth of Two Minds.* New York: Doubleday, 1987.

Benhabib, Seyla. "The Generalized and the Concrete Other." In *Feminism as*

Critique: On the Politics of Gender, eds. Seyla Benhabib and Drucilla Cornell, 77–95. Minneapolis: University of Minnesota Press, 1987.

Benjamin, Jessica. "A Desire of One's Own: Psychoanalytic Feminism and Intersubjective Space." In *Feminist Studies/Critical Studies,* ed. Teresa de Lauretis, 78–101. Bloomington: Indiana University Press, 1986.

Benvenuto, Bice, and Roger Kennedy. *The Works of Jacques Lacan.* London: Free Association Books, 1986.

Berenson, Bernard. "The Central Italian Painters of the Renaissance." In *The Bernard Berenson Treasury.* Edited by H. Kiel. New York: Simon and Schuster, 1962.

Berkeley, Bishop George. *Essay Toward a New Theory of Vision* (1709). In *Berkeley Selections.* Edited by Mary W. Calkins. New York: Charles Scribner's Sons, 1929.

Bertrand, Mireille. *The Behavioral Repertoire of the Stumptail Macaque.* Basel, Switzerland: Karger, 1969.

Bordo, Susan. "Feminism Reconceives the Body." Paper presented at Bates College, Lewiston, ME, April 1991.

———. "Reading the Male Body." *Michigan Quarterly Review* 32/4 (Fall 1993): 696–737.

———. *Unbearable Weight.* Berkeley: University of California Press, 1993.

Borgia, Gerald. "Sexual Selection in Bowerbirds." *Scientific American* 254 (1986): 92–100.

Boston Women's Health Book Collective. *Our Bodies, Ourselves.* New York: Simon and Schuster, 1971.

Bourque, Linda Brookover, ed. *Defining Rape.* Durham, NC: Duke University Press, 1989.

Bower, T. G. R. *Development in Infancy.* San Francisco: W. H. Freeman, 1974.

Bramblett, Claud A. *Patterns of Primate Behavior.* Palo Alto, CA: Mayfield, 1976.

Brodsky, Carroll M. "Rape at Work." In *Sexual Assault,* eds. Marcia J. Walker and Stanley L. Brodsky, 35–51. Lexington, MA: Lexington Books, 1976.

Brown, Jason W. *The Life of the Mind.* Hillsdale, NJ: Lawrence Erlbaum, 1988.

Brownmiller, Susan. *Against Our Will.* New York: Simon and Schuster, 1975.

Bunch, Charlotte. "A Broom of One's Own." In *Passionate Politics.* New York: St. Martin's Press, 1987.

Burgess, Ann W., and Lynda L. Holmstrom. "Rape Trauma Syndrome." In *Forcible Rape: The Crime, the Victim, and the Offender,* eds. Duncan Chappell, Robley Geis, and Gilbert Geis, 315–328. New York: Columbia University Press, 1977.

Butler, Judith. "Gendering the Body: Beauvoir's Philosophical Contribution." In *Women, Knowledge, and Reality: Explorations in Feminist Philosophy,* eds. Ann Garry and Marilyn Pearsall, 253–273. Boston: Unwin Hyman, 1989.

———. *Gender Trouble.* New York: Routledge, 1990.

Butterworth, George. "Structure of the Mind in Human Infancy." In *Advances in Infancy Research,* vol. 2, eds. Lewis P. Lipsitt and Carolyn K. Rovee-Collier, 1–29. Norwood, NJ: Ablex Publishing Corporation, 1983.

Calhoun, J. B. "A Comparative Study of the Social Behavior of Two Inbred Strains of House Mice." *Ecological Monographs* 26 (1956): 81–103.

———. "A Behavioral Sink." In *Roots of Behavior,* ed. Eugene L. Bliss, 295–315. New York: Harper and Bros., 1962.

———. "Population Density and Social Pathology, 139–148." *Scientific American* 206 (1962).

Chappell, Duncan, Robley Geis, and Gilbert Geis, eds. *Forcible Rape: The Crime, the Victim, and the Offender.* New York: Columbia University Press, 1977.

Chevalier-Skolnikoff, Suzanne. "Heterosexual Copulatory Patterns in Stumptail Macaques (*Macaca arctoides*) and in Other Macaque Species." *Archives of Sexual Behavior* 4/2 (1975): 192–220.

Christian, Barbara. "The Race for Theory." In *Gender and Theory,* ed. Linda Kauffman, 225–237. New York: Basil Blackwell, 1989.

The Chronicle of Higher Education, Vol. 40/12 (November 10, 1992).

Clark, Lorenne M. G., and Debra J. Lewis. *Rape: The Price of Coercive Sexuality.* Toronto: The Women's Press, 1977.

Clément, Catherine. *The Lives and Legends of Jacques Lacan.* Translated by Arthur Goldhammer. New York: Columbia University Press, 1983.

Code, Lorraine. *What Can She Know?* Ithaca: Cornell University Press, 1991.

Collins, Margery, and Christine Pierce. "Holes and Slime: Sexism in Sartre's Psychoanalysis." *Philosophical Forum* 5/1–2 (1973–74): 112–27.

Crews, Frederick. "The Unknown Freud." *The New York Review of Books* 40/19 (18 November 1993): 55–66.

Crook, J. H. "Sexual Selection, Dimorphism, and Social Organization in Primates." In *Sexual Selection and the Descent of Man, 1871–1971,* ed. B. Campbell, 231–281. Chicago: Aldine, 1972.

Daily Emerald. University of Oregon, 28 May 1991, p. 8.

Dallery, Arleen B. "The Politics of Writing (the) Body: *Écriture Feminine.* In *Gender/Body/Knowledge,* eds. Alison M. Jaggar and Susan R. Bordo, 52–67. New Brunswick: Rutgers University Press, 1989.

Daly, Martin, and Margo Wilson. *Sex, Evolution, and Behavior,* 2nd ed. Belmont, CA: Wadsworth Publishing, 1983.

Darwin, Charles. *Charles Darwin's Notebooks, 1836–1844.* Edited by Paul H. Barrett et al. Ithaca: Cornell University Press, 1987.

———. *The Origin of Species* (1859). Edited by J. W. Burrow. Middlesex: Penguin, 1968.

———. *The Descent of Man and Selection in Relation to Sex* (1871). Princeton: Princeton University Press, 1981.

————. *The Expression of the Emotions in Man and Animals* (1872). Chicago: University of Chicago Press, 1965.

de Waal, Frans. *Chimpanzee Politics: Power and Sex Among the Apes.* New York: Harper Colophon Books, 1982.

Dennett, Daniel C. "Intentional Systems in Cognitive Ethology: The 'Panglossian Paradigm' Defended." *The Behavioral and Brain Sciences* 6 (1983): 343–90.

Derrida, Jacques. *Of Grammatology.* Translated by Gayatri Spivak. Baltimore: Johns Hopkins University Press, 1976.

————. "Force et signification." In *Writing and Difference.* Translated by Alan Bass, 3–30. Chicago: University of Chicago Press, 1978.

————. "La parole soufflée." In *Writing and Difference.* Translated by Alan Bass, 169–95. Chicago: University of Chicago Press, 1978.

————. *"Geschlecht:* Sexual Difference, Ontological Difference." *Research in Phenomenology* 13 (1983): 65–83.

————. *"Geschlecht II."* In *Deconstruction and Philosophy.* Edited by John Sallis, 161–96. Chicago: University of Chicago Press, 1987.

Descartes, René. *Meditations on First Philosophy.* Translated by John Cottingham, Robert Stoothoff, and Dugald Murdoch. New York: Cambridge University Press, 1984.

Doane, Mary Ann. "Veiling Over Desire: Close-ups of the Woman." In *Feminism and Psychoanalysis,* eds. Richard Feldstein and Judith Roof. 105–41. Ithaca: Cornell University Press, 1989.

Dolhinow, Phyllis. "The North Indian Langur." In *Primate Patterns,* ed. Phyllis Dolhinow 181–238. New York: Holt, Rinehart and Winston, 1972), pp. 223–24.

Douglas, Mary. *Natural Symbols,* 2nd ed. London: Barrie & Jenkins, 1973.

Dreyfus, Hubert L., and Paul Rabinow. *Michel Foucault: Beyond Structuralism and Hermeneutics,* 2nd ed. Chicago: University of Chicago Press, 1983.

Dundes, Alan, ed. *The Evil Eye: A Folklore Casebook.* New York: Garland Publishing, 1981.

Eberhard, William G. *Sexual Selection and Animal Genitalia.* Cambridge: Harvard University Press, 1985.

Eibl-Eibesfeldt, Irenaus. "Similarities and Differences Between Cultures in Expressive Movements." In *Nonverbal Communication,* ed. R. A. Hinde, 297–314. New York: Cambridge University Press, 1972.

————. *Ethology,* 2nd ed. New York: Holt, Rinehart and Winston, 1975.

Eimerl, Sarel, and Irven DeVore. *The Primates.* New York: Times, Inc., 1965.

Ekman, Paul. "Facial Expression." In *Nonverbal Behavior and Communication,* eds. Aron W. Siegman and Stanley Feldstein, 97–116. Hillsdale, NJ: Lawrence Erlbaum, 1978.

————. "The Argument and Evidence about Universals in Facial Expressions of

Emotions," in *Handbook of Social Psychophysiology,* eds. H. Wagner and A. Manstead, 143–64. New York: John Wiley and Sons, 1989.

Ekman, Paul, W. V. Friesen, M. O'Sullivan, A. Chan, I. Diacoyanni-Tarlatzis, K. Heider, R. Krause, W. A. LeCompte, T. Pitcairn, P. E. Ricci-Bitti, K. R. Scherer, M. Tomita, and A. Tzavaras. "Universals and Cultural Differences in the Judgements of Facial Expressions of Emotion." *Journal of Personality and Social Psychology* 53 (1987): 712–17.

Ellis, Lee. *Theories of Rape.* New York: Hemisphere Publishing, 1989.

Enomoto, Tomas. "The Sexual Behavior of Japanese Monkeys." *Journal of Human Evolution* 3 (1974): 351–72.

Fausto-Sterling, Anne. *Myths of Gender.* New York: Basic Books, 1985.

Feder, H. H. "Specificity of Steroid Hormone Activation of Sexual Behaviour in Rodents." In *Biological Determinants of Sexual Behaviour,* ed. J. B. Hutchison, 395–424. New York: John Wiley & Sons, 1978.

Fedigan, Linda Marie. *Primate Paradigms.* Montreal: Eden Press, 1982.

Firth, Raymond. "Postures and Gestures of Respect." In *The Body Reader,* ed. Ted Polhemus, 88–108. New York: Pantheon Books, 1978.

Fisher, Ronald A. *The Genetical Theory of Natural Selection,* 2nd ed. New York: Dover, 1958.

Flax, Jane. *Thinking Fragments.* Berkeley: University of California Press, 1990.

Ford, Clellan S., and Frank A. Beach. *Patterns of Sexual Behavior.* New York: Harper Torchbooks, 1951.

Foucault, Michel. *Discipline and Punish.* Translated by Alan Sheridan. New York: Vintage Books, 1979.

———. *The History of Sexuality,* vol. 1. Translated by Robert Hurley. New York: Vintage Books, 1980.

———. *The Use of Pleasure* (vol. 2 of *The History of Sexuality*). Translated by Robert Hurley. New York: Vintage Books, 1986.

Fraser, Nancy. "Foucault's Body-Language." *Salmagundi* 61 (Fall 1983): 55–70.

Freud, Sigmund. "The Taboo of Virginity." *Standard Edition of Complete Works,* vol. XI. Translated by James Strachey, 193–208. London: The Hogarth Press, 1957.

———. "The Ego and the Id." *Standard Edition of Complete Works,* vol. XIX. Translated by James Strachey, 13–59. London: Hogarth Press, 1955.

———. "The Infantile Genital Organization." *Standard Edition of Complete Works,* vol. XIX. Translated by James Strachey, 141–45. London: The Hogarth Press, 1961.

———. "The Question of Lay Analysis." *Standard Edition of Complete Works,* vol. XX. Translated by James Strachey, 183–250. London: The Hogarth Press, 1959.

———. "The Dissection of the Psychical Personality." In "New Introductory Lectures on Psycho-Analysis." *Standard Edition of Complete Works,* vol. XXII. Translated by James Strachey, 57–80. London: The Hogarth Press, 1964.

————. "The Unconscious." *Standard Edition of Complete Works,* vol. XXII. Translated by James Strachey, 166–215. London: The Hogarth Press, 1964.

————. "Analysis Terminable and Interminable." *Standard Edition of Complete Works,* vol. XXIII. Translated by James Strachey, 216–253. London: The Hogarth Press, 1961.

————. "Observations on 'Wild' Psycho-analysis." *Collected Papers,* vol. 2. Translated by Joan Rivière. Edited by Ernest Jones, 297–304. London: The Hogarth Press, 1924.

————. "Analysis Terminable and Interminable." *Collected Papers,* vol. 5. Translated by Joan Rivière. Edited by James Strachey, 316–357. London: The Hogarth Press, 1957.

————. "The Acquisition of Power over Fire." *Collected Papers,* vol. 5. Translated by Joan Rivière. Edited by James Strachey, 288–94. London: The Hogarth Press, 1957.

Fuss, Diana J. "'Essentially Speaking': Luce Irigaray's Language of Essence." *Hypatia* 3/3 (Winter 1989): 62–80.

Gajdusek, Carleton. "Physiological and Psychological Characteristics of Stone Age Man." In *Engineering and Science:* Symposium on Biological Bases of Human Behavior (Pasadena: California Institute of Technology, Alumni Association, March 1970): 26–33; 56–62.

Galdikas, Birute M. F. "Orangutan Adaptation in Tanjung Puting Reserve: Mating and Ecology." In *The Great Apes,* eds. David A. Hamburg and Elizabeth R. McCown, 195–233. Menlo Park, CA: Benjamin/Cummings Publishing, 1979.

Galen. *On the Usefulness of the Parts of the Body,* 2 vols. Edited and translated by Margaret T. May. Ithaca: Cornell University Press, 1968.

Gallop, Jane. *The Daughter's Seduction: Feminism and Psychoanalysis.* Ithaca: Cornell University Press, 1982.

————. *Reading Lacan.* Ithaca: Cornell University Press, 1985.

Gibson, J. J. *The Senses Considered as Perceptual Systems.* Boston: Houghton Mifflin, 1966.

————. *The Ecological Approach to Visual Perception.* Boston: Houghton Mifflin, 1979.

Givens, David. "Greeting a Stranger: Some Commonly Used Nonverbal Signals of Aversiveness." In *Nonverbal Communication, Interaction, and Gesture,* ed. Adam Kendon, 219–50. *Approaches to Semiotics,* No. 41. The Hague: Mouton, 1981.

Gordon, Margaret T., and Stephanie Riger. *The Female Fear.* New York: The Free Press, 1989.

Gould, Carol C. "The Woman Question: Philosophy of Liberation and the Liberation of Philosophy." In *Philosophy of Woman,* ed. Mary B. Mahowald, 415–452. Indianapolis: Hackett Publishing, 1983.

Gould, Stephen Jay, and Richard Lewontin. "The Spandrels of San Marco and the

Panglossian Paradigm: A Critique of the Adaptationist Programme." *Proceedings of the Royal Society of London,* Series B., Biological Science 205 (1979): 581–98.

Gregor, Thomas. *Anxious Pleasures: The Sexual Lives of an Amazonian People.* Chicago: University of Chicago Press, 1985.

Grossman, William I., and Walter A. Stewart. "Penis Envy: From Childhood Wish to Developmental Metaphor." In *Female Psychology: Contemporary Psychoanalytic Views,* ed. Harold P. Blum, 193–212. New York: International Universities Press, 1977.

Grosz, Elizabeth. *Sexual Subversions.* Boston: Allen & Unwin, 1989.

———. *Jacques Lacan: A Feminist Introduction.* New York: Routledge, 1990.

Groth, A. Nicolas. *Men Who Rape.* New York: Plenum Press, 1979.

Hall, Edward T. *The Hidden Dimension.* New York: Doubleday, 1966.

Hall, K. R. L. "Social Vigilance Behaviour of the Chacma Baboon (*Papio ursinus*)." *Behavior* 16 (1960): 261–94.

Hall, K. R. L., and Irven DeVore. "Baboon Social Behavior." In *Primate Patterns,* ed. Phyllis Dolhinow. New York: Holt, Rinehart and Winston, 1972.

Hanby, J. "Sociosexual Development in Primates." In *Perspectives in Ethology,* vol. 2, eds. P. P. G. Bateson and Peter H. Klopfer, 1–67. New York: Plenum Press, 1976.

Hanisch, Carol. "The Personal is Political." In *Notes from the Second Year: Women's Liberation, Major Writings of the Radical Feminists,* eds. Shulamith Firestone and Anne Koedt, 76–78. New York: Radical Feminists, 1970.

Haraway, Donna. "A Manifesto for Cyborgs: Science, Technology, and Socialist Feminism in the 1980s." *Socialist Review* 15/2 (1985): 65–107.

———. "Situated Knowledges: The Science Question in Feminism and the Privilege of Partial Perspective." *Feminist Studies* 14/3 (1988): 575–599.

———. *Primate Visions.* New York: Routledge, 1989.

Harding, Sandra. "The Instability of the Analytical Categories." In *Sex and Scientific Inquiry,* eds. Sandra Harding and Jean F. O'Barr, 282–302. Chicago: University of Chicago Press, 1987.

Harding, Cheryl F. "Sociobiological Hypotheses About Rape: A Critical Look at the Data behind the Hypotheses." In *Violence Against Women,* eds. Suzanne R. Sunday and Ethel Tobach, 23–58. New York: Gordian Press, 1985.

Hass, Hans. *The Human Animal.* New York: G. P. Putnam's Sons, 1970.

Hediger, H. *Studies of the Psychology and Behavior of Captive Animals in Zoos and Circuses.* London: Butterworths, 1955.

———. "The Evolution of Territorial Behavior." In *Social Life of Early Man,* ed. Sherwood L. Washburn, 34–57. New York: Viking Fund Publications in Anthropology No. 31, 1961.

Heider, Karl G. *Grand Valley Dani.* New York: Holt, Rinehart and Winston, 1979.

Hewes, Gordon W. "World Distribution of Certain Postural Habits." *American Anthropologist* (n.s.) 57/2 (1955): 231–44.

―――. "The Anthropology of Posture." *Scientific American* 196 (1957): 123–132.

Hubbard, Ruth. *The Politics of Women's Biology.* New Brunswick: Rutgers University Press, 1990.

Husserl, Edmund. *The Crisis of European Sciences and Transcendental Phenomenology.* Translated by David Carr. Evanston: Northwestern University Press, 1970.

―――. *Cartesian Meditations.* Translated by Dorion Cairns. The Hague: Martinus Nijhoff, 1973.

―――. *Ideas Pertaining to Pure Phenomenology and to a Phenomenological Philosophy, Third Book: Phenomenology and the Foundation of the Sciences.* Translated by T. E. Klein and W. E. Pohl. Boston: Martinus Nijhoff, 1980.

―――. "The Origin of Geometry." Translated by David Carr. In *Husserl, Shorter Works,* eds. Peter McCormick and Frederick A. Elliston, 255–270. Notre Dame, IN: Notre Dame Press, 1981.

―――. *Ideas Pertaining to a Pure Phenomenology and to a Phenomenological Philosophy, Second Book: Studies in the Phenomenology of Constitution.* Translated by R. Rojcewicz and A. Schuwer. Boston: Kluwer Academic, 1989.

Irigaray, Luce. *This Sex Which Is Not One.* Translated by Catherine Porter. Ithaca: Cornell University Press, 1985.

Jaggar, Alison M. "Love and Knowledge: Emotion in Feminist Epistemology." In *Gender/Body/Knowledge,* eds. Alison M. Jaggar and Susan R. Bordo, 145–171. New Brunswick: Rutgers University Press, 1989.

James, William. *Principles of Psychology,* vol. 1. New York: Dover Publications, 1950.

Johnston, Jill. *Marmalade Me.* New York: E. P. Dutton & Co., 1971.

Johnstone, Albert A. *Rationalized Epistemology: Taking Solipsism Seriously.* Albany: State University of New York Press, 1991.

Jones, Ivor H., and Dorothy Frei. "Exhibitionism—A Biological Hypothesis." *British Journal of Medical Psychology* 52 (1979): 63–70.

Jung, Carl G. *On the Nature of the Psyche.* Translated by R. F. C. Hull. Princeton: Princeton University Press, 1960.

―――. *Aion: Researches into the Phenomenology of the Self,* 2nd ed. Translated by R. F. C. Hull (Princeton: Princeton University Press, 1968).

―――. "The Transcendent Function." Translated by R. F. C. Hull. In *The Portable Jung.* Edited by Joseph Campbell, 273–300. New York: Viking Press, 1971.

―――. *Memories, Dreams, Reflections,* rev. ed. Edited by A. Jaffe. Translated by R. and C. Winston. New York: Pantheon Books, 1973.

―――. *Psychological Types.* Translated by H. G. Baynes, revised by R. F. C. Hull. Princeton: Princeton University Press, 1976.

―――. *The Archetypes and the Collective Unconscious,* 2nd ed. Translated by R. F. C. Hull. Princeton: Princeton University Press, 1990.

————. "The Tavistock Lectures: On the Theory and Practice of Analytical Psychology." *Collected Works,* vol. 18. Translated by R. F. C. Hull, 5–182. Princeton: Princeton University Press, 1980.

Kant, Immanuel. *Critique of Pure Reason.* Translated by Norman K. Smith. London: Macmillan, 1953.

————. *Metaphysical Foundations of Natural Science.* Translated by J. Ellington. Indianapolis: Bobbs-Merrill, 1970.

Kaufmann, John H. "Social Relations of Adult Males in a Free-Ranging Band of Rhesus Monkeys." In *Social Communication Among Primates,* ed. Stuart A. Altmann, 73–98. Chicago: University of Chicago Press, 1967.

Keller, Evelyn Fox. "Feminism as an Analytical Tool for the Study of Science." *Academe* 69/5 (September–October 1983): 15–21.

————. "Making Gender Visible in the Pursuit of Nature's Secrets." *Feminist Studies/Critical Studies,* ed. Teresa de Lauretis, 66–77. Bloomington: Indiana University Press, 1986.

————. "The Gender/Science System; or Is Sex to Gender as Nature is to Science?" *Hypatia* 2/3 (1987): 37–49.

————. "Gender and Science." In *Discovering Reality,* eds. Sandra Harding and Merrill Hintikka, 187–205. Boston: D. Reidel, 1983.

Kevles, Bettyann. *Females of the Species.* Cambridge: Harvard University Press, 1986.

Kleeman, James A. "A Boy Discovers His Penis." *The Psychoanalytic Study of the Child* 20 (1965): 239–66.

Koss, Mary, and Mary Harvey. *The Rape Victim.* Lexington, MA: Lexington Books, 1976.

Kugelmann, Robert. "Life Under Stress: From Management to Mourning." In *Giving the Body Its Due,* ed. Maxine Sheets-Johnstone, 109–31. Albany: State University of New York Press, 1992.

Kummer, Hans. *Primate Societies.* Chicago: Aldine, 1971.

Lacan, Jacques. "Some Reflections on the Ego." *International Journal of Psycho-Analysis* 34 (1953, Part I): 11–17.

————. *Écrits.* Paris: Éditions du Seuil, 1966.

————. *Écrits: A Selection.* Translated by Alan Sheridan. New York: W. W. Norton, 1977.

————. *The Four Fundamental Concepts of Psycho-Analysis.* Edited by Jacques-Alain Miller. Translated by Alan Sheridan. New York: W. W. Norton, 1978.

————. *Le Séminaire de Jacques Lacan, Livre Trois, Les Psychoses.* Paris: Éditions du Seuil, 1981.

————. *Feminine Sexuality.* Edited by Juliet Mitchell and Jacqueline Rose. New York: Pantheon Books, 1982.

La Fontaine, J. S. "Person and Individual: Some Anthropological Reflections." In *The Category of the Person: Anthropology, Philosophy, History,* eds. Michael

Carrithers, Steven Collins, and Steven Lukes, 123–40. New York: Cambridge University Press, 1985.

Lancaster, Jane. *Primate Behavior and the Emergence of Human Culture.* New York: Holt, Rinehart and Winston, 1975.

Langer, Susanne. *Feeling and Form.* New York: Charles Scribner's Sons, 1953.

Leonardo da Vinci. *Philosophical Diary.* Translated by Wade Baskin. New York: Wisdom Library, 1959.

Lerner, Harriet E. "Parental Mislabeling of Female Genitals as a Determinant of Penis Envy and Learning Inhibitions in Women." In *Female Psychology: Contemporary Psychoanalytic Views,* ed. Harold P. Blum, 269–84. New York: International Universities Press, 1977.

Leroi-Gourhan, André. *Le geste et la parole,* vol. 1. Paris: Albin Michel, 1964.

Liebow, Elliott. *Tally's Corner.* Boston: Little, Brown, 1967.

———. *Tell Them Who I Am.* New York: Free Press, 1993.

Lindeman, E. "Symptomatology and Management of Acute Grief." *American Journal of Psychiatry* 101 (1944): 141–48.

Linden, Eugene. "A Curious Kinship: Apes and Humans." *National Geographic* 181/3 (March 1992): 2–45.

Longino, Helen. *Science as Social Knowledge.* Princeton: Princeton University Press, 1990.

Lorenz, Konrad. *On Aggression.* Translated by Marjorie K. Wilson. New York: Bantam, 1967.

Lowe, Marion. "Social Bodies: The Interaction of Culture and Women's Biology." In *Biological Woman—The Convenient Myth,* eds. Ruth Hubbard, Mary Sue Henifer, and Barbara Fried, 91–116. Cambridge, MA: Schenkman, 1982.

Lyons, William. *The Disappearance of Introspection.* Cambridge: MIT Press, 1986.

Macey, David. *Lacan in Contexts.* New York: Verso, 1988.

MacKinnon, Catherine A. "Feminism, Marxism, Method and State: An Agenda for Theory." *Signs* 7 (Spring 1982): 514–44.

MacKinnon, John. "Reproductive Behavior in Wild Orangutan Populations." In *The Great Apes,* eds. David A. Hamburg and Elizabeth R. McCown, 257–273. Menlo Park, CA: Benjamin/Cummings Publishing, 1979.

Maclean, Paul D. "The Triune Brain, Emotion, and Scientific Bias." In *The Neurosciences: Second Study Program,* ed. F. O. Schmitt, 336–349. New York: Rockefeller University Press, 1970.

Maclean, Paul D., and D. W. Ploog. "Cerebral Representation of Penile Erection." *Journal of Neurophysiology* 25 (1962): 29–55.

Maloney, Clarence, ed. *The Evil Eye.* New York: Columbia University Press, 1976.

Martin, Emily. *The Woman in the Body.* Boston: Beacon Press, 1987.

Masters, William H., Virginia E. Johnson, Robert C. Kolodny. *Human Sexuality.* Boston: Little, Brown and Company, 1982.

Maybury-Lewis, David, and Uri Almagor, eds. *The Attraction of Opposites: Thought and Society in the Dualistic Mode.* Ann Arbor: University of Michigan Press, 1989.

McKinney, F., S. R. Derrickson, and P. Mineau. "Forced Copulation in Waterfowl." *Behaviour* 86/3–4 (1983): 250–94.

McKnight, David. "Men, Women, and Other Animals: Taboo and Purification among the Wik-mungkan." In *The Interpretation of Symbolism,* ed. Roy Willis, 77–97. London: Malaby Press, 1975.

McWhorter, Ladelle. "Culture or Nature? The Function of the Term 'Body' in the Work of Michel Foucault." *Journal of Philosophy* 86/11 (November 1989): 608–14.

Meltzoff, Andrew N. "Imitation, Intermodal Co-ordination and Representation in Early Infancy." In *Infancy and Epistemology: An Evaluation of Piaget's Theory,* ed. George Butterworth, 85–114. Brighton, Sussex [England]: Harvester Press Ltd., 1981.

Merleau-Ponty, Maurice. *Phenomenology of Perception.* Translated by Colin Smith. London: Routledge & Kegan Paul, 1962.

———. "The Child's Relations with Others." Translated by William Cobb. In *The Primacy of Perception.* Edited by James M. Edie. Evanston: Northwestern University Press, 1964.

Metzger, Deena. "It Is Always the Woman Who Is Raped." In *The Rape Victim,* ed. Deanna R. Nass. Dubuque, IA: Kendall/Hunt Publishing, 1977.

Miller, David B. Letter dated 4 August 1983. In *Violence Against Women,* eds. Suzanne R. Sunday and Ethel Tobach. New York: Gordian Press, 1985.

Mitchell, G. *Human Sex Differences: A Primatologist's Perspective.* New York: Van Nostrand Reinhold, 1981.

Mitchell, Juliet, and Jacqueline Rose, eds. *Feminine Sexuality.* New York: Pantheon Books, 1982.

Money, John, and Patricia Tucker. *Sexual Signatures.* Boston: Little, Brown and Company, 1975.

Monick, Eugene. *Phallos: Sacred Image of the Masculine.* Toronto: Inner City Books, 1987.

Muller, John P., and William J. Richardson. *Lacan and Language.* New York: International Universities Press, 1985.

Murdoch, George Peter. "The Common Denominator of Cultures." In *Readings in Introductory Anthropology,* vol. 1, ed. Richard G. Emerick, 323–26. Berkeley: McCutcheon Publishing, 1969.

Nadelson, Leslee. "Pigs, Women, and the Men's House in Amozonia [sic]: An Analysis of Six Mundurucu Myths." In *Sexual Meanings: The Cultural Construction of Gender and Sexuality,* eds. Sherry B. Ortner and Harriet Whitehead, 240–272. New York: Cambridge, 1981.

Napier, J. R., and P. H. Napier. *A Handbook of Living Primates.* New York: Academic Press, 1967.

———. *The Natural History of the Primates.* London: British Museum of Natural History, 1985.

Nass, Deanna R., ed. *The Rape Victim.* Dubuque, IA: Kendall/Hunt Publishing, 1977.

Needham, Rodney. *Counterpoints.* Berkeley: University of California Press, 1987.

Ochs, Elinor, and Bambi B. Schieffelin. "Language Acquisition and Socialization." In *Culture Theory: Essays in Mind, Self, and Emotion,* eds. Richard A. Shweder and Robert A. LeVine, 276–320. Cambridge: Cambridge University Press, 1984.

Ong, Walter J. *Interfaces of the Word.* Ithaca: Cornell University Press, 1977.

Pierce, Christine. "Philosophy." *Signs* 1/2 (1975): 487–503.

Ploog, D. W., J. Blita, and F. Ploog. "Studies on Social and Sexual Behavior of the Squirrel Monkey *(Saimiri sciureus).*" *Folia Primatologica* 1 (1963): 29–66.

Poirier, Frank E. "Colobine Aggression: A Review." In *Primate Aggression, Territoriality, and Xenophobia,* ed. Ralph L. Holloway, 123–157. New York: Academic Press, 1974.

Poovey, Mary. "Feminism and Deconstruction." *Feminist Studies* 14/1 (Spring 1988): 51–65.

Portmann, Adolph. *Animal Forms and Patterns.* Translated by Hella Czech. New York: Schocken Books, 1967.

Radin, Paul. *The Trickster: A Study in American Indian Mythology.* New York: Shocken Books, 1972.

Rancour-Laferrière, Daniel. *Signs of the Flesh: An Essay on the Evolution of Hominid Sexuality.* New York: Mouton de Gruyter, 1985.

Rodman, Peter S. "Individual Activity Patterns and the Solitary Nature of Orangutans." In *The Great Apes,* eds. David A. Hamburg and Elizabeth R. McCown, 235–255. Menlo Park, CA: Benjamin/Cummings Publishing, 1979.

Rogers, C. M. "Implications of Primate Early Rearing Experiment for the Concept of Culture." In *Precultural Primate Behavior,* ed. Emil Menzel, 185–191. Basel, Switzerland: Karger, 1973.

Roustang, François. *The Lacanian Delusion.* Translated by Greg Sims. New York: Oxford University Press, 1990.

Rowell, Thelma. *The Social Behaviour of Monkeys.* Middlesex: Penguin, 1974.

Sade, Donald Stone. "Determinants of Dominance in a Group of Free-Ranging Rhesus Monkeys." In *Social Communication Among Primates,* ed. Stuart A. Altmann, 99–114. Chicago: University of Chicago Press, 1967.

Sanday, Peggy Reeves. "Rape and the Silencing of the Feminine." In *Rape,* ed. Sylvana Tomaselli and Roy Porter, 84–101. New York: Basil Blackwell, 1986.

Saroyan, William. *The Time of Your Life.* In *Famous American Plays of the Nineteen Thirties,* ed. Harold Clurman. New York: Dell Publishing, 1980.

Sartre, Jean-Paul. *Being and Nothingness.* Translated by Hazel E. Barnes (New York: Philosophical Library, 1956).

———. *Huis clos.* Paris: Gallimard, 1945.

Schafer, Roy. "Problems in Freud's Psychology of Women." In *Female Psychology: Contemporary Psychoanalytic Views,* ed. Harold P. Blum, 331–360. New York: International Universities Press, 1977.

Schilder, Paul. *The Image and Appearance of the Human Body.* New York: International Universities Press, 1950.

Scott, Joan W. "Deconstructing Equality-Versus-Difference: Or, the Uses of Poststructuralist Theory for Feminism." *Feminist Studies* 14/1 (Spring 1988): 33–50.

Sheets-Johnstone, Maxine. "Evolutionary Residues and Uniquenesses in Human Movement." *Evolutionary Theory* 6 (April 1983): 205–209.

———. *The Roots of Thinking.* Philadelphia: Temple University Press, 1990.

———. "The Materialization of the Body: A History of Western Medicine, A History in Process." In *Giving the Body Its Due,* ed. Maxine Sheets-Johnstone, 132–158. Albany: State University of New York Press, 1992.

———. "Taking Evolution Seriously." *American Philosophical Quarterly* 29/4 (October 1992): 343–352.

———. "The Body as Cultural Object/The Body as Pan-Cultural Universal." In *Phenomenology of the Cultural Disciplines,* eds. Mano Daniel and Lester Embree, 85–114. Boston: Kluwer Academic, 1994.

Shields, W. M., and L. M. Shields. "Forcible Rape: An Evolutionary Perspective." *Ethology and Sociobiology* 4 (1983): 115–36.

Shirek-Ellefson, Judith. "Social Communication in Some Old World Monkeys and Gibbons." In *Primate Patterns,* ed. Phyllis Dolhinow, 297–311. New York: Holt, Rinehart and Winston, 1972.

Smith, John W., Julia Chase, and Anna Katz Lieblich. "Tongue Showing: A Facil [sic] Display of Humans and Other Primate Species." In *Nonverbal Communication, Interaction, and Gesture,* ed. Adam Kendon, 501–548. *Approaches to Semiotics,* No. 41. The Hague: Mouton, 1981.

St. Augustine. *The City of God.* Books XIII and XIV. New York: Fathers of the Church, 1952.

Steele, Robert S. *Freud and Jung.* Boston: Routledge & Kegan Paul, 1982.

Stern, Daniel. *The Interpersonal World of the Infant.* New York: Basic Books, 1985.

———. *Diary of a Baby.* New York: Basic Books, 1990.

Stoller, Robert J. "Sexual Deviations." In *Human Sexuality in Four Perspectives,* ed. Frank A. Beach, 190–214. Baltimore: Johns Hopkins University Press, 1976.

————. *Sexual Excitement: Dynamics of Erotic Life.* New York: Pantheon Books, 1979.

————. "Facts and Fancies: An Examination of Freud's Concept of Bisexuality." In *Women and Analysis,* ed. Jean Strouse, 343–64. Boston: G. K. Hall, 1985.

Sugiyama, Yukimaru. "Social Behavior of Chimpanzees in the Budongo Forest, Uganda." *Primates* 10 (1969): 197–225.

Sunday, Suzanne R., and Ethel Tobach, eds. *Violence Against Women.* New York: Gordian Press, 1985.

Symons, Donald. *The Evolution of Human Sexuality.* New York: Oxford University Press, 1979.

Teleki, Geza. "Chimpanzee Subsistence Technology: Materials and Skills." *Journal of Human Evolution* 3 (1974): 575–594.

Theweleit, Klaus. *Male Fantasies,* vol. 1 and 2. Translated by Stephen Conway. Minneapolis: University of Minnesota Press, 1987.

Thiessen, Delbert. "Rape as a Reproductive Strategy: Our Evolutionary Legacy." American Psychological Association Fellow's Address, 1983.

Thompson-Handler, N., R. K. Malenky, and N. Badrian. "Sexual Behavior of *Pan paniscus* under Natural Conditions in the Lomako Forest, Equateur, Aire." In *The Pygmy Chimpanzee,* ed. R. L. Susman, 347–368. New York: Plenum Press, 1984.

Thornhill, Randy, and Nancy W. Thornhill. "Human Rape: An Evolutionary Analysis." *Ethology and Sociobiology* 4 (1983): 137–173.

Thornhill, Randy, Nancy W. Thornhill, and Gerard Dizinno. "The Biology of Rape." In *Rape,* eds. Sylvana Tomaselli and Roy Porter, 102–121. New York: Basil Blackwell, 1986.

Tutin, C. E. G., and P. R. McGinnis. "Chimpanzee Reproduction in the Wild." In *Reproductive Biology of the Great Apes,* ed. C. E. Graham, 239–264. New York: Academic Press, 1981.

van Hooff, J. A. R. A. M. "The Facial Displays of the Catarrhine Monkeys and Apes." In *Primate Ethology,* ed. Desmond Morris, 9–88. Garden City, NY: Anchor Books, 1969.

van Lawick-Goodall, Jane. "The Behaviour of Free-Living Chimpanzees in the Gombe Stream Reserve." *Animal Behaviour Monographs,* vol. 1, part 3 (1968).

————. "A Preliminary Report on Expressive Movements and Communication in the Gombe Stream Chimpanzees." In *Primate Patterns,* ed. Phyllis Dolhinow, 25–84. New York: Holt, Rinehart and Winston, 1972.

Washburn, Sherwood. "The Analysis of Primate Evolution with Particular Reference to the Origin of Man." *Cold Spring Harbor Symposia on Quantitative Biology* 15 (1950): 57–78.

Watson, O. Michael. *Proxemic Behavior: A Cross-Cultural Study.* The Hague: Mouton & Co., 1970.

Weedon, Chris. *Feminist Practice and Poststructuralist Theory.* New York: Basil Blackwell Ltd., 1987.

Weiss, Rick. "Historic Priapism Pegged to Frog Legs." *Science News* 139/1 (January 5, 1991): 6.

Whiten, A., and R. W. Bryne. "Tactical Deception in Primates." *The Behavioral and Brain Sciences* 11 (1988): 233–273.

Wickler, Wolfgang. "Socio-sexual Signals and Their Intra-specific Imitation Among Primates." *Primate Ethology,* ed. Desmond Morris. Garden City, NY: Anchor Books, 1969.

Winkler, Cathy. "Rape as Social Murder." *Anthropology Today* 7/3 (June 1991): 12–14.

Winnicott, D. W. "Birth Memories, Birth Trauma, and Anxiety." *Collected Papers: Through Paediatrics to Psycho-Analysis,* 174–193. New York: Basic Books, 1958.

Winson, Jonathan. "The Meaning of Dreams." *Scientific American* 263/5 (November 1990): 89–96.

Zilboorg, Gregory. "Masculine and Feminine: Some Biological and Cultural Aspects." *Psychiatry* 7 (1944): 257–296.

Zillmann, Dolf. *Connections Between Sex and Aggression.* Hillsdale, NJ: Lawrence Erlbaum Associates, 1984.

INDEX